To

DAD and HEIDI

Energy Econo
Demand Manag
and Conserva
Policy

Energy Economics, Demand Management and Conservation Policy

Mohan Munasinghe

Senior Energy Advisor
to the President of Sri Lanka,
Colombo, Sri Lanka

Gunter Schramm

Professor of Resource Economics
School of Natural Resources
The University of Michigan
Ann Arbor, Michigan

Foreword by Marcel Boiteux

VNR VAN NOSTRAND REINHOLD COMPANY
NEW YORK CINCINNATI TORONTO LONDON MELBOURNE

Library of Congress Catalog Card Number: 83-3493
ISBN: 0-442-25838-0

Manufactured in the United States of America

Published by Van Nostrand Reinhold Company Inc.
135 West 50th Street, New York, N.Y. 10020

Van Nostrand Reinhold
480 Latrobe Street
Melbourne, Victoria 3000, Australia

Van Nostrand Reinhold Company Limited
Molly Millars Lane
Wokingham, Berkshire, England

15 14 13 12 11 10 9 8 7 6 5 4 3 2 1

Library of Congress Cataloging in Publication Data

Munasinghe, Mohan, 1945–
  Energy economics, demand management, and conservation policy

  Includes bibliographical references and index.
  1. Energy policy. 2. Energy consumption. 3. Energy conservation. 4. Energy development. 5. Energy industries. I. Schramm, Gunter. II. Title.
HD9502.A2M845   1983   333.79   83-3493
ISBN 0-442-25838-0

FOREWORD

Two successive oil crises, in 1973 and 1979, revealed the energy vulnerability of a great many countries, both rich and poor. With their economies long organized in relation to low oil prices, these countries now find, under the weight of excessive and sometimes unbearable oil import bills, that their energy supply strategies are failing them. Only hindsight can justify talk of improvidence.

The only alternative is for these countries to endeavor to redesign their energy economies within the shortest possible time, focusing on resources and systems that cost less and are more reliable. In the process of doing so, however, they find themselves coming up against forces of inertia within the energy sector itself. Although it is supremely capable of reducing unit costs by combining economies of scale with standardization, this sector lacks flexibility when it comes to converting to entirely new processes, on both the supply and demand sides. It takes time to optimize performance in fresh fields—time to bring innovation to maturity, to convert energy-generation and supply systems, to instill new behavioral attitudes, and to write off existing energy-inefficient equipment. These are all fairly irreducible time lags, each of which delays the substitution process.

In this context, the least malleable link in the energy chain is to be found on the demand side, which involves the greater part of the capital invested in the existing system, although it is less obtrusive as a result of being spread among an enormous and varied mass of energy end-users. The ability to channel the trend of demand in such a manner that both the individual and the general interest are safeguarded as far as possible is thus a matter of great importance.

The foremost merit of this book by Mohan Munasinghe and Gunter Schramm is that it provokes a fundamental reappraisal of all the aspects of this question—first, by identifying the demand factor within the overall view of the energy economy (Chapter 2); second, by widening the field of investigation to take in the other major links in the energy chain represented by reserves and production facilities, and last, by analyzing demand within the framework of overall energy policy (Chapter 3).

The authors stress that pricing policy is the main instrument of demand management (Chapters 4 and 5). They then proceed to explain how long-run marginal cost is the practical pricing basis most conducive to reconciling individual consumer options with the optimization of net benefits to society as a whole. Cost transparency, a clear distinction between economic efficiency and technical constraints on the one hand and social objectives on the other, and compatible management and planning policies are some of the undoubted advantages to be derived from this approach—an approach now being adopted by many countries.

Furthermore, the oil crisis made it plain that priority could and should be awarded to energy-saving measures—the most logical way to check any rise in consumption. When assessing achievements in this regard, however, due allowance must be made for the negative effects of a global economic recession on energy consumption, which have tended to reinforce conservation-oriented policies. We must also remember that reduced consumption in terms of calories alone cannot be a policy target and that the very concept of energy saving has different weight and meaning depending on whether the countries we consider are industrialized or developing (Chapter 6).

What then remains is to place all these factors in a proper perspective, evaluate as accurately as possible the economic and financial consequences of errors of calculation and projection, and highlight the close interrelationship between supply and demand (Chapters 7 and 8). Such considerations make it especially welcome that the authors devote the second part of their book to a review of country case studies that tests the theory against actual experience (Chapters 9 to 13).

Mohan Munasinghe and Gunter Schramm unfold before us a wide and promising field of reflection. They have combined the intellectual rigor of the economist with a rich store of personal experience to produce a most lucid, rewarding, and up-to-date exposition of their subject.

M. BOITEUX
President
Electricité de France
Paris

PREFACE

The content of this book reflects our cumulative experiences and involvement with energy policy issues in many parts of the world, spanning a period of almost two decades—from the balmy pre-oil crisis days of falling real energy prices and emerging environmental concerns, to the massive dislocations following the 1973 and 1979/80 oil price shocks and their aftermaths. As this volume goes to press, once again world energy markets are in disequilibrium, with gleeful predictions by some that oil prices may fall by as much as a third or more compared to 1982 price levels. Whatever actually happens in the short run, however—and the short run may well extend for several years—it is our view that the real price of petroleum resources will tend to increase over time, gradually approaching the levels determined by the costs of replacements such as shale oil, heavy oil and tar sands extraction, or coal liquefaction. Whether these levels will be reached well before or shortly after the turn of this century is uncertain, but it is inevitable that this situation will eventually occur, given a gradually expanding world economy and limited and decreasing petroleum resources. Hence, in our view, short-term fluctuations in prices should not be taken as a reliable guideline for long-term policy decisions.

During the past decade of turbulent energy prices, many studies on energy policy have been published. Although early ones focused mainly on regulatory issues, later ones reflected the increasing concerns about environmental problems. The first oil price shock in 1973 resulted in a plethora of doomsday scenarios that equated the perceived finiteness of depletable resources with the unsustainability and decline of modern civilization as we know it. In these forecasts, energy use and alleged misuse played a prominent part. Energy consumption patterns were initially analyzed in terms of potential energy savings on a strictly technical basis, usually invoking the first and second law of thermodynamics, but forgetting that conservation measures, too, involve costs in terms of other economic resources. Other studies attempted to analyze existing consumption patterns on the basis of statistical comparisons of aggregate energy to GDP ratios and similar input-output relationships. From them, for a while, emerged a tendency to summarily condemn those countries whose ratios were higher than those of others as "energy wasters." What these studies conveniently overlooked, however, was that energy is but one input among others and that relative costs and prices in the past have shaped the interrelationship between different inputs and outputs and have also determined the location of energy intensive activities and industries. Instead, what is needed for drawing conclusions about energy "waste" are studies of these inter-relationships at the micro-level. At the same time, macro-economic changes that could have had a major effect on the structure and use of energy were not given sufficient attention in these early studies. Some of them concentrated on the analysis of energy supply, often grossly neglecting other aspects, especially potential changes in demand. In our view, most of them made at least some contribution to the overall issues of energy planning and policy analysis even if many suffered from the partial nature of their analysis and a lack of integration of all relevant economic, technological, physical, and socio-political factors. This lack of integration is something that we have tried to avoid in our analysis.

The book has been written for the use of a broad range of persons involved in energy sector work, including energy planners and policymakers, economists, other specialists, consultants, teachers, and researchers, and advanced undergraduates as well as graduate-level students. In order to address such a wide audience, we have made a deliberate attempt to avoid the use of economic jargon as much as possible. Where appropriate, mathematical derivations have been placed in technical appendices and the results of our findings explained as simply as possible in the main text, often using diagrammatic expositions.

The first eight chapters present the necessary background information and analytical base. The following five consist of a series of case studies that apply the principles developed to real-world situations. A special characteristic of all of these case studies is that one or both of us were closely involved in the original fieldwork and subsequent analysis.

The writing of this book has spanned several years. Although this imposed a special burden on both of us and our families, it had the advantage that we could observe the dynamics of energy supply-demand interactions over time, in a given location, and for a given use. This ongoing process of observation has provided us with important insights into the dynamics of change and the effects of bottlenecks caused by lack of consumer comprehension, lock-in effects, and delayed policy responses.

In developing our analysis and writing this book, we have benefited from the advice of many of our friends and colleagues around the world. We are grateful to Marcel Boiteux, distinguished 'father' of modern marginal cost pricing, for having contributed the Foreword. Special thanks are due to Romesh Dias-Bandaranaike of IFC, Inc., and Martin J. Beckmann of Brown University, who helped develop, respectively, the mathematical models of electricity demand and the optimal user cost that are contained in the Annex to Chapter 13 and Annex 4.2. A close friend and colleague, Jeremy Warford, was instrumental in stimulating our interest in the issues of energy pricing and demand analysis. Without his encouragement, this book would not have been undertaken. Michael G. Webb and Robert Pindyck read earlier versions of the manuscript and made many valuable suggestions for its improvement. At the World Bank, we want to thank a large number of colleagues and friends whose insights and comments have contributed much to our own understanding of energy issues. Among them are Yves Rovani, Richard Sheehan, Robert Sadove, D. C. Rao, James Fish, Richard Dosik, DeAnne Julius, David Hughart, Trevor Byer, Anwer Malik, Dennis Anderson, Karl Jechoutek, Ibrahim Elwan, Ernie Terrado, Fernando Manibog, Edwin Moore, Pierre Moulin, and Edward Minnig.

Since its inception in 1978, several hundred developing-country officials and fellow lecturers who participated in the USAID-sponsored energy management training program at Stonybrook have been exposed to and helped to clarify many of the concepts presented here. Among them, we would like to mention Robert Nathans, Peter Meier, Romir Chatterjee, Gerhard Tschannerl, David Jhirad, Vinod Mubayi, and Owew Carroll. At the University of Michigan, Kenneth Shapiro, Richard Porter, and Harvey Brazer all contributed, directly or indirectly, to the basic ideas incorporated in this volume. The members of three successive graduate seminars on energy economics at the University of Michigan worked their way through preliminary drafts and made many useful suggestions for improvements and clarification. Special thanks are due to Dale Avery, Theodore Graham-Tomasi, Enrique Crousillat, and Bruce DenUyl for their insightful comments and suggestions. Others we wish to thank include Jean-Romain Frisch, Blair T. Bower, Marc Ross, Roy J. Piggott, Robert Bakely, Jayant Madhab, Barin Ganguli, Shyam Rungta, V. V. Desai, Corazon Siddayao, Gail R. Wilensky, Tilak Siyambalapitiya, and K. D. P. Gunatilaka. Betty Manoulian typed many successive drafts of the volume and remained cheerful as always. Last, but not least, we want to thank our respective families—Sria, Anusha, and Ranjiva Munasinghe and Heidi, Eileen, and Barbara Schramm—for their patience, understanding and unstinting support.

Mohan Munasinghe
Gunter Schramm

CONTENTS

Part A
Theory and Methodology

1
Introduction and Overview

The rapid increases in energy prices in the 1970s have clearly indicated that the era of cheap and abundant energy, especially oil, is over. The cost of energy is now becoming increasingly significant relative to the costs of other factor inputs such as capital, labor, and land. The critical dependence of modern economies on energy in various forms underlines the need for effective development and use of scarce energy resources. The intricate links between energy and the different sectors of the economy, as well as the interactions among the various forms of energy, need to be carefully examined within a disaggregate but integrated framework. The application of rational economic analysis will improve the efficiency of investment decisions and pricing policy in the energy area. Even small improvements can have a significant beneficial impact, given the large investment needs and consumer expenditures on energy.

Energy use permeates every sector of the economy. Governments generally seek to ensure that the cheapest energy sources are available in sufficient quantities to meet the future needs of the economy and maximize the welfare of citizens. The energy crises of the 1970s have shown that when demand for energy exceeds supply, rapid price rises and economic dislocations occur, especially in the short run when the economy is unable to adjust smoothly to the new situation. Energy planning has therefore emerged as an important discipline that seeks to achieve the best balance between energy and other inputs such as capital and labor, and to match the uses and availabilities of different types of energy, such as oil, coal, and electricity in various uses, including industry, transportation, and households.

Actual or potential imbalances that result when energy demand exceeds supply can be avoided by either augmenting supply or by reducing demand or both. These two options may be broadly labeled supply and demand management respectively. More emphasis has been placed in the past on increasing supply, especially through the development of new and higher cost energy sources or by the application of more advanced technologies to existing resources. This book seeks to examine the policy options available to manage demand. The achievement of desirable energy conservation goals fall within this category. Supply management and investment planning issues are also treated here to the extent that they influence demand, since demand and supply are often closely interrelated.

There are, of course, many uncertainties concerning the future availability and prices of different fuels, worldwide. However, the decade of the 1970s has shown that future energy costs will be higher relative to the costs of other goods and services than in earlier years. A major objective of all countries in the next decade will be the adjustment of their economies to higher energy prices. Thus demand management and conservation will be an important theme for some time to come, and there is good evidence to indicate that additional resources devoted to this effort are likely to be more cost effective, at least initially, than funds spent on merely increasing energy supply.

Chapter 1 continues with a summary of the other chapters in this book followed by a brief history of commercial energy use. It then explores the macroeconomic impacts of increased energy costs including the direct effects via the balance of payments, and reduced real incomes, as well as dynamic effects such as inflation, recession, and reduced GNP, investment, and employment. A special discussion of the developing country situation is provided.

1.1 SUMMARY OF THE BOOK

The rest of Chapter 1 begins with a brief history of energy use. In ancient times, human beings progressed from the use of their own muscle power to the harnessing of the energy of other animals such as horses and cattle. Since the industrial revolution, the use of machines and direct conversion of inanimate forms of energy for production has developed rapidly and resulted in high rates of economic growth and output.

Wood is the earliest and most common form of inanimate energy used by man; the practice dates back to the discovery of fire in prehistoric times. Fuelwood has also been used in manufacturing for many thousands of years. Next, the dominance of coal in industry developed during the industrial revolution in the 18th century, starting in Britain. Since the early 20th century, petroleum products have become increasingly important, particularly for transportation. Electricity, whether produced from hydro, oil, coal, or nuclear has also become a dominant and convenient form of energy in the modern economies of the 20th century.

Chapter 1 continues with an explanation of why modern industrialized or developing societies could not function without the use of large quantities of inanimate energy. In fact, the history of economic development is to a very large degree the history of the increasing use and substitution of inaminate for human or animal energy. The correlation between per capita energy use and real income per capita throughout the world is striking. On an aggregate basis, higher energy use implies higher income and vice versa. It is only at a much more disaggregated level that significant variations in this general relationship become apparent.

While large-scale energy use is absolutely essential for the functioning of modern economies, the overall costs of energy relative to the total value of output is still modest, rarely exceeding 10% of total GNP even in the most energy-intensive countries of the world. Analyzed on an activity by activity basis, the cost of energy relative to the value of output of a given industry or process is less, usually amounting to only about 1–3% of the value of output, although energy costs may be in excess of 50% for a few energy-intensive production processes.

However, while the costs of energy relative to the value of output are small, in terms of foreign exchange balances the drastic increases in world prices for petroleum products in the 1970s have resulted in major dislocations for most petroleum importing countries. In 1978, for example, (prior to the more than 100% increase in world petroleum prices in 1979/80) a number of developed as well as developing countries spent close to 50% of their total income from exports on energy imports alone. If relative prices for petroleum products keep rising, by 1990 some of them may be forced to spend close to 100%. These increased costs of energy imports have led to a dangerous increase in the external debt burden of many energy-importing countries, and it is forcing some of them to curtail other essential imports.

Generally, the increases in energy costs have had adverse impacts on the economies of practically all countries, except those of the oil exporters, in terms of reduced growth rates and output, employment, and personal incomes. The direct effects of the energy crisis can be analyzed in an essentially static framework, and include the worsening of balance-of-payments mentioned earlier, as well as a fall in real income. Even if the level of gross national product is unaffected, more resources will have to be diverted to pay for energy and therefore less output will be available for domestic consumption and investment. These effects will be milder, the greater and more rapid the substitution possibilities between energy and other inputs to production. The dynamic impact of energy price increase occurs when the economy tries to adjust to a price shock. Rigidities in the economy and resistance to the necessary fall in real wages and prices of most other goods can lead to spiraling wage-price inflation. As workers fight to maintain real wages, employment will decline with a consequent loss of output. Finally, investment and economic growth will also fall off following a major energy price increase, because of uncertainty in the market regarding costs, interest rates, and profits, and the complementarity between capital and energy in the short run.

Such effects are evident in the performance of most economies following the 1973 oil price increase. Thus the OECD group of industrialized countries experienced sudden spiraling wage-price inflation (except for Germany), reduced output, and recession during the 1973–1975 period.[1] In the oil importing developing countries (OIDCs), there has been a tendency to borrow heavily from external sources to tide over the initial shock. This is reasonable since in the short-run the economy is most inflexible to the energy price increase and foreign exchange is scarcest. Early borrowing is therefore made against subsequent periods when the economy has more time to adjust through energy substitution and other energy management policies. These macroeconomic effects of an increase in the costs of energy are analyzed more fully later in this chapter.

A major problem for the OIDCs is the increasing foreign debt service burden and their declining credit-worthiness as oil prices continue to increase. Improved mechanisms for recycling petrodollars and providing financial resources to these countries from the oil exporters, developed countries and multilateral aid agencies will have to be developed for the OIDCs to meet their increasing energy needs for economic growth and development. In general, the oil importing countries will need to finance massive investments for developing alternative energy sources, and for implementing energy conservation policies, to prevent catastrophic increases in their oil import bills.

Chapter 2 seeks to introduce the reader to the principal sources and uses of energy. While it has become customary to talk about energy as if it represented a homogenous commodity, this is inaccurate. Energy supplies come in many different forms. In most cases they have to be tailored precisely to specific uses. Petroleum products, for example, are divided into many different products, ranging from liquid petroleum gases (LPG) to gasoline, diesel fuels, kerosene, jet fuels, and various grades of fuel oils. All of these are designed for rather specific uses, and are not easily interchangeable with each other. Natural gas and coal are largely used as boiler and furnace fuels. Electricity is the major source of lighting, of mechanical energy for stationary uses, and of telecommunication equipment, as well as the energy source for electrochemical and electrosmelting processes. Electricity itself, in turn, is produced from other energy sources such as petroleum, natural gas, coal, or nuclear through the intermediary of high pressure steam turbines, gas turbines, internal combustion engines, or water-driven hydraulic turbines. Natural gas, petroleum products, coal, wood, charcoal, and biomass residues are the major sources of household and commercial energy for heating, cooking, and cooling. The lighter petroleum products are the almost exclusive fuels for independent transport equipment (e.g., airplanes, cars, trucks, ships, and railroad engines). Wood and other

---

[1]The 24 member countries of the OECD include most of the industrialized market economies of the world; they are: Australia, Austria, Belgium, Canada, Denmark, Finland, France, West Germany, Greece, Iceland, Ireland, Italy, Japan, Luxemburg, Netherlands, New Zealand, Portugal, Norway, Spain, Sweden, Switzerland, Turkey, U.K., and U.S.

traditional fuels are particularly important in the developing countries, which contain the bulk of the world's population. In some regions, these fuels are in critically short supply.

The major characteristics of all of these diverse energy sources are that they are designed to be used in specific applications with rather specific equipment. Only in a few applications, such as specially designed boilers, furnaces, or kilns, can a variety of fuels be used interchangeably. Most energy-using equipment is energy-source specific. Hence energy substitution is usually difficult. Given the relatively long life expectancies of most equipment and appliances (from four years for a truck to up to 50 years for a smelter or industrial furnace) change-overs from one source of energy to another are not easily accomplished. Usually, the equipment has to be replaced first and this may not be economical until it has outlived its usefulness. Furthermore, existing equipment operates at specific energy efficiency rates. These can rarely be changed without replacement of the equipment itself. Both of these factors, the long-life of equipment and its built-in energy efficiency make it difficult to change energy consumption rates in the short-run in response to substantial changes in costs. In the long-run, however, significant substitutions can be accomplished by switching from higher-cost to lower-cost energy sources, or by increasing energy-use efficiencies of equipment through substitutions by more sophisticated designs and/or lower-cost labor or capital.

However, in order to bring such changes about, appropriate energy demand management policies are needed that provide users with the appropriate signals (either through higher prices, persuasion, or direct intervention). This is why demand management is so important for the overall task of energy management.

Chapter 3 begins with a discussion of the functioning of ideal energy markets in an economically efficient manner, and then goes on to identify the real world problems that give rise to market distortions such as monopoly power, externalities, indivisibilities, and so on. Therefore, some government intervention is usually necessary, ranging from mild oversight, guidance, and decentralized incentives in a market economy, to more direct control of practically all aspects of energy sector activities in a centrally planned economy.

The scope of energy planning is defined, and the specific role of demand management is described next in Chapter 3. It is generally accepted that the broad rationale underlying modern energy management and planning is to make the best use of available energy resources for promoting economic development, and improving social welfare, and the quality of life. Therefore, energy planning is an essential part of the overall management of the national economy, and should be carried out in close coordination with the latter. However, in energy management and planning, the principal emphasis is on the comprehensive and disaggregate analysis of the energy sector, with due regard for the main interactions with the rest of the economy, and among the different owned, energy subsectors themselves. The efficient management of government owned energy related corporations, and where necessary, the provision of correct investment and price signals to the private sector, are an integral part of successfully implementing national energy policies.

In a strictly technical sense, the best strategy might be to seek the least-cost method of meeting future energy requirements. However, energy planning also includes a variety of other and often conflicting objectives, such as reducing dependence on foreign sources, supplying basic energy needs of the poor, reducing the trade and foreign exchange deficits, priority development of special regions or sectors of the economy, raising sufficient revenues to finance energy sector development (at least partially), ensuring continuity of supply, maintaining price stability, preserving the environment, and so on.

In general, energy planning requires analysis at the following three hierarchical levels in relation to fundamental national objectives: (1) links between the energy sector and the rest of the economy, (2) interactions between different subsectors within the energy sector, and (3) activities in each individual energy subsector. The steps involved in the planning procedure usually include energy supply and demand analyses and forecasting, energy balancing, policy formulation, and impact analysis, to meet short-, medium-, and long-range goals. Implementation of the results of this analysis could be considered within the framework of a formal national energy master plan (EMP), or a more decentralized policy package that relies on voluntary responses of private energy producers and consumers to market prices. This will generally require the coordinated use of a number of interrelated policy tools such as: (1) physical controls and legislation, (2) technical methods (including research and development), (3) direct investments or investment inducing policies, (4) education and propaganda, and (5) pricing.

Energy planning may be carried out initially at a relatively simple level, but, as data and analytical capabilities improve, more sophisticated techniques, including computer modeling, may be used. The institutional structure could also be rationalized by setting up a central energy authority or ministry of energy, whose principal focus should be on energy planning and policy-making. Some central guidance and coordination of the many policy tools, energy supplying institutions, and consuming sectors is necessary even in countries in which energy supply activities are dominated by the private sector. The influence of government actions in the various energy subsectors is quite pervasive in all countries. However, regardless of the degree of centralization of planning functions, the execution of policy, and day-to-day operations, would remain the responsibility of government institutions or private firms such as electricity utilities or petroleum corporations that already exist.

Demand management includes that group of elements in a national energy policy package which seeks to establish optimal or desirable levels, patterns, and mixes of energy use in the economy. This book seeks to focus on demand management, but also deals with specific related issues in the complementary area of supply management and energy resource development. It is clear that the hard policy tools for demand management such as physical controls, technical methods, and directed investments, are more effective in the short run, while the soft policy tools including pricing, financial incentives, and education and propaganda, have a greater impact in the long run. For best results the use of both soft and hard techniques of demand management should be carefully coordinated.

Chapters 4 and 5 introduce and develop the basic concepts of integrated energy pricing. An integrated pricing framework must begin with a clear statement of national objectives, and must provide a method for trading off among mutually contradictory goals.

Energy pricing structures, disaggregated by energy subsector, are derived in two stages. First we seek to satisfy the economic efficiency objective. The shadow-priced marginal opportunity cost (MOC) or long run marginal cost (LRMC) of a given form of energy is determined, based essentially on supply-side considerations. For a tradable form of energy, an appropriate measure of MOC would be the marginal cost of imports or export earnings foregone, with adjustments for local transport and handling costs. For nontraded fuels, MOC would be the marginal supply cost, plus a user cost component (in the case of nonrenewable resources). Demand-side effects including distortions in the prices of other goods, especially substitute fuels, are used to derive from the MOC the strictly efficient energy price level p_e. In practice, this basic theoretical framework may be extended to cover dynamic effects relating to both supply and demand, price feedback effects, capital indivisibilities, problems of joint cost allocation, supply and demand uncertainty, shortage costs, externalities, promotional pricing, and so on.

In the second stage of the pricing procedure, the efficient price (p_e) is adjusted to yield a realistic pricing structure that meets social-subsidy or income distributional considerations, sector financial requirements, and other practical constraints such as the need to change prices as gradually as possible, simplicity of price structure for metering and billing, and so on.

In developing countries, special attention must be paid to the hitherto neglected area of traditional energy. Direct pricing policies are usually inapplicable in the traditional fuels subsector, due to the lack of well-developed markets for these forms of energy. Therefore, indirect methods—including augmentation of supply, the appropriate pricing of substitute fuels, improvements in the efficiency of woodfuel energy conversion, and punitive measures for excessive use—must be used in close coordination to manage demand.

Energy policy-makers in developing countries face several special difficulties, such as high levels of market distortion, shortages of foreign exchange and investment funds, large numbers of poor consumers whose basic energy needs must be met, and relatively greater usage of traditional fuels, in addition to the energy related problems found in industrialized countries. Because of such distortions, important linkages between the energy sector and the rest of the economy, as well as interactions between and activities within different energy subsectors, must, in general, be analyzed by using shadow prices, essentially within a partial equilibrium framework. For consistency, the shadow pricing methodology used for pricing energy sector outputs must be the same as the one used to make investment decisions.

The topic of energy conservation is addressed in Chapter 6. Conservation broadly defined includes reduced consumption of energy by changing usage patterns or by energy saving investments, and substitution of cheaper energy sources for more expensive ones. Because of the potential to realize significant energy savings, conservation is an important element of demand management. While the concept of energy conservation is intuitively appealing, in a period of increased costs and scarcity of energy, the desirability of specific energy conservation policies must be verified by applying specific tests that compare their costs and benefits.

Two purely technical measures of the efficiency of energy use may be derived from the first and second laws of thermodynamics. The principal weakness of these measures is that they focus almost exclusively on the amount of energy used in a particular process and do not take into consideration the necessary inputs of other scarce resources such as capital, labor, and land. Also, the first and second law criteria may not always be consistent.

A more comprehensive and unambiguous test of specific energy conservation policies is based on the concept of economic efficiency and cost-benefit analysis. Generally, a conservation measure gives rise to cost saving benefits B due to reduced consumption or substitution of energy, and to additional costs C_1 of implementing the conservation policy (including hardware costs), and to costs C_2 representing benefits foregone due to energy consumption. If B exceeds the sum of C_1 and C_2, the conservation measure is desirable. In some cases, however, increasing energy consumption may improve overall economic benefits, e.g., if the price of an energy product is well above its MOC, the optimal price should be lowered towards MOC, with a consequent increase in demand.

These economic costs and benefits are evaluated on a life cycle basis, in terms of present discounted values. Appropriate measures of the opportunity costs of goods and services are used to verify whether the given policy will increase economic welfare from the national viewpoint. If so, the same calculation must be repeated using market prices to check whether private individuals will actually find it profitable to adopt this measure. Often, since market prices diverge from opportunity costs, it may be necessary to change taxes, prices, and legislation to promote a desirable conservation policy.

Chapter 6 ends with a summary of the principal practical possibilities for energy conservation, in the near term, in several selected energy producing and consuming sectors of the economy: transportation, buildings, industry, and electric power. It is also shown that energy conservation measures can have both adverse and favorable environmental consequences, depending on the specific case.

The importance of energy demand analysis and forecasting are described in Chapter 7. Errors in demand projections lead to shortages of energy which may have serious repercussions on economic growth and development. The methodologies for demand forecasting include, historical trend analysis, sector-specific econometric multiple-correlation methods, macroeconomic or input-output methods, and empirical surveys. In general, the relatively sophisticated econometric and input-output approaches are easier to apply in the developed countries, while data and manpower constraints indicate that the simpler techniques will be more effective in the developing countries. There is no universally superior method of demand forecasting, and the use of several different methods is recommended to cross-check the final result.

Chapter 7 continues with a discussion of the problems of verifying patterns of energy consumption. Price subsidies and anomalies in the availability of energy products lead to leakages and use of fuels in unexpected and undesirable activities. The energy use patterns for traditional fuels are also difficult to determine and project, because well-established markets do not exist for these products. Chapter 7 concludes with a detailed description of the determinants of demand in the various energy subsectors including: electric power, natural gas, coal, petroleum products, traditional fuels, and nonconventional sources.

Chapter 8 addresses selected issues in investment planning and supply management that bear a direct relationship to demand management and pricing issues. Two important ways by which supply affects demand is through availability and quality of supply. Also, because of the long life expectancy of capital-intensive energy supply systems, choices in energy use—such as types of fuels utilized in given locations—are often times predetermined for long periods of time, thus "locking-in" energy users to particular alternatives.

The basic cost-benefit principles underlying energy investment decisions are briefly presented, followed by a discussion of the role of shadow pricing for efficiency as well as income distribution or equity objectives.

The iterative nature and interaction of supply and demand decisions are explored, and a simplified example is used to discuss the differences in making social versus private investment decisions, how this affects demand, and the possible ways in which the two viewpoints could be reconciled with each other. Finally, a separate technical Annex is provided, elaborating some of the formal concepts of shadow pricing.

Chapters 9 to 13 consist of several case studies in Sri Lanka, Thailand, Bangladesh, the United States, Costa Rica, and Brazil, which illustrate the potential applications of the theory and methodology presented in the first eight chapters. The overall role of energy planning in the macroeconomic context, problems arising from energy supply-demand forecasting and balancing, coordinated use of the tools of demand management, and selected issues in investment planning and financial resource mobilization are analyzed and discussed. In particular, the determination of optimal pricing strategies, possibilities for energy substitution both on the supply side (e.g., coal versus natural gas versus hydro in electric power generation) and on the demand side (e.g., kerosene versus fuelwood/charcoal versus liquified petroleum gas (LPG) for domestic cooking), as well as energy conservation policies are highlighted.

1.2   HISTORICAL PERSPECTIVE

Living organisms have made use of energy since life emerged on earth. The direct absorption of energy or the ingestion of substances in the environment and the conversion of energy stored in them into other forms of energy that support activities such as maintaining metabolism or locomotion is a well known aspect of plant and animal life. All energy available on the earth is ultimately dependent on the sun.[2]

Inanimate energy provides the essential underpinning for the functioning of modern economies. But the use of nonhuman, nonanimal energy is as old as human history. Wood, charcoal, crop residues, and dung provided the energy for cooking without which most of our food would have been indigestible. The same sources provided warmth in cold climates. Even today close to half of the world's population, and the vast majority of people living in the rural areas of the developing countries, depend for cooking and heating on these traditional fuels. Wood and charcoal traditionally provided the energy for making bricks and ceramic ware, and for melting and forming metals. Even today, several large steel works (e.g., in Brazil) use charcoal as their principal fuel and reducing agent.

The Industrial Revolution of the seventeenth and eighteenth centuries provided the take-off for modern economic development and growth. It would have been impossible without a manifold increase in the use of energy. Steam engines, fired initially by wood, but later mainly by coal, provided the necessary mechanical energy for mining, water pumping, manufacturing, and the propulsion of ships and locomotives. Coal, converted into coke, became the main

---

[2]Strictly speaking, the renewable forms of energy derived from water, wind, solar radiation, and biomass, as well as fossil fuels such as coal and oil, are directly attributable to the sun. Others such as geothermal and nuclear energy are more related to the original formation of the earth itself, but since the dynamics of the whole solar system is dominated by the sun, these forms of energy also have a "solar" origin.

source of energy for steel production, replacing dwindling and increasingly costly charcoal resources. The drastic cost reduction in steelmaking which followed brought about an explosion in the use of steel for machinery, transport equipment, and structures.

Towards the end of the nineteenth century, electricity produced from high pressure steam or the force of falling water came into widespread use. At the same time, the internal combustion engine made its appearance, revolutionizing the transport industry. Large oil finds in the U.S. in the latter part of the Nineteenth Century had made kerosene the favorite fuel for lighting fixtures, replacing dwindling resources of whale oil and inferior wax candles. Low btu. city gas, a by-product of the conversion of coal into coke, became an alternative source of lighting, but more often replaced coal as a convenient fuel for cooking purposes.[3]

The superior quality of incandescent light bulbs compared to other sources of artificial illumination, and the convenience of electricity as a source of mechanical power in industrial operations led to an explosive expansion of electric utility systems throughout the world. While electricity became available on a commercial basis only in 1887,[4] by the first decade of the twentieth century most of the larger urban centers around the globe had at least some limited electricity services installed.[5]

Electricity also became the exclusive energy source for the growing telecommunication industry. Telephone and telegraph systems, and more recently, movies, radio, and television all depend on the use of electricity as their source of energy. Modern computers could not function without it. Another major user of electricity became the electrochemical and electrosmelting industry. Without electricity many products such as aluminum, magnesium, titanium, high-grade zinc and copper, caustic soda, and phosphoric acid, to name but a few, could not be produced or produced only at prohibitive costs.

Similar claims can be made for petroleum products. The invention of the internal combustion engine, both in the form of spark-plug and diesel engines, and, later, gas-turbines, would not have been possible without the availability of highly refined petroleum products. Their use revolutionized the transport industry, making mechanized land transport economically feasible and providing the high-grade, high energy-density fuel needed for powering airplanes. So great are the advantages of petroleum products in transportation, that the once-dominant steam engine has all but disappeared.

---

[3]For example, city gas was used for decades for street lighting in the Upper Silesian coal mining areas in Poland until well after the end of World War II.

[4]Electricity was first produced on a commercial basis in 1887 by the Rochester Light and Power Company.

[5]For example, Port-au-Prince, capital city of Haiti, a low income developing country, had its first electric utility installed as early as 1912.

Overall, what we can observe is that inanimate energy consumption has grown rapidly over the course of the last century or so. This is apparent from Figure 1.1, which traces energy consumption in the United States from 1850 to the present. This rapid growth was accompanied by a fundamental shift in the use of different energy resources. This can be seen from Figure 1.2. In 1850 more than 90% of inanimate energy in the U.S. was supplied by wood. By 1900, coal had taken over the lead accounting for 71% of energy consumed, with wood still amounting to 20%. In 1950 oil and gas dominated with 51%, and coal had fallen off to 41% of energy supply. Hydroelectric energy has supplied a steady 3–4% since 1900, while nuclear has increased from zero in the early 1960s to 2% of total energy supply by 1975, and woodfuel use is now negligible.

Without the use and ready availability of these various sources of energy the majority of products that we use and consume today would not be available. There would be no plastics, no light metals, and no fresh fruits and vegetables during the winter months in cold climatic zones. There would be no airplanes, automobiles, trucks, or self-powered ocean transports. Without the all-pervasive use of these inanimate energy sources, humanity would have to return to a subsistence existence, using biomass fuels for heat, and animal and human muscle power as mechanical energy to sustain itself, a lifestyle that prevailed for all but the last few centuries.

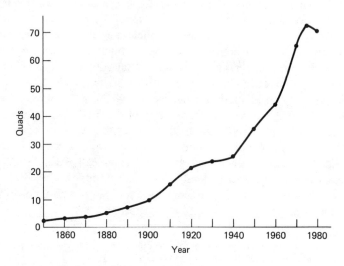

Figure 1.1.   Historic growth of U.S. energy consumption.
Sources: Sam H. Schurr, *Energy in The American Economy,* Johns Hopkins Press, Baltimore, 1960, table 1
National Academy of Sciences, *Energy in Transition 1985–2010* W. H. Freeman & Sons, San Francisco, 1979, Fig. 1-3.

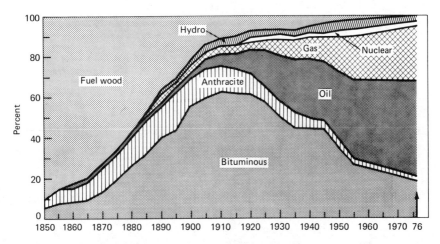

Figure 1.2.   Specific energy resources as percentages of aggregate U.S. energy consumption 1850–1978.
Sources: Sam H. Schurr et al, *Energy in the American Economy*, Johns Hopkins, Baltimore, 1969, fig. 10
Sam H. Schurr et al, *Energy in America's Future*, Johns Hopkins, Baltimore, 1979, table 2.1.

1.3   MACROECONOMIC RELATIONSHIPS

Analyzed on an aggregate basis there appears to exist a strong correlation between energy consumption and income per capita. This can be seen from Figure 1.3 which shows this relationship for low-income, middle-income, and high income (industrialized market) countries. As per capita income increases energy consumption apparently increases at a constant rate.

It is only at a more disaggregate, country-by-country level that significant variations from this general trend become apparent. This can be seen from Table 1.1 which, in addition to the grouped averages, shows the countries with the highest and lowest energy/gross national product (GNP) ratios within each income group.[6] Within the low-income countries, for example, the energy/ GNP ratios range from a low of 0.11 for Nepal to a high of 3.21 for China, and in the middle-income countries from a low of 0.17 in Yemen Arab Republic to a high of 1.62 in Singapore. However, among the industrialized countries the variation is much smaller, ranging from a low of 0.65 for Italy to a high of 1.40 for Canada.

One important explanation for the wide variations among the developing countries is that the data cover only commercial energy sources, but exclude

---

[6]For detailed country-by-country data see Appendix 1.1, Table 1. While this section analyzes energy/GNP ratios, the related concept of energy demand elasticity with respect to GNP (i.e., percentage change in energy use relative to change in GNP) is discussed in Chapter 7.

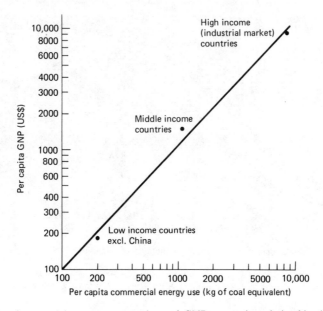

Figure 1.3. Commercial energy consumption and GNP per capita relationships by country groups.

traditional ones.[7] The latter, however, account for as much as 75–90% of total energy consumption in some of them.[8] Another reason is the importance of a few, relatively energy-intensive activities in a number of them. For example, China has a very high ratio of 3.21 because of a substantial industrial base, even though the country's average per capita income is very low. Similar reasons underlie India's and Pakistan's relatively high per capita use. For example, if China and India are excluded, the weighted average energy/GNP ratio of all other low-income countries falls to 0.54. Indonesia's high consumption is due to its large oil and gas industry and its low domestic energy prices. Mauritania's and Zambia's relatively high consumption are a function of their substantial mining industries (see Annex Table A.1.1.).

---

[7]Commercial energy sources are generally those that are traded in the market and produced by modern industrial methods, i.e., petroleum products, hydropower, natural gas, coal, and electricity, while traditional sources comprise wood, charcoal, crop residues, and dung.

[8]According to a recent study, traditional energy resources in 1978 accounted for about 73% of total energy consumption in Afghanistan and Bangladesh and as much as 96% in Nepal, while in some of the middle income countries, such as the Philippines and Thailand, their share was a still significant but much lower 34% and 23%, respectively. From T. L. Sankar and G. Schramm, *Asian Energy Problems* Praeger, New York, 1982, Annex 2.

Table 1.1. Average and High/Low Per Capita Commercial Energy Consumption
Rates (1979 Data).

COUNTRY OR COUNTRᵢ GROUP	GNP/CAPITA (dollars)	ENERGY CONSUMPTION PER CAPITA (kilograms of coal equivalent)	ENERGY/GNP RATIO
LOW INCOME COUNTRIES	*230*	*463*	*2.01*
High: China	260	835	3.21
Low: Nepal	130	14	0.11
MIDDLE INCOME COUNTRIES	*1,420*	*1,225*	*0.86*
High: Singapore	3,830	6,211	1.62
Low: Yemen, Arab Rep.	420	73	0.17
INDUSTRALIZED COUNTRIES	*9,440*	*7,892*	*0.84*
High: Canada	9,640	13,453	1.40
Low: Italy	5,250	3,438	0.65
CAPITAL-SURPLUS, OIL EXPORTING COUNTRIES	*5,470*	*1,458*	*0.27*
High: Kuweit	17,100	6,348	0.29
Low: Iraq	2,410	692	0.96

Source: (see anne A.1.1. of Chapter 1.

There are, of course, many other reasons why significant variations in
energy/GDP ratios should exist. One is climate. Less energy is needed for
space conditioning in more moderate climatic zones.[9] Another important reason is provided by differences in types of economic activities. Countries with a
substantial mineral extraction industry will tend to use more energy per unit
of output than countries with similar per capita incomes generated largely by
agricultural activities. Major differences will exist between countries that have
large electroprocess industries, such as the United States, Canada, and Norway
and others that do not, such as Japan, West Germany, and Sweden. The very
reason why the former have attracted large, high-energy consuming industries
in the first place is, of course, that they originally had cheap and abundant
sources of energy available.[10] It is noteworthy, for example, that in the United
States five industrial activity groups, which accounted for about 26% of total

----

[9]Although hot weather in high-income countries (such as the United States) may bring about
substantial energy consumption for air-conditioning.
[10]For an analysis of the relationships between low cost energy and industrial location see: Gunter
Schramm, "The Effects of Low-Cost Hydropower on Industrial Location," *Canadian Journal of
Economics,* Vol. 2, No. 2, May 1969, pp. 210–229.

industrial value added, consumed as much as 70% of the energy used by the whole manufacturing sector.[11]

Significant variations in energy use may also result from factors such as population density, urban structures, types of housing, and types of transportation systems. Some societies, such as the United States and Canada, have opted for many decades for a combination of freeways, single-family dwellings, large lot sizes, and strict separations between industrial/commercial activities and residential areas. They necessarily have to use much more energy for personal transportation as well as space conditioning than others that have developed higher-density housing-industrial-commercial complexes. Multistory apartment buildings, the predominant form of housing in Europe, Japan, or Hong Kong, are inherently more energy-efficient than unattached single-family homes. High housing densities also allow the maintenance of efficient and convenient public transport systems. These spatial differences explain much of the differences in personal energy consumption patterns between North America and regions such as Western Europe and Japan. Such structural differences cannot easily be changed in response to new energy-cost relationships. Decades are required.[12]

Given the many factors that can explain differences in per capita energy use, it is perhaps remarkable that among the developed, industrialized countries, energy use relative to income varies as little as it does. This is but another confirmation of the fact that high per capita income levels depend on high levels of modern-sector activities, which in turn vitally depend on the use of substantial quantities of inanimate energy.

It is equally important to note, however, that for each given activity, or energy-dependent end use, there exists a significant range of energy inputs that, in combination with varying inputs of capital and labor, could produce comparable levels of services. The general issue of identifying and implementing policies that will induce consumers to choose the most desirable or optimal combinations of energy and other inputs for different activities is analyzed at greater length in Chapter 6 under the topic of energy conservation.

---

[11]The industries are paper and allied products, chemical and allied products, stone, clay and glass products, and the primary metals industry. From: John G. Myers et al., *Energy Consumption in Manufacturing,* Ballinger Publishing Company, Cambridge, MA 1974, tables 1–4. For a detailed investigation of the underlying factors that cause differences in energy consumption rates see: Joel Darmstadler, Joy Dunkerley, and Jack Altman, *How Industrial Societies Use Energy,* Johns Hopkins Univ. Press, Baltimore, 1977.

[12]The low-density, urban developments typical for North America are, of course, as much the result of low land prices as of low energy prices. While the latter have changed rather drastically, the fact remains that land prices in North America generally are much lower than those in Europe or Japan. If it can be assumed that low-density housing provides more attractive living conditions, it is likely that the North American life-style will persist (or remain more prevalent) even if energy costs reach the higher European levels.

Table 1.2. Primary Energy Consumption in North America, Western Europe, 25 Asian Countries, and the World. (in million tons of oil equivalent (toe) and percent)

ENERGY SOURCE	NORTH AMERICA		WESTERN EUROPE		JAPAN		25 ASIAN COUNTRIES		WORLD	
	TOE	%	TOE	%	TOE	%	TOE	%	TOE	%
Oil	894	45	697	58	255	74	97	74	2,693	43
Natural gas	580	29	175	14	11	3	10	8	1,182	19
Coal	365	18	235	19	52	15	20	15	1,907	30
Hydro	113	6	74	6	21	6	4	3	1,744	28
Nuclear	46	2	27	2	7	2				
Total	19998	100	1,208	100	346	100	131	100	6,268	100

Note: Percentages may not add to 100% due to rounding.
Sources: Ferdinand E. Banks, *The Political Economy of Oil,* Lexington Books, Lexington, MA, 1980, Table 5-1 (for North America, Western Europe, and Japan, (1976 data); T. L. Sankar and G. Schramm, *Asian Energy Problems,* Asian Development Bank, Praeger. New York, 1981, Annex 2; (1978 data for 25 Asian countries and the world). The Asian countries included the following: Indonesia, Malaysia, Afghanistan, Bangladesh, Burma, Cambodia, Laos, Nepal, Cook Islands, Miribati, Maldives, Solomon Islands, Tonga, Western Samoa, Pakistan, Sri Lanka, Vietnam, Papua New Guinea, Philippines, Thailand, Taiwan, Fiji, Hongkong, Korea, Singapore.

1.4   BALANCE OF PAYMENTS EFFECTS

The most important characteristic of post-World War II energy consumption trends was the rapid ascendency of petroleum which became the dominant source of commercial energy throughout the world. This can be seen from the data in Table 1.2, which show that petroleum products accounted for some 43% of total world energy consumption in 1978; 45% in North America, 58% in Western Europe, and as much as 74% in Japan and 25 Asian countries. Among the latter, oil dependency was below 50% only in Pakistan, while 15 of them depended on oil for more than 90% of their commercial energy requirements.[13]

Petroleum products also completely dominate the international trade in energy resources. In 1970, they accounted for some 7.5% and 9.8% of the value of total imports of the developing countries and the developed countries, respectively. Because of the oil price increases, by 1977 these shares had increased to 15.0% and 22.2%.[14] Other energy commodities traded internationally are

---

[13]T. L. Sankar and G. Schramm, *Asian Energy Problems,* Asian Development Bank, Praeger, New York, 1982, Table 2.7.
[14]World Bank, *Commodity Trade and Price Trends* (1979 Edition), Washington, D.C., p. 5.

largely limited to metalurgical coal, natural gas (through pipelines or in the form of LNG), and electricity. However, compared to the trade in oil, their quantities are insignificant. This pattern will change during the 1980s, however, with much greater quantities of natural gas and thermal coal competing with petroleum products. There are a variety of reasons for the preference for petroleum products. Among them are convenience in use and ease of handling, high energy density per unit of weight, fewer adverse environmental effects, and low costs relative to other sources of energy, at least prior to 1973/74. Petroleum prices in real terms were actually declining until that time. This can be seen from the upper graph in Figure 1.4 which shows average crude oil prices between 1950 and 1980 in both current and constant dollars. By 1973/74, when prices quadrupled, most of the world's commercial energy consumption patterns had become firmly dependent on the use of oil as their major input.

Two important problems arising from this dependence on petroleum are that oil deposits are highly unevenly distributed and that oil represents a depletable asset whose known reserves are small relative to consumption rates. This can

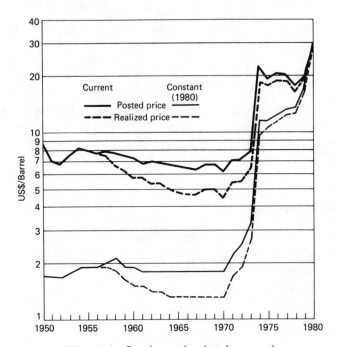

Figure 1.4. Petroleum prices (yearly average).

Table 1.3. Oil and Natural Gas Liquids Proved Reserves and Estimated Resources in 1979.

CRUDE OIL	PROVED RECOVERABLE RESERVES		ADDITIONAL RESOURCES IN PLACE	ESTIMATED ADDITIONAL RESOURCES		
				IN PLACE		RECOVERABLE
	10^6t	%	10^6t	10^6t	%	10^6t
Africa	8040	9	2180	85000	16	34000
North America	4480	5	42100	60000	11	24000
Latin America	7770	9	15800	30000	6	12000
Far East/Pacific	2390	3	7740	30000	6	12000
Middle East	51040	57	600	130000	24	52000
Western Europe	2710	3	7770	25000	5	10000
USSR, China, Eastern Europe	12700	14		160000	30	64000
Antarctic				10000	2	4000
Total	89140	100		530000	100	212000

NATURAL GAS LIQUIDS	PROVED RECOVERABLE RESERVES		ADDITIONAL RESOURCES IN PLACE	ESTIMATED PROVED REC. RESERVES		ESTIMATED ADD. REC. RESOURCES	
	10^6t	%	10^6t	10^6t	%	10^6t	%
Africa	454	33		470	7	450	4
North America	510	34	1980	690	10	1100	10
Latin America	382	25	133	390	6	250	2
Far East/Pacific	110	7	577	310	5	850	7
Middle East				1890	29	2100	18
Western Europe	9	1	12	360	5	550	5
USSR, China, Eastern Europe				2470	38	6300	54
Total	1505	100		6580	100	11600	100

Source: World Energy Conference, *Survey of Energy Resources,* Fed. Institute for Geosciences and Natural Resources, Munich, Sept. 1980. Part A, Table 2.6

be seen from the data in Tables 1.3 and 1.4. Some 46% of known recoverable oil reserves are owned by members of OPEC and another 26% by centrally planned economies. Only 14% of world reserves are located in OECD countries and the remaining 14% in other market economies. This high concentration of reserves, in combination with the generally much higher costs for direct petroleum substitutes such as synfuels, provides the fundamental economic underpinning for the success of OPEC.

Table 1.4. Static Lifetime of Proved, Recoverable Oil Resources.

	STATIC LIFE-TIME IN YEARS	PROPORTION OF THE WORLD PRODUCTION IN %	PROPORTION OF THE WORLD CONSUMPTION IN %
Africa	27	10	2
North Africa	9	16	32
Latin America	31	8	6
Far East/Pacific	17	5	12
Middle East	48	35	3
Western Europe	31	3	23
USSR, China, Eastern Europe	18	23	22
OECD	13	20	63
Countries with centrally planned economy	18	23	21
OPEC	41	50	4
Other	31	7	12

Source: World Energy Conference, *Survey of Energy Resources,* Fed. Institute for Geosciences and Natural Resources, Munich, Sept. 1980. Part A, Table 2.8.

The drastic changes in the price of oil in the 1970s had major financial and balance-of-payments repercussions on the oil importing countries. Total energy imports as a percent of total merchandise exports in 1960 had amounted to less than 11% worldwide. In 1977 energy import costs had increased to between 16 and 40% of exports for the low-income non-oil producing countries, to 24% for the middle-income countries, and to 20% for the industrialized countries (see Table 1.5). However, these average percentages hide rates of dependency that are much higher for a number of individual countries. As can be seen from the data in Annex 1.1, some of them, such as Jordan, Pakistan, Panama, Syria, and Turkey spent between 40 and 90% of export earnings for oil imports, and the United States and Japan about 31% and 32% respectively. Given the more than 100% oil price increases in 1979/80 these percentages are likely to have increased substantially since then.

Table 1.5. Energy Imports as a Percentage of Merchandise Exports.

	1960	1978
Low Income Countries	8	16–40
Middle Income Countries	13	24
Industralized Countries	11	20

Source: Annex to Chapter 1, Table A.1.1

Projections made for 1990 for member countries of the Asian Development Bank indicate that energy imports may account for between 55 and 120% of export earnings in low-income countries and between 17 and 50% for the middle-income country group, in spite of the fact that the projections also foresee a significant substitution of other energy resources for oil.[15]

These increased foreign exchange requirements for oil imports were a major contributor to a significant deterioration in the balance of payments of many of the oil importing countries. As a result their external debt load increased sharply. This is particularly true for developing countries. Since much of their oil expenditures will have to be financed through foreign exchange borrowings rather than increased exports, the accumulation of debt and declining international credit worthiness of many importing developing countries are major unresolved issues. The special problems of the third world are discussed in greater detail later.

We conclude this section by noting that a reduction in oil demand will not only reduce the import bill but also help dampen increases in future oil prices. The latter argument is particularly true for large importers like the industrialized market economies. This point may be demonstrated using a simple static model. The cost of oil imports for a given country are given by: $C = pQ$ where p and Q are the world market price and the quantity of oil imported, respectively. Differentiating gives:

$$(dC/dQ) = p + Q(dp/dQ) \tag{1.1}$$

For a reduction ΔQ in oil use; we get

$$(dC/dQ)\Delta Q = p.\Delta Q + Q(dp/dQ)\Delta Q \tag{1.2}$$

A recent study estimates that by 1990, the world oil price (although higher than the 1981 value) will tend to drop by between $0.9 and $2.40 per barrel in 1981 constant dollars, for each 1 million barrels per day (MMBD) reduction in world oil demand.[16] This is represented by the (dp/dQ) component, but note that the latter is not constant and depends on the size of the reduction ΔQ. Thus oil importers have a significant incentive to limit oil demand, acting either individually or collectively.

This implies that even for an oil importer acting alone, the value of a barrel of conserved oil could be significantly greater than the nominal market price p. For example, if a conservation program reduces oil imports from 5 to 4.5

---

[15]T. L. Sankar and G. Schramm, *op. cit.* Table 9.9.
[16]Energy Modeling Forum, "World Oil", EM6-Summary Report, Stanford University, CA94305, Dec. 1981.

MMBD, assuming $(dp/dQ) = 1$, equation (1.1) indicates that the marginal value of the conserved oil is given by: $(dC/dQ) = p + 5$, or \$5 more than the original import price.

1.5   DIRECT MACROECONOMIC EFFECTS

One of the most important direct effects of the energy price increases of the 1970s was the resulting fall in real incomes of oil importing countries. World oil prices rose by some 350% in 1973–74 and by some 140% in 1979–80. These increases occurred in response to relatively small reductions in available world supplies proving that the short-run demand and supply elasticities for oil are low. In both periods, panic buying reinforced the upward pressures on prices with the result that spot market prices (which account for between 5 to 15% of world trade) rose even faster than the announced contractual prices. Given the uneven distribution of oil producing and importing countries these drastic price rises resulted in major real income transfers from importing to producing countries. The consequence was, of course, that real income in the importing countries had to fall.

Suppose the share or fraction of energy costs in GNP is ES. Then, in the worst case where the price-elasticity of energy demand is zero and no energy substitution is possible, a simple and essentially static analysis shows that an increase in energy prices of p percent will reduce real national income at most by $[p \times ES/(1 - ES)]$ percent.[17] Assuming a typical value of 0.1 for ES, real income will fall by a maximum of 1.1% following a 10% rise in energy costs, neglecting the dynamic effects of the energy price shock to be discussed later. Thus if the energy cost share, ES, is small, this will reduce the adverse impact on national income, basically because the economy is less dependent on energy and therefore less vulnerable to the price increase. However, for a given value of ES, the greater the possibilities of substituting other relatively cheaper inputs for energy in various productive activities (i.e., a higher value for price elasticity), the smaller the decline in real national income. A recent study estimates that a 10% increase in energy prices will reduce real income by 0.8 percent in the group of OECD countries, assuming an overall price elasticity of energy demand of -0.6.[18]

The direct effects will be the same whether the price of imported energy rises or domestic energy sources become more expensive to exploit. Only countries

---

[17]The price elasticity of demand for energy is essentially the percentage change in energy demand divided by the percentage change in energy price, keeping other variables constant. See Chapter 7 on Demand Forecasting for further details.

[18]Robert E. Hall and Robert S. Pindyck, "Oil Shocks and Western Equilibrium," *Technology Review, Vol. 83,* May/June 1981, pp. 32–40.

with significant amounts of cheap (and preferably exportable) energy resources will not be adversely affected. It is important to realize the distinction between national income and national output. Thus, if there is little substitution of other inputs such as capital and labor for energy, real output or GNP will be relatively unaffected, but since more of this product is used to pay for expensive energy, less will be available as wages and profits, i.e., real incomes and the standard of living will decline. It has been pointed out that, paradoxically, the greater the substitution of other inputs for energy in production. the larger the fall in real GNP, but the smaller the reduction in real national income.[19]

Dynamic Effects

The dynamic effects of an energy price increase occur as the economy seeks to adjust to this sudden shock, encumbered by wage and price rigidities and constraints. These impacts are more severe, the sharper and more unexpected the increase. As discussed in the previous section, the increased share of output claimed by energy costs implies reduced incomes for other factors of production such as capital and labor. However, in the post-1973 period various economic agents in the oil importing countries sought to maintain their real incomes by raising wages and prices of other goods. These actions simply boosted money incomes (with little effects on real incomes). and triggered off a spiral of wage-price inflation. Among the OECD nations, only Germany was able to avoid a sharp increase in the rate of inflation from 1970 to 1976, e.g., the relevant inflation rates were 6% for 1970–73 and 6.5% for 1973–75.[20]

In addition to the inflationary effects mentioned above, the adjustment of the economy will also cause real output to fall. All the well-known adverse effects such as increased unemployment, decrease in GNP growth rate, and general recession were felt in oil importing countries after the 1973 and 1979–80 oil price increases. In many cases, the tendency of workers to resist

---

[19]Robert S. Pindyck, "Energy, Productivity, and the New U.S. Industrial Policy." Energy Laboratory Working Paper No. MIT-EL81-006WP, MIT, Cambridge, MA, February 1981.

[20]As an exasperated Canadian finance minister aptly described the underlying process: "Put simply, energy price increases have to be absorbed by the economy—if we are serious about reducing inflation they cannot be allowed to spread to other sectors. Energy prices have to rise not only in absolute terms, but in relation to other prices as well. Any attempt to compensate for every blip in the energy component of the consumer price index (CPI) through higher income demands will only result in higher costs elsewhere in the economy. Since prices respond mainly to costs which are just somebody else's prices, the final result can only be higher inflation. We are all too familiar with these endless catch ups and leap frogs that only make matters worse with time. We end up tossing a hot potato from one to another, and nobody wants to be holding the thing when the game ends. But I assure you we will all end up taking the energy price increases sooner or later." Allan J. MacEachen, an Address to the Conference Board of Canada, Toronto, May 6, 1981, p. 3.

decreases in real wage levels and increases in other input costs resulted in reduced profits. This led to layoffs and decreases in output. At the same time uncertainty regarding prices, interest rates and profits resulted in reduced investment. Consumer demand also reacted negatively to reduced future real incomes. Thus all the OECD nations went into recession during the period 1973–75, with the uniformity and synchronization of these responses being enhanced by the tight linkage due to large trade flows among them.

Studies undertaken by the Organization for Economic Cooperation and Development (OECD) have shown that the 1973–74 price rises were a major contributor to the 1974–75 world recession. although restrictive economic policies and other factors played a part as well. As the studies show, attempts to raise money incomes in response to the oil price rise led to wage/price spirals, followed by a profit squeeze that sharply reduced investments in later years.[21] For the U.S., a detailed study concluded that the 1973/74 oil price increased inflation by 4 and 2 percentage points in 1974 and 1975 respectively. Real output decreased by 30 billion in 1974, 66 billion in 1975, and around 50 billion in each of the following three years. while unemployment was estimated to have increased by 1 percentage point in 1974, 1.9% in 1975, tapering off thereafter.[22]

Simulation studies were also undertaken to predict the effects of the 1979/80 price rises on the economies of the OECD member countries. The results are summarized in Table 1.6. As can be seen, all indicators of real growth, disposable income, private consumption, total demand, imports, exports, and GNP were negatively affected. The cumulative real effect on GNP over the 1979–81 period was estimated at 5.4%. Consumer prices were projected to be higher by 1.6%, 5%, and 3.6% in the three-year period than they would have been in the absence of the oil price increase, while export and import prices were affected even more.

The overall effects of the two oil price shocks on the seven major OECD member countries are shown in Figure 1.5. As can be seen, in 1973–74 there was an almost immediate and sharp recessionary effect on GNP, accompanied by a sharp rise in inflation. In 1979/81 both the recessionary impact and the inflationary response appear to have been more muted and drawn-out over a longer period.

These immediate, macroeconomic effects of the income transfers from oil importers to oil exporters will probably diminish, however, as price-induced substitutions by other energy resources and conservation measures take hold. In the long-run the higher oil prices will lead to a decline in the use of oil, even

---

[21]From: "The impact of oil on the world economy," *OECD Observer*, No. 105, July 1980, p. 7.
[22]From: Knut Anton Mark and Robert E. Hall, "Energy Prices, Inflation, and Recession, 1974–75," *Energy Journal*, Vol. 1,33,July 1980, pp. 31–63.

Table 1.6. Simulated Effects on OECD Area of Real Oil Price Increase Since the End of 1978.

	1979	1980	1981
Increments to volume growth rates			
Disposable income	−1.2	−3.2	−1.0
Private consumption	−0.7	−2.4	−1.3
Total domestic demand	−1.0	−3.4	−1.9
Imports[a,b]	−1.4	−4.5	−2.7
Exports[a,b]	−0.9	−2.2	−1.0
GNP	−0.9	−3.0	−1.5
Cumulative effect on:			
GNP	−0.9	−3.8	−5.4
Savings ratio	−0.5	−1.4	−1.1
Increments to growth rates			
Consumer prices	1.6	5.0	3.6
Export prices[a]	3.0	5.4	3.1
Import prices[a]	6.9	12.2	2.6
GNP deflator[a]	0.9	3.7	3.8
Increment to foreign balance relative to baseline[a]			
($ billion)	−39.1	−93.4	−61.5

a) National accounts basis; trade includes services: prices are implicit deflators.
b) Includes OECD area trade.
Source: "The Impact of Oil on the World Economy," OECD Observer, No. 105.

though it is not clear at this point in time if world consumption rates will fall in absolute terms or only relative to rates that would have prevailed at lower prices. In the industralized world, oil consumption in 1980 has actually declined relative to 1978, the year of peak consumption. In the developing countries, however, oil consumption is likely to rise further in order to support the essential process of economic development and growth, and to switch over from traditional to commercial energy-dependent, modern methods of production as discussed in the next section.

However, oil-substitution itself will generally require higher initial investment costs to provide the same level of service that would have been obtained with the use of oil. Hence this substitution process is not costless. One of the advantages of the use of oil is that most oil-using equipment or facilities generally have low capital costs relative to alternative energy supply systems. For example, the costs of a coal-fired power plant per kilowatt of capacity in the United States may be as much as 60 to 80% higher than the costs of a comparable oil-fired one. Costs for an atomic plant may be as much as four times higher. Although the average costs of electric energy will be lower for the atomic or coal fired alternatives over the life expectancies of the respective plants, the problem is that higher initial capital investments are needed per

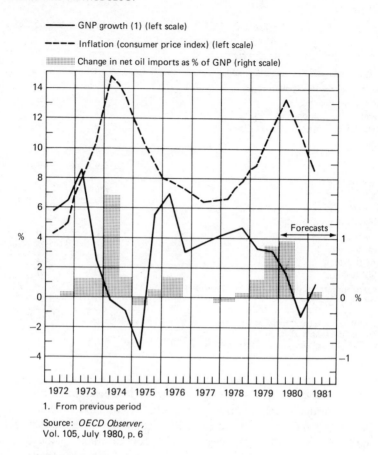

——— GNP growth (1) (left scale)

– – – – Inflation (consumer price index) (left scale)

░░░░░ Change in net oil imports as % of GNP (right scale)

1. From previous period

Source: *OECD Observer,*
Vol. 105, July 1980, p. 6

Figure 1.5.   Effects of the oil price hikes; comparison of 1973/74 with 1979/81 projections. Major seven OECD countries seasonally adjusted annual rates. Source: *OECD Observer* vol. 105, p. 6, July 1980.

unit of output. To bring about such substitution, therefore, will require a significant increase in the amount of investments relative to the amount of useful output produced. Higher savings and lower consumption rates will be needed to finance these shifts in the capital/output ratios of energy-producing and energy-consuming equipment and facilities.[23] These increases in capital requirements put a major strain on world capital markets.

---

[23]Capital deepening is also needed for new energy conservation measures ranging from better heat insulation and industrial retrofits to more fuel efficient transportation.

Because there is less time for dynamic adjustment, sudden increases in the world oil price are more detrimental to the economies of oil importers than a gradual price rise. While there is general agreement that oil prices will continue to rise (because it is a depleting resource), the rate of increase is less certain. Recent estimates range between 1 and 3% per year, on average, over the next two decades. However, greater conservation efforts by oil importers will help to limit world oil price increases (as described in the next section), whereas unforeseen supply disruption in oil producing areas could lead to sharp price rises. For large oil users like the U.S., several recent studies have explored the merits of a number of measures including oil stockpiling and high tariffs on imported oil to minimize the shock-like effects of future supply shortages.[24]

There will also have to be considerable adjustment in the oil refining sectors of many countries.[25] As more excess refining capacity comes on-stream in the OPEC countries during the 1980s, many less efficient refineries in other countries will have to be phased out. Also, as heavier crudes that yield relatively more of the heavier refinery fractions predominate over lighter crudes in the future, and demand for fuel oil drops relative to the lighter products, refineries will have to adapt, e.g., by installing cracking facilities (See Section 2.1).

1.6   DIFFICULTIES FACED BY THE OIL IMPORTING DEVELOPING COUNTRIES

The problems encountered by the oil importing developing countries (OIDCs) require special mention because these nations are the least well equipped to deal with the energy crisis.[26] Most of the oil importing developing countries such as Brazil and Korea financed their increased energy import bills by external borrowings following the 1973 oil price increase because they were unable to boost exports sufficiently to balance their foreign trade. A few like Taiwan sought to cut back sharply on domestic demand for oil. On the whole, the OIDCs succeeded in cushioning their economies from the oil price shock, and posted a 4.5 percent growth rate of real GNP, while the OECD countries were deep in recession. However, the increasing debt service burden of the OIDCs makes it unlikely that continued foreign financing of energy needs can be sustained indefinitely into the future. Between 1970 and 1978 their debt service

---

[24]Robert S. Pindyck, and Julio S. Rotenberg, "Energy Shocks and the Macroeconomy", Energy Laboratory Working Paper No. MIT-EL 82-043WP, MIT, Cambridge, Mass., July 1982; and Henry S. Rowen and John P. Weyant, "Reducing the Economic Impact of Oil Supply Interruptions: An International Perspective", *Energy Journal, vol. 3*, January 1982, pp. 1–34.

[25]Fereidun Fesharaki and David J. Isaak, *OPEC, the Gulf and the World Petroleum Market: A Study in Government Policy and Downstream Operation,* Westview Press, Colorado, 1982.

[26]For a list of these countries, see *World Development Report 1980,* The World Bank, Washington, D.C., August 1980, pp. 110–11.

as a percentage of GNP increased from 1.2 to 2.7% for low income and from 1.5 to 2.2% for middle income countries.[27] The future looks even bleaker for a number of them. For 12 major energy importing Asian countries, an Asian Development Bank energy survey predicts that energy imports as a percent of GNP will amount to almost 15% by 1990, as against a range of 3 to 7½% in 1978.[28]

The more detailed pattern of foreign exchange borrowings by OIDCs shown in Table 1.7 has been analysed in a recent paper.[29] Clearly, there was a sharp increase in borrowing after the 1973 oil price increase which gradually decreased up to 1978, only to rise again following the volatility of international oil markets during 1979–80. Many of these loans were obtained from private commercial banks, and therefore not under concessionary terms usually provided by international aid agencies. These rapid increases in short term borrowings have increased the debt service burden of many OIDCs to dangerous levels. For example, by 1985 the more industrialized middle income OIDC's are projected to spent about 29 percent of their export earnings to meet interest and amortization payments on their foreign debt.

Increased borrowing immediately after a sudden oil price increase is generally in the interest of the OIDCs. An economy is usually most rigid in the short run, with few possibilities for energy substitution in production and consumption. Therefore it makes sense to borrow more heavily in the initial period following the oil price increase, assuming that energy substitution and conservation programs will provide greater flexibility in later years. However, it is clear that the ability to withstand repeated energy shocks will be diminished if the magnitude and frequency of these price increases is too great to permit economies to adjust to new equilibria and allow debt service ratios to fall to acceptable levels. While the demand for foreign loans may not be very sensitive to the cost of borrowing (i.e., interest rates and terms of repayment), the supply of funds, especially from the commercial banks, will depend critically on creditworthiness. Already, preliminary results indicate that the OIDCs will not be able to cushion their economies from the effects of the most recent oil price shocks as well as they did in the post-1973 period.

Both the other direct and dynamic effects of higher energy costs discussed earlier will continue to reduce already low real incomes and living standards, decrease output and employment levels, and fuel inflation in the OIDCs. In the

---

[27]World Bank, *World Development Report* 1980, Table 13.
[28]T. L. Sankar and G. Schramm, *Asian Energy Problems,* Asian Development Bank, Praeger, New York, 1982, tables 9.2 and 9.9. The projections assumed an average annual increase of the price of oil of 3% up to 1990.
[29]Ricardo Martin and Marcelo Selowsky, "Energy Prices, Substitution, and Optimal Borrowing in the Short Run: An Analysis of Adjustment in Oil Importing Developing Countries.", Staff Working Paper No. 466, World Bank, Washington, D.C., July 1981.

Table 1.7. Borrowing of Oil Importing Developing Countries (as % of GNP) and Oil Prices.[a]

	1970–73	1975	1978	1980
Low income	1.9	3.8	2.7	3.6
Middle income	1.7	5.3	2.2	4.0
Total	1.7	5.1	2.3	3.9
OPEC petroleum prices in 1977 dollars (per barrel):	4.7	12.0	11.1	19.3

[a]From *World Development Report 1980* and *Price Prospects for Major Primary Commodities* (World Bank, January 1980).

next decade or so, energy substitution and conservation policies will help to alleviate these problems. However, the third world will still need increasing amounts of energy to sustain its drive for economic growth and industrialization. Thus, according to the World Bank, total commercial energy consumption in oil importing developing countries will increase from 12.4 million barrels a day of oil equivalent in 1980 to 22.8 million barrels by 1990.[30] Not all of this increase, of course, will be in the form of oil consumption. For example, for 19 Asian countries oil dependence is projected to decrease from some 78% in 1978 to 65% by 1990, even though in absolute terms oil consumption in these countries is expected to double.[31] As oil dependence declines relative to total GNP, the negative effects of high and rising oil prices will become less significant for income and employment.

As mentioned earlier, energy related investments in developing countries for developing new energy sources and for conservation programs will also be substantial.

According to estimates by the World Bank, for example, the energy-related investment needs of the developing countries as a whole are projected to rise from some 34.4 billion U.S. dollars in 1980 to approximately 82.2 billion dollars per year by 1990 (in 1980 U.S. dollars). This represents an average annual increase of 12.3% over the decade.[32]

In summary, solving the energy problems of the third world will be a formidable challenge. In the past, commercial banks have been willing to act as intermediaries by holding loans from OIDCs and issuing their own internationally recognized instruments to finance oil payments and energy related investments. Since this cannot continue indefinitely, greater direct flows of

[30]World Bank, *Energy in the Developing Countries,* Washington, D.C. August 1980, Table 1.1.
[31]T. L. Sankar and G. Schramm, p. XXIX.
[32]World Bank, *Energy in the Developing Countries,* Table 3.

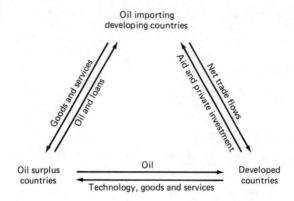

Figure 1.6.   Energy and economic links between oil surplus, developed and oil importing developing countries.

petrodollars from the oil-surplus countries to the OIDCs in the form of concessionary loans for investments, etc. may be required. More aid from both multilateral and bilateral sources as well as higher levels of private foreign investment will be required. The developed countries should also be willing to run greater trade deficits with OIDCs. The broad outlines of a possible triangular relation between oil exporters, OIDCs, and developed countries is given in Figure 1.6.

We conclude by noting that any evolution of international economic relationships will depend on many difficult and unknown political factors. Energy will be only one issue, albeit an important one, to be thrashed out within some broader framework such as the North-South dialogue and the new international economic order (NIEO) discussions.[33] It is very clear, however, that whatever the outcome of international developments, the more efficient use of energy, and demand management and conservation policies pursued within individual countries, will play an important role in reducing the adverse effects of the energy crisis. Most of this book is dedicated to discussing exactly how such practical policies may be developed and applied.

---

[33]See for example: Roger D. Hansen, "North South Policy—What's the Problem?" *Foreign Affairs, Vol. 58,* Summer 1980, pp. 1104–1128; and Mahbub ul Haq, "Negotiating the Future," *Foreign Affairs, Vol. 59,* Winter 1980/81, pp. 398–417.

Table A.1.1. World Tables: Population, GNP/Capita, Energy Consumption and Imports, 1960–1978.

	POPULATION (millions) MID-1979	GNP PER CAPITA DOLLARS 1979	GNP PER CAPITA AVERAGE ANNUAL GROWTH (percent) 1960–79	AVERAGE ANNUAL RATE OF INFLATION (percent) 1960–70[a]	AVERAGE ANNUAL RATE OF INFLATION (percent) 1970–79	ENERGY PRODUCTION 1960–74[a]	ENERGY PRODUCTION 1974–79	ENERGY CONSUMPTION 1960–74	ENERGY CONSUMPTION 1974–79	ENERGY CONSUMPTION PER CAPITA (kilograms of coal equivalent) 1960	1979	ENERGY IMPORTS AS A PERCENTAGE OF MERCHANDISE EXPORTS 1960[b]	1978[c]
Low-income countries	**2,260. t**	**230 w**	**1.6 w**	**3.0 m**	**10.8 m**	**5.2 w**	**8.4 w**	**4.4 w**	**8.1 w**	**356 w**	**463 w**	**8 w**	**..**
China and India	**1,623.7 t**	**230 w**	**..**	**..**	**..**	**4.5 w**	**8.8 w**	**4.3 w**	**8.5 w**	**439 w**	**594 w**	**..**	**..**
Other low-Income	**636.5 t**	**240 w**	**1.8 w**	**3.0 m**	**10.9 m**	**9.5 w**	**6.8 w**	**6.4 w**	**4.5 w**	**86 w**	**129 w**	**7 w**	**..**
1 Kampuchea, Dem.	..	..	..	3.8	..	..	..	−0.7	−38.9	32	..	9	..
2 Lao PDR	3.3	..	..	..	..	..	16.1	13.7	13.6	17	102	..	..
3 Bhutan	1.3	80	−0.1	..	..	..	..	..	..	..	..	..	..
4 Bangladesh	88.9	90	−0.1	3.7	15.8	..	10.1	..	6.3	..	41	..	35
5 Chad	4.4	110	−1.4	4.6	7.9	..	..	7.5	4.6	8	24	23	..
6 Ethiopia	30.9	130	1.3	2.1	4.3	14.1	2.3	14.0	−5.3	9	20	11	20
7 Nepal	14.0	130	0.2	7.7	8.7	26.8	4.6	12.4	2.3	5	14	..	..
8 Somalia	3.8	..	−0.5	4.5	11.3	..	..	8.7	13.0	17	78	4	..
9 Mali	6.8	140	1.1	5.0	9.7	..	8.3	5.6	5.3	15	30	13	..
10 Burma	32.9	160	1.1	2.7	12.1	5.6	12.4	3.7	5.6	58	72	4	..
11 Afghanistan	15.5	170	0.5	11.9	4.4	38.8	−2.8	10.1	6.6	24	90	12	12
12 Viet Nam	52.9	..	..	..	..	..	7.6	11.2	−4.0	99	140	..	..
13 Burundi	4.0	180	2.1	2.8	11.2	..	22.0	..	6.9	..	17	..	..
14 Upper Volta	5.6	180	0.3	1.3	9.8	..	..	7.7	10.2	5	29	38	..
15 India	659.2	190	1.4	7.1	7.8	4.9	9.1	5.1	8.3	111	242	11	27
16 Malawi	5.8	200	2.9	2.4	9.1	..	6.9	..	5.7	..	70	..	22
17 Rwanda	4.9	200	1.5	13.1	14.6	..	3.5	..	10.4	..	30	..	..
18 Sri Lanka	14.5	230	2.2	1.8	12.3	10.1	8.2	3.9	3.8	114	140	8	18
19 Benin	3.4	250	0.6	1.9	9.2	..	..	9.5	−0.6	40	68	16	..
20 Mozambique	10.2	250	0.1	2.8	11.0	3.2	60.0	5.2	1.1	113	139	11	..

Table A.1.1. World Tables: Population, GNP/Capita, Energy Consumption and Imports, 1960–1978. (Continued)

	POPULATION (millions) MID-1979	GNP PER CAPITA DOLLARS 1979	GNP PER CAPITA AVERAGE ANNUAL GROWTH (percent) 1960-79	AVERAGE ANNUAL RATE OF INFLATION (percent) 1960-70[a]	AVERAGE ANNUAL RATE OF INFLATION (percent) 1970-79	ENERGY PRODUCTION 1960-74[a]	ENERGY PRODUCTION 1974-79	ENERGY CONSUMPTION 1960-74	ENERGY CONSUMPTION 1974-79	ENERGY CONSUMPTION PER CAPITA (kilograms of coal equivalent) 1960	ENERGY CONSUMPTION PER CAPITA 1979	ENERGY IMPORTS AS A PERCENTAGE OF MERCHANDISE EXPORTS 1960[b]	ENERGY IMPORTS 1978[c]
21 Sierra Leone	3.4	250	0.4	2.9	11.3	..	..	9.0	-1.1	31	89	11	..
22 China	964.5	260	..			4.4	8.7	4.1	8.5	650	835	..	..
23 Haiti	4.9	260	0.3	4.1	10.9	..	13.7	1.4	20.8	36	66	..	*16*
24 Pakistan	79.7	260	2.9	3.3	13.9	9.3	7.5	5.3	5.0	136	218	*17*	40
25 Tanzania	18.0	260	2.3	1.8	13.0	10.6	10.4	9.4	-2.9	43	53	..	..
26 Zaire	27.5	260	0.7	29.9	31.4	3.0	18.1	3.8	0.4	98	103	3	..
27 Niger	5.2	270	-1.3	2.1	10.8	*16.0*	(.)	14.8	12.8	6	48	6	..
28 Guinea	5.3	280	0.3	1.5	4.4			3.2	1.6	67	87	7	..
29 Central African Rep.	2.0	290	0.7	4.1	9.1	14.1	4.1	7.6	8.5	38	55	12	1
30 Madagascar	8.5	290	-0.4	3.2	10.1	6.7	4.1	9.0	3.9	40	94	9	16
31 Uganda	12.8	290	-0.2	3.0	28.3	5.2	-4.4	9.1	-8.2	43	39	5	..
32 Mauritania	1.6	320	1.9	1.6	10.1	..	..	21.1	5.5	18	185	39	..
33 Lesotho	1.3	340	6.0	2.5	11.6					..		..	..
34 Togo	2.4	350	3.6	1.1	10.3	..	22.3	12.7	11.8	23	117	10	*13*
35 Indonesia	142.9	370	4.1	..	20.1	8.5	6.5	3.8	10.1	130	237	3	5
36 Sudan	17.9	370	0.6	3.7	6.8	..	13.7	13.1	-0.9	54	141	8	*24*

Middle-income countries	985.0 t	1,420 w	3.8 w	3.0 m	13.3 m	12.7 w	−0.5 w	8.4 w	6.3 w	509 w	1,225 w	10 w	20 w
Oil exporters	324.8t	1,120w	3.1w	3.0m	14.0m	15.0w	−2.1w	9.0w	6.1w	362w	893w	5w	10w
Oil importers	660 2t	1,550w	4.1w	3.0m	12.2m	6.5w	3.8w	8.2w	6.4w	576w	1,388w	13w	24w
37 Kenya	15.3	380	2.7	1.5	11.1	9.6	17.6	3.3	3.5	150	180	18	30
38 Ghana	11.3	400	−0.8	7.6	32.4	..	2.6	12.2	2.3	105	265	7	19
39 Yemen Arab Rep.	5.7	420	10.9	..	17.8	..	..	12.8	15.8	7	73	..	..
40 Senegal	5.5	430	−0.2	1.7	7.6	..	..	4.7	12.4	110	266	8	..
41 Angola	6.9	440	−2.1	3.3	21.6	35.5	−2.4	10.3	1.1	90	208	6	..
42 Zimbabwe	7.1	470	0.8	1.3	8.4	2.5	−3.1	2.4	−0.3	1,346	791	..	..
43 Egypt	38.9	480	3.4	2.7	8.0	9.4	27.1	3.6	10.3	299	565	12	6
44 Yemen, PDR	1.9	480	11.8	..	..	..	..	7.6	7.0	237	545	..	..
45 Liberia	1.8	500	1.6	1.9	9.4	31.8	−1.3	18.9	−0.9	88	448	3	17
46 Zambia	5.6	500	0.8	7.6	6.8	..	5.1	..	5.2	..	858	..	11
47 Honduras	3.6	530	1.1	2.9	8.4	29.4	6.4	7.7	1.5	157	248	10	14
48 Bolivia	5.4	550	2.2	3.5	32.4	17.1	−3.0	6.8	9.3	185	470	4	1
49 Cameroon	8.2	560	2.5	4.2	10.3	1.1	45.3	6.2	7.8	87	148	7	9
50 Thailand	45.5	590	4.6	1.8	9.5	28.2	0.8	16.2	7.6	63	376	12	28
51 Philippines	46.7	600	2.6	5.8	13.3	2.4	24.9	8.3	5.6	159	356	9	32
52 Congo. People's Rep.	1.5	630	0.9	5.4	10.9	15.8	5.1	5.3	7.0	125	213	25	1
53 Nicaragua	2.6	660	1.6	1.9	12.2	26.4	−16.3	10.3	2.7	183	455	12	14
54 Papua New Guinea	2.9	660	2.8	3.6	9.5	12.3	16.2	16.4	4.9	51	299	7	..
55 El Salvador	4.4	670	2.0	0.5	10.8	5.1	15.6	7.7	8.3	150	351	6	13
56 Nigeria	82.6	670	3.7	2.6	19.0	36.6	1.0	9.4	1.4	29	83	7	2
57 Peru	17.1	730	1.7	10.4	26.8	3.5	18.5	6.5	2.7	436	737	4	20
58 Morocco	19.5	740	2.6	2.0	7.3	2.0	4.7	6.4	6.4	169	315	9	28
59 Mongolia	1.6	780	3.0	..	..	10.4	14.6	7.3	13.1	553	1,667	..	..
60 Albania	2.7	840	4.2	..	..	9.7	5.0	11.3	8.6	327	1,103	..	..
61 Dominican Rep.	5.3	990	3.4	2.1	8.4	1.8	−5.1	14.4	−1.0	164	515	..	32

Table A.1.1. World Tables: Population, GNP/Capita, Energy Consumption and Imports, 1960–1978. (Continued)

| | POPULATION (millions) MID-1979 | GNP PER CAPITA | | AVERAGE ANNUAL RATE OF INFLATION (percent) | | AVERAGE ANNUAL GROWTH RATE (percent) | | | | ENERGY CONSUMPTION PER CAPITA (kilograms of coal equivalent) | | ENERGY IMPORTS AS A PERCENTAGE OF MERCHANDISE EXPORTS | |
| | | DOLLARS 1979 | AVERAGE ANNUAL GROWTH (percent) 1960–79 | 1960–70[a] | 1970–79 | ENERGY PRODUCTION | | ENERGY CONSUMPTION | | 1960 | 1979 | 1960[b] | 1978[c] |
						1960–74[a]	1974–79	1960–74	1974–79				
62 Colombia	26.1	1,010	3.0	11.9	21.5	3.5	2.0	5.7	7.0	510	938	3	7
63 Guatemala	6.8	1,020	2.9	0.1	10.6	9.9	2.5	6.2	1.6	175	251	12	14
64 Syrian Arab Rep.	8.6	1,030	4.0	1.9	12.7	86.2	7.5	7.5	15.2	323	971	16	..
65 Ivory Coast	8.2	1,040	2.4	2.8	13.5	9.7	−12.2	14.3	5.5	75	234	5	10
66 Ecuador	8.1	1,050	4.3	..	14.7	19.4	5.0	8.7	14.9	208	654	2	1
67 Paraguay	3.0	1,070	2.8	3.1	9.3	..	6.7	8.2	10.7	85	251	..	..
68 Tunisia	6.2	1,120	4.8	3.7	7.5	72.1	5.5	8.7	10.8	173	618	15	21
69 Korea, Dem. Rep.	17.5	1,130	3.5	..	..	9.4	3.0	9.3	3.6	1,193	2,846	..	..
70 Jordan	3.1	1,180	5.6	..	..	..	..	5.9	13.3	197	552	79	52
71 Lebanon	2.7	..	..	1.4	..	12.7	0.5	8.6	−3.7	567	1,083	68	..
72 Jamaica	2.2	1,260	1.7	3.9	17.4	−0.7	−2.0	11.0	−5.4	446	1,390	11	14
73 Turkey	44.2	1,330	3.8	5.6	24.6	7.6	3.1	9.8	7.0	254	807	16	63
74 Malaysia	13.1	1,370	4.0	−0.3	7.3	37.3	27.2	10.5	4.1	253	767	2	9
75 Panama	1.8	1,400	3.1	1.6	7.4	14.7	35.9	9.0	4.3	438	947	..	91
76 Cuba	9.8	1,410	4.4	..	..	21.2	5.6	4.5	6.0	896	1,148	..	..
77 Korea, Rep. of	37.8	1,480	7.1	17.5	19.5	6.3	4.2	13.0	11.4	261	1,642	70	19
78 Algeria	18.2	1,590	2.4	2.3	13.3	11.1	6.5	7.1	12.3	277	671	14	2
79 Mexico	65.5	1,640	2.7	3.6	18.3	5.8	15.5	7.7	7.8	769	1,673	3	4
80 Chile	10.9	1,690	1.2	32.9	242.6	3.9	0.1	6.1	0.7	824	1,193	10	18
81 South Africa	28.5	1,720	2.3	3.0	11.8	3.8	8.1	5.0	4.4	2,320	3,479	9	..
82 Brazil	116.5	1,780	4.8	46.1	32.4	8.2	7.5	8.2	7.7	392	1,062	21	39
83 Costa Rica	2.2	1,820	3.4	1.9	15.4	9.5	3.5	10.1	7.6	315	842	7	13
84 Romania	22.1	1,900	9.2	−0.2	0.8	5.8	3.1	8.2	6.9	1,469	4,810	..	..

85 Uruguay	2.9	2,100	0.9	51.1	64.0	3.7	8.5	2.8	3.4	895	1,274	35	34
86 Iran	37.0	..	..	-0.5	..	14.6	-9.1	15.5	1.4	270	1,214	1	..
87 Portugal	9.8	2,180	5.5	3.0	16.1	4.4	11.7	7.4	6.0	473	1,496	17	34
88 Argentina	27.3	2,230	2.4	21.7	128.2	6.5	3.7	5.5	3.1	1,110	2,038	14	17
89 Yugoslavia	22.1	2,430	5.4	12.6	17.8	4.9	4.1	7.2	5.2	875	2,400	8	25
90 Venezuela	14.5	3,120	2.7	1.3	10.4	1.1	-3.3	7.0	5.4	1,615	3,055	1	22
91 Trinidad and Tobago	1.2	3,390	2.4	3.2	19.5	2.8	3.9	10.2	5.8	1,747	5,037	35	39
92 Hong Kong	5.0	3,760	7.0	2.4	7.9	..	..	9.6	16.7	468	2,401	5	6
93 Singapore	2.4	3,830	7.4	1.1	5.5	..	..	13.4	17.1	518	6,211	17	31
94 Greece	9.3	3,960	5.9	3.2	14.1	14.3	19.1	12.8	9.6	424	2,841	26	42
95 Israel	3.8	4,150	4.0	6.2	34.3	41.8	-62.3	11.6	4.7	1,270	3,643	17	20
96 Spain	37.0	4,380	4.7	8.2	15.9	2.6	6.0	8.8	3.8	892	2,822	22	40
Industrial market economies	671.2 t	9,440 w	4.0 w	4.3 m	9.4 m	4.1 w	2.3 w	5.3 w	2.5 w	4,486 w	7,892 w	11 w	20 w
97 Ireland	3.3	4,210	3.2	5.2	14.6	0.1	-1.2	4.9	4.3	1,922	3,819	17	13
98 Italy	56.8	5,250	3.6	4.4	15.6	2.3	0.9	7.8	1.4	1,317	3,438	18	24
99 New Zealand	3.2	5,930	1.9	3.3	12.3	5.7	5.6	6.0	1.7	2,699	4,891	7	13
100 United Kingdom	55.9	6,320	2.2	4.1	13.9	-1.0	13.5	2.0	1.0	4,489	5,637	14	13
101 Finland	4.8	8,160	4.1	5.6	12.9	3.3	2.9	8.7	2.4	1,925	6,259	11	20
102 Austria	7.5	8,630	4.1	3.7	6.5	1.4	0.4	5.0	2.6	2,523	5,206	12	14
103 Japan	115.7	8,810	9.4	4.9	8.2	-1.7	3.4	9.7	3.0	1,333	4,260	18	32
104 Australia	14.3	9,120	2.8	3.1	11.7	10.9	4.9	5.6	2.8	3,935	6,975	12	9
105 Canada	23.7	9,640	3.5	3.1	9.1	8.7	1.7	6.2	3.1	7,087	13,453	9	9
106 France	53.4	9,950	4.0	4.2	9.6	-1.3	2.9	5.5	2.3	2,674	4,995	16	21
107 Netherlands	14.0	10,230	3.4	5.4	8.3	16.1	0.3	9.0	2.7	2,500	6,745	15	16
108 United States	223.6	10,630	2.4	2.8	6.9	3.5	1.0	4.4	2.3	8,228	12,350	8	31
109 Norway	4.1	10,700	3.5	4.3	8.2	6.8	22.1	5.8	5.1	4,938	11,919	15	13
110 Belgium	9.8	10,920	3.9	3.6	8.1	-7.2	5.2	4.2	2.0	3,846	6,745	11	13
111 Germany, Fed. Rep.	61.2	11,730	3.3	3.2	5.3	0.3	4.9	6.0	4.3	2,711	6,627	7	14

Table A.1.1. World Tables: Population, GNP/Capita, Energy Consumption and Imports, 1960–1978. (Continued)

	POPULATION (millions) MID-1979	GNP PER CAPITA		AVERAGE ANNUAL RATE OF INFLATION (percent)		AVERAGE ANNUAL GROWTH RATE (percent)				ENERGY CONSUMPTION PER CAPITA (kilograms of coal equivalent)		ENERGY IMPORTS AS A PERCENTAGE OF MERCHANDISE EXPORTS	
		DOLLARS 1979	AVERAGE ANNUAL GROWTH (percent) 1960-79	1960-70[a]	1970-79	ENERGY PRODUCTION		ENERGY CONSUMPTION		1960	1979	1960[b]	1978[c]
						1960-74[a]	1974-79	1960-74	1974-79				
112 Denmark	5.1	11,900	3.4	5.5	9.8	-20.4	39.5	8.1	0.8	2,767	5,978	15	20
113 Sweden	8.3	11,930	2.4	4.4	9.8	3.6	6.0	4.7	2.5	4,599	8,502	16	15
114 Switzerland	6.5	13,920	2.1	4.4	5.4	4.2	2.7	5.5	1.9	2,762	5,138	10	8
Capital-surplus oil exporters	**25.4 t**	**5,470 w**	**5.0 w**	**1.7 m**	**18.2 m**	**12.7 w**	**4.0 w**	**7.6 w**	**10.4 w**	**771 w**	**1,458 w**	**..**	**(·) w**
115 Iraq	12.6	2,410	4.6	1.7	14.1	5.0	9.2	6.0	2.6	494	692	(·)	(·)
116 Saudi Arabia	8.6	7,280	6.3	..	25.2	14.0	3.6	9.3	14.3	741	1,554	..	(·)
117 Libya	2.9	8,170	5.8	5.2	18.7	29.1	6.9	16.7	27.2	251	2,360	83	(·)
118 Kuwait	1.3	17,100	-1.6	0.6	17.7	4.5	-0.2	4.0	9.2	10,584	6,348	..	(·)
Nonmarket industrial economies	**351.2t**	**4,230w**	**4.3w**	**..**	**..**	**5.3w**	**4.7w**	**5.2w**	**3.9w**	**2,990w**	**6,164w**	**..**	**..**
119 Bulgaria	9.0	3,690	5.6	..	..	3.3	2.0	9.7	4.1	1,366	5,403	..	..
120 Poland	35.4	3,830	5.2	..	..	3.9	4.2	4.5	2.6	3,115	5,803	7	..
121 Hungary	10.7	3,850	4.8	..	..	2.6	3.7	4.7	4.8	1,732	4,073	13	14
122 USSR	264.1	4,110	4.1	..	..	5.9	5.2	5.2	4.4	2,866	6,122	4	..
123 Czechoslovakia	15.2	5,290	4.1	..	..	1.4	-3.3	3.2	-0.4	4,509	6,830	..	18
124 German Dem. Rep.	16.8	6,430	4.7	..	..	0.6	5.3	6.0	4.7	4,579	8,718	..	..

Source: World Bank, World Development Report 1981, Washington, D.C., Aug. 1981, Tables 1 and 7.
a. Figures in italics are for 1961-74, not 1960-74. b. Figures in italics are for 1961, not 1960. c. Figures in italics are for 1977, not 1978; t = totals; w = weighted averages.

Table A.1.2.
WORLD PETROLEUM PRICES
(US$/barrel)

YEAR	SAUDI ARABIAN (POSTED PRICE)[1]		SAUDI ARABIAN (REALIZED PRICE)[2]	
	CURRENT $	1980 CONSTANT $	CURRENT $	1980 CONSTANT $
1950	1.71	8.30	1.71	8.30
1951	1.71	6.95	1.71	6.95
1952	1.71	6.79	1.71	6.79
1953	1.84	7.67	1.84	7.67
1954	1.93	8.18	1.93	8.18
1955	1.93	8.04	1.93	8.04
1956	1.93	7.85	1.93	7.85
1957	2.01	7.82	1.86	7.24
1958	2.08	7.68	1.83	6.75
1959	1.92	7.44	1.56	6.05
1960	1.86	7.05	1.50	5.68
1961	1.80	6.79	1.45	5.47
1962	1.80	6.87	1.42	5.42
1963	1.80	6.82	1.40	5.30
1964	1.80	6.72	1.33	4.96
1965	1.80	6.55	1.33	4.84
1966	1.80	6.41	1.33	4.73
1967	1.80	6.32	1.33	4.67
1968	1.80	6.74	1.30	4.87
1969	1.80	6.72	1.28	4.78
1970	1.80	6.06	1.30	4.38
1971	2.21	6.88	1.65	5.14
1972	2.47	7.00	1.90	5.38
1973	3.30	7.84	2.70	6.41
1974	11.59	22.20	9.78	18.74
1975	11.53	19.28	10.72	17.93
1976	12.38	20.36	11.51	18.93
1977	13.33	20.20	12.40	18.79
1978	13.66	17.49	12.70	16.26
1979	17.28	19.31	17.00	18.99
1980	31.03	31.03	28.50	28.50
1981 Jan.–June	34.41		32.00	

[1]Light Crude oil, 43°–34.9° API gravity, f.o.b. Ras Tanura.
[2]Light Crude oil, 34°–34.9° API gravity, average realized price, f.o.b. Ras Tanura.
Source: World Bank, *Commodity Trade and Price Trends (1981 ed.)*. Washington, D.C., Aug. 1981.

2
Energy Sources, Uses, and Substitution Possibilities

Chapter 2 contains an overview of the most significant sources of energy, their most important uses, and an analysis of the reasons for the differences in energy use intensities among different countries and different uses. It concludes with an examination of the short-run and long-run potentials for energy substitutions, either by other energy sources, by labor, by capital, or by changed habits and consumption patterns.

Inanimate energy is utilized in three basic forms: as heat, as mechanical energy, and directly as electric energy.[1] Most of the energy used in the world today is the result of some form of combustion process, i.e., the burning of a fuel and the subsequent utilization of the resulting heat or pressure or both. In the United States, for example, combustion processes supplied some 93.2% of all energy uses, followed by hydropower with 4.1% and nuclear energy with 2.7% (1976 data). Other sources of energy, such as solar and wind, are quantitatively insignificant at the present time.

But while combustion processes form the backbone of our energy sources, such processes consist of many different forms that are not freely interchangeable with each other. Wood, coal, or some liquid hydrocarbon fuel can be burned interchangeably in an open fire to cook a meal, but even in a simple enclosed furnace these fuels can no longer be substituted for each other in the absence of special grates for solid fuels and burners for liquid ones. Most energy using equipment and appliances, such as furnaces, boilers, or heat engines are fuel specific. They are designed for use with a specialized fuel only and usually cannot be easily converted to utilize another. Hence, the management of energy resources and energy uses must be broken into fuel-specific and use-specific components, and considerable attention must be given to the limits of practical fuel substitutions for given tasks.

Our main fuel sources are petroleum products, coal, natural gas, and renewable biomass (alcohol, wood, dung). Nonhydrocarbon based energy sources are

---

[1] For electrolytic processing, lighting, telecommunications, etc.

nuclear fuels, hydro, geothermal, solar, wind, tidal power, ocean currents, and ocean temperature gradients. Of the latter, only hydro, nuclear, and geothermal are of some quantitative significance at the present.

2.1   PETROLEUM PRODUCTS

Over the course of this century petroleum products have become the most important sources of commercial energy throughout the world. This is the result of the significant advantages they offer in many uses compared to other sources of energy: Their energy density per unit of volume is higher than those of any other fuel except nuclear (see Table 2.1); they can be easily transported in nonpressurized containers; their handling and transshipping cost are low because they can be easily pumped from one container to another or moved by pipelines; they do not deteriorate when exposed to open air; when "used" they can be metered and injected with great precision into a firing chamber or a furnace to achieve optimal combustion conditions. For stationary uses (e.g., power production or industrial furnaces) oil creates fewer pollution problems than coal, and the pollution abatement costs of oil burning furnaces and appliances are lower. Until 1973/74 the costs of petroleum products that competed directly with other fuels such as coal were quite favorable. For example, the costs of heavy fuel oils at the U.S. East Coast ranged between $3 and $4 per million kcalories, about equal to the average cost of coal delivered to the same destination.

Almost all of the petroleum products utilized today originate from crude oil deposits which are found in sedimentary rocks of various depths. Presently known deposits are highly unevenly distributed, with close to 60% of them located in the Middle East (see Table 1.3).

Crude oil is usually composed of a wide range of paraffins, olefins, and aromatic hydrocarbons.[2] While several thousand different hydrocarbons are known to exist, crude deposits generally are composed of varying mixtures of a few dozen main components ranging from dissolved methane (a gas) with a boiling point of $-150°C$ through octane to heavy paraffin-wax and tar-like substances that are normally solid at room temperature. In addition, most deposits contain various contaminants such as sulfur, heavy metals, and others. These may have to be removed before processing. Admixtures of gaseous components have to be separated out prior to shipping via pipeline or other means.

In deposits that contain large amounts of dissolved gas (i.e., associated gas) the separated gas must either be removed by pressurized pipeline to suitable

---

[2]Paraffins are saturated, and olefins unsaturated fatty hydrocarbons. Aromatic hydrocarbons have ring-shaped molecules derived from the benzene ring C_6H_6.

Table 2.1. Heat Values of Different Fuels Per Unit Weight and/or Volume.

FUEL	HEAT VALUE (kcal/unit)	kcal/liter	kcal/kg
Natural gas (methane)	252/scf.	9.1	—
Hydro-Electric Energy	2,979/kWh.	—	—
Crude Oil	—	8,050	9,100
L.P.G.	—	6,358	11,154
Gasoline	—	8,398	11,197
Diesel	—	9,371	11,025
Kerosene	—	9,280	11,457
Jet Fuel (J.P.4)	—	8,354	10,440
Light Fuel Oil	—	9,371	10,185
Anthracite	—	—	7,950
Heavy Fuel Oil	—	9,826	9,925
Bituminous Coal (avg.)	—	—	6,950
Sub-Bituminous Coal (avg.)	—	—	5,900
Lignite	—	—	3,500
Fuel Wood[1]	405,720/m^3	—	680
Charcoal[2]	1,363,320/m^3	—	6,000
Paddy Husk	1,401,120/metric ton	—	1,401
Bagasse	1,910,000/metric ton	—	1,910
Peat	1,890/kg	—	1,890
Coke	6,690/kg	—	6,690

Sources: David Crabbe and Richard McBride, The World Energy Book, MIT Press, Cambridge, MA 1979; Oil and Thailand 1980, National Energy Administration, Bangkok, Thailand, 1981.
[1]Assuming 600 kg of roundwood per m^3; [2] Assuming 227 kg per m^3.

markets, reinjected into the substrata for pressure maintenance of the deposit, or be burned off. The amount of gas thus burned is still very large; worldwide it amounted to 40% of total associated gas production in 1978 (see Table 2.2).

Before being used, crude oils are broken down or refined into their various major constituents. The basic refining process consists of fractioned distillation which is performed by simply heating the crude oil in a large still and drawing off the various components as they vaporize at various temperatures. The distilling process can only single out the different fractions contained in the crude oil, without being able to influence the percentage in which they occur naturally.[3]

Crude oil compositions and existing demands for the various components are rarely matched. Hence it is desirable to change the composition of the outputs. This can be done by cracking. Cracking is a process that splits the heavier

---

[3]For further details, see, for example: Hans Thirring, Energy for Man, Harper and Row, New York, 1976, Chapter 9.

Table 2.2. World Associated Gas Production, 1978 (million barrels of oil equivalent).

	PRODUCTION	FLARED	PERCENTAGE OF PRODUCTION FLARED
Africa	330	265	80
Asia/Middle East	1,300	770	59
Latin America	390	120	31
Eastern Europe	620	130	21
North America	770	40	5
Western Europe	80	60	75
Total	3,490	1,385	40

Source: World Bank, Energy In The Developing Countries, Washington, D.C., August 1980, Table 15.

molecules contained in the mixture into lighter ones. Three basic cracking processes are used: thermal cracking, which essentially consists of heating under high pressure of up to 98 atm and temperatures of 482° to 538°C. This process yields mostly additional gasoline fractions. Catalytic cracking consists of the addition of a catalyst (such as silicon and aluminum oxides) to the process. Catalytic cracking has the advantage over thermal cracking in that it not only yields more gasoline but also gasoline with higher octane ratings, which improves engine performance. This allows the use of engines with higher compression ratios that are more fuel efficient.

Finally, a third process consists of hydro-cracking, a process that adds hydrogen under high temperatures (400°C) and pressures (150 atm).[4] Hydro-cracking not only produces light gasoline fractions but also heavier middle distillates such as diesel, kerosene, jet fuels, etc., which are in particular demand in many of the developing nations. Hydro-cracking at U.S. $5,000 per daily barrel of capacity, is far more costly than thermal cracking, which requires investments of about $1,000 per daily barrel.[5]

Cracking processes consume considerably more energy than simple distillation processes. The latter require about 2% of the energy content of the fuel. Thermal cracking requires about 13% and catalytic cracking 18% or more.[6]

Other nonconventional sources of petroleum products are heavy oils, tar sands, shale oils, and synthetic gasoline from the hydrogenation of coal. Basic

---

[4]David Crabbe and Richard McBride, p. 97.
[5]World Bank, Energy in Developing Countries, p. 25.
[6]Hans Thirring, p. 203.

deposits of these resources are very large—considerably larger than the known deposits of crude oil. This is apparent from the data in Table 2.3 which show that the so-called proved, recoverable reserves of oil shales and bituminous sands contain some 86,300 million tons of hydrocarbons, or about the same amount as the existing conventional crude oil deposits (see Table 1.3). However, if estimated additional resources are taken into account, nonconventional hydrocarbon resources of some 353,000 million tons are substantially larger than the estimated 234,000 million tons of conventional reserves. Furthermore, as one recent survey points out: "When one attempts to estimate the total quantities, i.e., the sum of proved reserves and additional resources, one should taken into account the fact that the figures obtained from the questionnaire are very incomplete. According to an inquiry made by Depraires, the potential is more than 700×10^9 tons and is thus several times greater than the expectations from conventional resources."[7]

The major problem confronting the use of these unconventional oil resources is that of costs. While at today's market prices oil recovery and processing from Alberta's tar sands appears economically feasible, it must be remembered that production costs from conventional resources range between less than \$1/bl to \$8/bl in expensive off-shore locations, while tar-sand recovery and processing costs are well in excess of \$20/bl, in 1980 constant prices. Similar cost ranges apply to heavy oil recovery processes and to shale oil production. At present, conventional known oil resources are sufficient to supply world demand for another 25–30 years. Any serious inroads by nonconventional production, therefore, may simply tend to reduce the selling price of existing conventional producers. This, probably, is the major factor explaining the cautious development of nonconventional resources until now.

At present, heavy oils are recovered through steam injection processes in California and Saskatchewan. Tar sands production in Canada is about 160,000 bls daily; the construction of two or three additional plants with 100 to 150 thousand bls/day each is planned. Construction of the first, commercial, 50,000 bl/day shale oil plant in the United States has just been announced. In the USSR crushed oil shale is used directly as a power plant fuel. Venezuela plans to produce some 5,000 bls/day of heavy oils by 1985 and about 100,000 bls/day by the year 2,000.[8] South Africa produces about 50,000 bls/day from coal in its Sassol plants. Additional capacity is being added at present. New Zealand has a gasoline plant under construction that will use off-shore natural gas as its raw material base. The ·plant capacity is 50,000 bls/day. Compared to the approximate daily production of 51 million barrels/day from conven-

---

[7]World Energy Conference, *Survey of Energy Resources, op. cit.*, p. 156.
[8]World Energy Conference, *Survey of Energy Resources, 1980*, p. 158.

Table 2.3. The Availability of Oil from Oil Shales and Bituminous Sands. Figures in 10^6 Tons of Recoverable Oil.

	OIL SHALES				BITUMINOUS SANDS**		
	PROVED RECOV. RESERVES 1.1.1979	ADDITIONAL RESOURCES 1.1.1979	PRODUCTION 1978	PROVED RECOV. RESERVES 1.1.1979	ADDITIONAL RESOURCES 1.1.1979	PRODUCTION 1978
Canada				19300	16300	4
USA	28000	236000		1		
Australia		490				
New Zealand	1					
Thailand	2015					
USSR	6820	49180	37			
Argentina	*)					
Brazil	86		*)			
Venezuela				20000	50000	
Jordan	800	7000		700	10000	
Marocco	7400					
Zaire	*)					
Federal Republic of Germany	250			50		
Austria			*)			
Sweden	880					
Spain	12					
	46262	292670	37	40051	76300	4

*Figures under 1 Mton.
**_Turkey_ quotes 202 Mton recoverable reserves and 5196 Mton additional resources of Bitumen and Asphalt. In 1978 the production was 465,000 ton.
Venezuela quotes an additional 2000 Mton of proved recoverable reserves and 3000 Mton additional resources of Asphalt.
Source: World Energy Conference, _Survey of Energy Resources 1980_, p. 155.

tional resources in the early 1980s these existing and announced production plans from nonconventional resources are quite modest so far.

Refineries processing crude oils are usually built to satisfy existing and projected product demands in their general marketing area. The range and specific quantities of their outputs produced are a function of their capacity, the characteristics of their crude inputs and the type of processing equipment available (for distillation and/or cracking plus auxiliary equipment for sulfur removal,

Table 2.4. Typical Refinery Output Ranges: USA and a
Developing Country (1978–9).

PRODUCT	U.S. AVERAGE	DEVELOPING COUNTRY (THAILAND)	
		WITH CRACKING FACILITIES	WITHOUT CRACKING FACILITIES
LPG and gases	5	—	—
Gasoline	44	29	15
Diesel oil	23	36	18
Kerosene	2	3	4
Jet fuel	7	12	2
Fuel oil	13	16	60
Bitumen and coke	6	3	1

Sources: T. L. Sankar and G. Schramm, p. 73; and USDOE, *Energy Data Reports*, Washington, D.C., Dec. 1980.

etc.). Most refineries are designed to utilize source-specific types of crude.[9] Some refineries have limited processing capabilities only and have to ship semiprocessed outputs to others for finishing.

Refineries in the industrialized countries usually are equipped to produce a much larger gasoline fraction than those in developing countries, whose major demands are for middle and heavy distillates. This is apparent from the comparative data in Table 2.4 which compare the output ratios of U.S. refineries and two representative Thai installations. As can be seen, over 50% of U.S. output is in the form of gasoline, while in Thailand gasoline accounts for only 18 and 36% respectively. This difference in output mix is brought about both by refinery configurations and by the judicious choice of crudes processed. It is interesting to note, for example, that Malaysia exports most of its domestic, light-fraction, low-sulfur crude to Japan and the U.S.A. where it is sold at premium prices, while most of the country's domestic consumption comes from imported, high-sulfur, heavy Middle-East crudes.[10]

Refineries also are subject to significant economies of scale. This is apparent from the data in Table 2.5. It explains why in certain locations such as Sin-

---

[9]This, for example, made it impossible for most of California's refineries to process the high-sulfur, heavy crudes from the Alaska North Slope instead of the light, low-sulfur crudes from Indonesia and Malaysia. As a consequence, imports of the latter continued while Alaskan crude had to be shipped via the expensive Panama Canal-route to U.S. Gulf and East Coast refineries.
[10]Domestic Malaysian product requirements in the 1970s were about 38% fuel oil, 33% diesel, 10% kerosene and jet fuels, 17% gasoline, and 2% LPG. From T. L. Sankar and G. Schramm, p. 51.

Table 2.5. Representative Refinery Production Costs In Developed and
Developing Countries.

TECHNICAL CHARACTERISTICS	TYPICAL DEVELOPING COUNTRY REFINERY	WORLD-SCALE REFINERY
Distillation Capacity: 10^6 tons/year	1	6
Bls/day	20,000	120,000
Investment costs per Daily Barrel (1980 US$)	4,450	2,100
Unit Investment Costs per ton/year of Distillation Capacity (1980 US$)	89	42
Operating Costs per bl of Throughput (1980 US$)[1]	4.50	2.10

[1]Includes a capital charge of 25% of investment costs, but excludes oil costs.
Source: World Bank, *Energy in Developing Countries*, Table 13.

gapore, huge, completely export-oriented refinery complexes were established that supply many of the neighboring countries.[11]

Petroleum products are used directly or indirectly in almost all energy-using activities, ranging from transportation to industrial processing, commercial and household use, and the production of electricity. They completely dominate the energy requirements of the transport sector. Gasoline is used for small- and medium-sized engines in cars, boats, and airplanes. Diesel fuels are used for heavy road transports and buses, tractors, construction equipment, armored vehicles, locomotives, and ship engines. Jet fuel is used for turbine-powered airplanes as well as for heavy tanks. Fuel oil is utilized for powering heavy turbines in ships or in the form of heavy bunker oils for steam engines.

The use of petroleum products is pervasive in other sectors as well. Gasoline is the favorite fuel for small, stationary or portable, lightweight engines. Diesel fuels are used for water pumps and stationary equipment where electricity is not available. Low-speed diesel-powered engines drive heavy compressors and electric generating plants. Light fuel oil is used for heating purposes, and in industrial heat and boiler applications. Heavy fuel oil is used for industrial boilers, smelters, kilns, and electric power plants.

Overall it can be seen that the versatility and ease of handling, transporta-

[11]For example, Indonesia, a major oil exporter, buys substantial amounts of middle distillates, such as kerosene and diesel fuel, from Singapore, because its domestic refineries are unable to satisfy local demands. Bangladesh, which has a refinery of its own, gets about a third of its petroleum products from Singapore refineries, while it has to export surplus gasoline fractions and fuel oil from its own installation in Chittagong.

tion, and utilization of petroleum products have made them the most widely used fuels worldwide. No other type of energy source can compete with petroleum in this respect.

2.2 NATURAL GAS

Natural gas is composed of hydrocarbons that occur in gaseous states in underground reservoirs. Usually it consists of methane, containing various small admixtures of heavier hydrocarbon gases, such as butane, propane, ethylene, and gasoline fractions.[12] These can be liquified through refrigeration or compression and split off from the gas in the form of LPG and other products. Natural gas may occur alone (nonassociated gas) or in conjunction with crude oil (associated gas). Associated gas that occurs not as a gas cap above an oil pool but as solution gas mixed with oil can only be produced together with the oil. Natural gas is a less convenient fuel than oil. Transportation of gas requires pressurized pipelines from the well to the ultimate consumer. Because of the lower energy density of gas, its cost of transportation per BTU supplied is 5 to 8 times higher than the comparable cost for oil. Gas can also be liquified at low temperatures and high pressures and transported in specially equipped, refrigerated ships. This is quite expensive and may account for over 50% of the value of the gas at the point of delivery.

By far the largest known deposits of natural gas are in the USSR, China, and Eastern Europe, followed by the Middle East and North America. The proven recoverable resources of natural gas are almost $75,000 \times 10^9 \text{m}^3$. In terms of heating value, this represents about 40% of the currently proved reserves of all hydrocarbons.[13] In terms of estimated ultimately recoverable resources the USSR, China, and Eastern Europe are in first place, followed by North America and Africa (see Table 2.6).

Because of the high cost of transport and infrastructure required, nonassociated gas deposits in remote areas, far from suitably located markets, have rarely been developed in the past, while associated gas was usually flared. The steep rise in the costs of petroleum products, however, has led to the conversion of many of these remoter natural gas deposits into economically viable resources.[14] What is needed are substantial front-end capital investments in the development of these fields and in the necessary transport facilities such as

---

[12]Gas deposits usually contain various types of impurities in widely different percentages. The most common are nitrogen, carbon dioxide, various sulfur compounds, and helium.
[13]World Energy Conference, *Survey of Energy Resources,* p. 124.
[14]As, for example, the huge gas deposits in Northeastern Siberia, the Alaskan North Slope, the Mackenzie Valley, and the Arctic Islands of Canada, and the offshore locations in Malaysia, Indonesia, and Northern Australia.

Table 2.6. World Natural Gas Reserves and Estimated Ultimate Reserves.

$(1.000 \times 10^9 \text{ m}^3)$	CUMULATIVE PRODUCTION UP TO 1.1.79	PROVED REC. RESERVES ON 1.1.1979	ESTIMATED ADDITIONAL RESOURCES	"ULTIMATE RECOVERY"
Africa	0.1	7.3	26	33.4
North America	16.9	7.5	42	66.4
Latin America	1.8	4.7	10	16.5
Far East/Pacific	0.2	3.3	10	13.5
Middle East	1.1	20.5	30	51.6
Western Europe	1.5	3.9	6	11.4
USSR, China, and Eastern Europe	5.2	26.9	64	96.1
Antarctic			4	4.0
Total	26.8	74.1	192	292.9

Source: World Energy Conference. Survey of Energy Resources, Table 2.11.

pipelines and compressor stations or liquefaction facilities and special cryogenic vessels to bring the gas to suitable markets.

The utilization of gas may also be extended to nontraditional uses, such as transport fuels. Compressed natural gas is already being used as a fuel in some parts of Europe. Methanol, a chemical derivative of natural gas, is eminently suitable for spark-plug engines either as a straight fuel or as an admixture to gasoline. Costs of methanol conversion are estimated at US$225/ton to US$400/ton including gas costs from US$0.40/MCF to US$1.50/MCF (all in 1980 prices).[15] This compares to free-market ex-refinery prices for gasoline of about $320/per ton.[16] Natural gas can also be converted into gasoline, although the cost of conversion is high.[17]

One of the important characteristics of gas is that it is an excellent feedstock for a wide variety of petrochemicals and nitrogenous fertilizers. While in the industrial nations the use of gas for these purposes comprises only a small percentage of total gas consumption, in some of the developing countries fertilizer production may account for a very large percentage of total gas use. It should be noted, however, that chemical and fertilizer plants are themselves large consumers of energy. Up to 40% of the gas consumed by these installations may be used as an energy input, rather than as feedstock.

---

[15]World Bank, Energy in the Developing Countries, p. 37.
[16]Average Singapore spot market price, First Quarter 1980. From: The Economist Intelligence Unit Limited, Second Quarter 1980.
[17]$360 to $400/ton from methanol feedstock. From: World Bank, Energy in the Developing Countries, p. 37.

An important advantage of natural gas over competing fuels is its low polluting characteristics which make gas a premium fuel in heavily populated and industrialized areas. The major uses of natural gas are as a boiler and industrial fuel, as a reducing agent, as a fuel for space heating or cooling, cooking and other household energy needs, and as a chemical feedstock for petrochemicals and fertilizers. Table 2.7 shows major use categories of natural gas in the United States and in Bangladesh and Pakistan, two major gas producing countries. As can be seen, the domestic and commercial uses form an important part of gas consumption in the U.S.A., but are of minor importance in the other two countries. There are two major reasons for this difference: In the U.S.A. individual household energy requirements are relatively high given the need for space conditioning (heating and/or cooling) and hot water supplies. This reduces the unit costs of gas sufficiently to make it competitive with other energy supplies such as fuel oil and/or electricity.[18] In tropical countries gas is used mainly for cooking and hot water heating in households. The quantities used, therefore, are small and often do not justify the high connection costs.[19] By 1990 natural gas is expected to supply approximately 1.4 million barrels per day of oil-equivalent, or about 10–12% of the total commercial energy requirements of developing nations. About 36% of this quantity is projected to be used for power generation, some 50% for industrial uses including fertilizers and chemicals, and 14% for commercial and household uses.[20]

Given the substantial gas surpluses and petroleum deficits of a number of the developing countries, it may become attractive for them to use gas systematically as a replacement fuel for petroleum products in nonconventional uses such as the transport sector. Pilot projects to utilize compressed natural gas, for example, already are underway in a number of them (e.g., Egypt, Pakistan, Bangladesh). The use of methanol as a vehicle fuel is another option. Another possibility of gas utilization for transport is given by the potential extraction of liquid petroleum gases (LPG fractions) from suitable gas deposits (i.e., deposits with sufficient percentages of these types of gases). LPG is already used as a fuel in spark-ignition vehicle engines in Japan, Thailand, and Canada, for example (see also the Thailand Case study).

As described earlier, because of its nonpolluting characteristics, natural gas

---

[18]Another major reason has been the low regulated well-head and interstate pipeline price of natural gas in the United States which kept gas prices far below those of competing fuels on a heat-content basis. In 1979, for example, average wholesale prices of mainline interstate pipelines were $2.04/10^6 BTU while heavy fuel oils sold for $4.56/10^6 BTU and the wholesale price of light fuel oils for household uses was $5.70/10^6 BTU. *From:* U.S. Department of Energy, *Cost and Quality of Fuels for Electric Utility Plants,* November 1980, Washington, March 4, 1981; U.S. Department of Energy, *Mainline Natural Gas Sales to Industrial Users 1979,* Washington, February 1981; U.S. Department of Energy, *Energy Data Report.*

[19]See the case study on Bangladesh gas (chapter 11).

[20]World Bank, *Energy in Developing Countries,* p. 30.

Table 2.7. Major Use Categories of Natural Gas in
the USA, Pakistan, and Bangladesh.

Percent

USE CATEGORY	USA	PAKISTAN	BANGLADESH
Power generation	19	31	39
Fertilizer and Chemicals		21	38
Cement	40	13	5
Industrial		26	11
Commercial	14	3	2
Residential	27	6	5

Sources: Bangladesh Planning Commission, data refer to 1978/79; Government of Pakistan, *Energy Year Book,* 1979, p. 54, data refer to 1978/79; U.S. Dept. of Energy, *Natural Gas Production and Consumption Data,* Washington D.C., January 1981, data refer to 1979.

is considered a premium fuel in comparison to coal or fuel oil. Nevertheless, experience in Canada has shown that gas, even today, has to be sold at a discount relative to crude oil in order to maintain its market share.[21]

Overall, it can be expected that a significant increase of natural gas consumption will take place as a substitute for high-priced oil. This conclusion holds for Europe, where supplies from offshore Norway as well as from the USSR will increase gas supplies notably in the future. It also holds for the United States, Canada, and Mexico where on a continent-wide basis significant gas surpluses already exist. Japan, Korea, and Taiwan will become major importers of liquified natural gas. The gas-rich nations of South and Southeast Asia will be both net exporters and domestic consumers. In South America, natural gas from Columbia, Bolivia, Argentina, Trinidad and Tobago, and Venezuela will be connected via pipelines or LNG installations to growing markets. Gas from North Africa and the Middle East will be shipped via pipelines or in the form of LNG or chemicals to markets in Europe, North America, and Japan. Worldwide, it can be expected that expansion of natural gas use will be faster than that of any other energy source with the possible exception of coal.

---

[21]Canadian natural gas prices are officially regulated at 85% of the price of crude oil. However, this has led to such an erosion of gas sales relative to low-priced, surplus fuel oil in industrial markets, that gas now is generally sold at prices equivalent to about 65% of the fixed crude oil price. Also, the Canadian policy of pricing gas exports to the United States on a BTU-equivalent basis with the world crude oil price 90 days previously, minus a transportation adjustment, has led to sharp reductions of Canadian gas sales to the United States (the Border price in late 1980 was $4.36/MCF). Canadian sources believe that Canada will have a gas surplus till the end of this century and that urgently needed additional export sales to the USA will require discounts of 15–20% below contracted prices. *From:* John F. Helliwell. "The National Energy Conflict," *Canadian Public Policy,* VII.1., Winter 1981, pp. 15–23.

2.3   COAL

Until the first quarter of this century, coal was the most important source of commercial energy in the world, both for providing high grade heat and mechanical energy. However, today the use of coal as a source of mechanical energy either in transport or industry has almost disappeared. Only in a few countries, notably China and India, are coal-powered steam locomotives still in widespread use. Elsewhere, the use of coal is largely limited to the production of high-temperature heat; in the direct reduction of iron ore in the form of coke or in power plants and industrial boilers. Coal is also used as a fuel for cement, brick, or ceramic kilns, pulp and paper plants, and similar heat-energy intensive installations. In some countries or regions, coal is widely used as a household fuel, as, for example, in Korea, India, China, East and West Germany, and Poland.

The huge price increases for competing petroleum products have made coal once again an economically attractive fuel, particularly for electric power generation and for use in industrial boilers and kilns.

The most common classification of coal is made in terms of its calorific content. Table 2.8 presents representative values for various types of coal.

Coking coal commands premium prices substantially in excess of the prices of steam coal. Coking coal is used mainly for the direct reduction of iron ore. It has to have special characteristics, such as low ash and sulphur contents and high structural strength to withstand the high pressures in a blast furnace. Until recently coking coal was the only type of coal for which a significant international market existed. Major suppliers were Australia, Canada, the United States, and Poland.

Anthracite, bituminous coal, and subbituminous coal present the major groups of so-called thermal coal, which is used for furnaces and boiler installations. As can be seen from Table 2.9, large deposits of such coals are avail-

Table 2.8. Average Calorific Values of
Various Grades of Coal.

THERMAL COAL:	kcal/kg
Anthracite	7,950
Bituminous Coal	6,950
Subbituminous Coal	5,900
Brown Coal and Lignite	3,500
Peat	1,900
Coking Coal	6,650

Source: David Crabbe and Richard McBride, Appendix 2, Table 5.

ENERGY SOURCES, USES, AND SUBSTITUTION POSSIBILITIES   53

Table 2.9. Coal, Recoverable and Additional Resourses and Peat (in Gt ce)
According to Continents and Economic Political Groups.

PROVED RECOVERABLE RESERVES

CONTINENT OR ECONOMIC-POLITICAL GROUPS	BITUMINOUS COAL AND ANTHRACITE Gt ce	%	SUBBITUMINOUS COAL Gt ce	%	LIGNITE Gt ce	%	PEAT Gt ce	%	TOTAL Gt ce	%
Africa	32.5	6.7	0.1	0.1	0.0	—	—	—	32.6	4.7
America	111.4	22.8	75.3	67.4	13.2	15.0	0.3	5.2	200.2	28.9
Asia	113.9	23.4	0.8	0.7	1.4	1.6	—	—	116.1	16.7
USSR	104.0	21.3	32.9	29.4	28.7	32.6	3.6	62.0	169.1	24.4
Europe	100.5	20.6	1.3	1.2	35.1	39.8	1.9	32.8	138.8	20.0
Oceania/ Australia	25.4	5.2	1.3	1.2	9.7	11.0	—	—	36.4	5.3
Total	487.7	100.0	111.6	100.0	88.1	100.0	5.8	100.0	693.2	100.0
Common Market	70.0	14.4	0.0	—	10.6	12.0	0.5	8.6	81.1	11.7
OECD	205.9	42.2	74.6	66.8	34.9	39.6	2.1	36.2	317.5	45.8
COMECON	134.2	27.5	32.7	29.3	45.1	51.1	3.6	62.1	215.6	31.1
Developing countries	22.5	4.6	2.8	2.5	1.3	1.5	0.0	—	26.6	3.8
OPEC	0.4	0.1	0.2	0.2	0.2	0.2	—	—	0.8	0.1

Source: World Energy Conference, *Survey of Energy Resources,* Table 1.7.

ADDITIONAL RESOURCES IN SITU

BITUMINOUS COAL AND ANTHRACITE Gt ce	%	SUBBITUMINOUS COAL Gt ce	%	LIGNITE Gt ce	%	PEAT Gt ce	%	TOTAL Gt ce	%
144.4	2.3	0.8	0.0	0.0	—	1.3	1.3	146.5	1.5
1181.2	19.2	1334.1	44.5	406.8	48.0	47.9	47.3	2970.0	29.4
1423.2	23.1	2.2	0.1	19.5	2.3	9.2	9.0	1454.1	14.4
2480.0	40.2	1570.9	52.5	381.5	45.0	37.4	37.0	4469.8	44.2
429.5	7.0	1.1	0.0	12.1	1.4	5.5	5.4	448.2	4.4
503.1	8.2	82.3	2.8	28.4	3.3	0.0	—	613.8	6.1
6161.4	100.0	2991.4	100.0	848.3	100.0	101.3	100.0	10102.4	100.0
335.4	5.4	0.0	—	0.0	—	2.5	2.5	337.9	3.3
2007.3	32.6	1399.6	46.8	432.5	51.0	51.4	50.8	3890.8	38.5
2572.3	41.7	1570.9	52.5	391.4	46.1	39.5	39.0	4574.1	45.3
214.4	3.5	19.0	0.6	9.3	1.1	9.9	9.8	252.6	2.5
4.7	0.1	4.6	0.2	5.9	0.7	8.3	8.2	23.5	0.2

1 at cc = $+10^6$ tons of standard coal equivalent.

able worldwide. At mining costs that range from less than $10 a ton in South Africa to as much as $100 in subsidized mines in Central Europe, coal is considerably less costly than competing petroleum products, on a heat equivalent basis. While handling costs for coal are higher than those for competing fuel oils, the cost differential relative to oil is sufficient to assure a major and growing market for thermal coal as a substitute for fuel oil in power generation and industrial furnace uses.

Brown coals and lignites are low heat-value coals whose major uses are as fuels in mine-mouth power plants.

Peat is used in a few places as a power plant fuel as, for example, in Ireland. Locally, it is also used as a household fuel.

World resources of coal are very large. They would last for several hundred years even if most energy uses were switched to coal. However, as in the case of other energy resources, coal deposits are unevenly distributed. Furthermore, coal mining costs vary substantially. Over the next several decades it can be expected that world trade in thermal coal will grow very rapidly. Major exporters are likely to be the United States, Western Canada, Australia, South Africa, and Columbia.

2.4   ELECTRICITY

Electricity is a particularly versatile form of energy that can provide heat, light, and motive power at high rates of efficiency. All devices for generating and transforming electric current are based on electromagnetic induction. Discovered by Faraday in 1831, current is induced in a coil when it cuts the lines of force in a magnetic field. Modern power stations produce electricity by coupling the rotor of a generator, which is a large cylindrical electromagnet, to a rotating turbine. When the turbine and rotor rotate alternating current is induced and fed to a transformer where the voltage is raised to a suitable level for transmission. The scientific principle underlying the production of direct current in a dynamo is similar. Dynamos are generally used in automotive applications to provide the electric energy for spark-plugs and secondary electric circuits.

To produce electricity, therefore, a source of mechanical energy is needed. The major ones are superheated steam produced through the combustion of some fossil fuel or nuclear energy, the force of falling water, geothermal steam or hot water, the direct combustion of some fuel in a gas turbine or internal combustion engine, or solar energy in the form of wind, concentrated solar heat that produces steam, or direct conversion through the use of solar cells. The latter is the only form of electricity production that does not use electromagnetic induction. However, its costs are prohibitive except in remote locations or space applications where alternatives are even more expensive.

About 22 percent of the technically feasible hydropower potential worldwide has already been developed or is under construction, while another 11 percent are under active consideration. This can be seen from the data in Table 2.10 which provides a summary of the estimated annual potentials by continent. Table 2.11 provides a breakdown in terms of theoretical capacity, giving detailed data for the developing countries. Most of the technically and econom- ically feasible sites in the industrialized nations have already been developed except for a few sites that have been set aside permanently for aesthetic or environmental reasons. Only in some of the remoter areas such as Alaska and Northern Canada, for example, is there substantial undeveloped hydro poten- tial.[22] These have not been developed so far because of a lack of suitable power markets within economical transmission distance.[23]

Similar problems afflict many of the hydro sites throughout the world. For example, the power potential of the Lower Congo River in Zaire is estimated at over 60,000 MW, more than the total installed hydro capacity in the United States. Similarly, the potential of the Tsangpo River, the principal tributary of the Brahmaputra River in Assam, India, is estimated to be in excess of 30,000 MW, but its location makes development unlikely for several more decades to come.[24] The reasons why these and similar sites cannot be developed inspite of growing local power markets is that they have to be developed in sizes and with capacities that are far in excess of these demands; for technical reasons staged development of hydro sites is generally not feasible. The problematic nature of this mismatch between site capacities and available market demand is graph- ically illustrated by the case of Nepal, whose estimated hydro capacity, much of it with very attractive, low unit-costs, is about 80,000 MW (see Table 2.11), while total installed generating capacity was only 48 MW in 1978.[25] A pro- gram to install several microhydro generating plants with a total capacity of 2,650 kW (2.65 MW) was just announced. Average unit costs per kW are

---

[22]Alaska's undeveloped, low-cost hydro potential has been estimated at about 19,000 MW, or about 35% of the total installed hydro capacity in the United States. For a discussion of the economic feasibility of developing this potential see: Gunter Schramm, *The Role of Low-Cost Power in Economic Development,* Arno Press, New York, 1979.

[23]To be economical, generation plus transmission costs must be lower than the costs of alternative power at load centers. Approximate limits today for very low-cost sites are 1,500 to 2,000 miles with dc transmission lines of at least one million kW capacity and load factors close to 100 percent. For an analysis of the effects of long-distance transmission costs on the economic fea- sibility of hydro sites see: Gunter Schramm, *Analyzing Opportunity Costs—The Nelson River Development,* Nat. Res. Inst., Univ. of Manitoba, Winnipeg, Manitoba, 1976.

[24]For a description of this project see: Hans Thirring, *Energy for Man,* p. 251 ff.

[25]T. L. Sankar and G. Schramm, Annex 313.

Table 2.10. Annual Hydraulic Potentials.

REGION	THEORETICAL POTENTIAL 10^{12} kWh	TECHNICALLY USABLE POTENTIAL 10^{12} kWh	OPERATING POTENTIAL 10^{12} kWh	POTENTIAL UNDER CONSTRUCTION 10^{12} kWh	PLANNED POTENTIAL 10^{12} kWh
Africa	10.118	3.14	0.151	0.047	0.201
America (North)	6.15	3.12	1.129	0.303	0.342
America (Latin)	5.67	3.78	0.299	0.355	0.809
Asia (excluding USSR)	16.486	5.34	0.465	0.080	0.368
Oceania	1.5	0.39	0.059	0.020	0.032
Europe	4.36	1.43	0.842	0.094	0.197
USSR	3.94	2.19	0.265	0.191	0.17 estimated
TOTAL	44.28	19.39	3.207	1.090	2.12

Source: World Energy Conference, Supply of Energy Resources, Table 4.7.

US$5,716, or five to ten times higher than the costs of conventional hydro sites.[26] Many of the Andean countries in South America, various Central African Nations, and several of the island nations of Southeast Asia face similar problems.

Geothermal resources have been utilized for power generation in some countries (e.g., Italy and New Zealand) for several decades. However, the plants in operation are relatively small, and it is only since 1973/74 that much greater efforts have been made to develop existing potentials, particularly in countries that depend heavily on oil imports for power generation. Worldwide, total installed capacity as of 1978 was only 1,468 MW, with an additional 2,220 MW under construction for completion by 1982 and another 5,177 MW planned for 1985 (see Table 2.12). Some countries have more ambitious plans than those indicated in the table. In the Philippines, for example, the Government has recently announced a target of 1,261 MW by 1985.

Little is known about the ultimate geothermal capacity. This is apparent from the wide discrepanies quoted by different sources. For the Philippines, for example, an authorative source quotes an estimated potential of about 2,100 MW[27] while recent Government surveys claim an ultimate potential of over 25,000 MW.[28]

Geothermal power can be generated from either steam or steam/hot-water

---

[26]Asian Development Bank, Project Announcement, 1981.
[27]Joseph Kestin, ed. Sourcebook on the Production of Electricity from Geothermal Energy, US Dept. of Energy, Washington, D.C., March 1981, p. 931.
[28]T. L. Sankar and G. Schramm, p. 91.

Table 2.11. World Hydroelectric Potential[a] (Estimated gross theoretical capacity, in megawatts).

DEVELOPED MARKET ECONOMIES			*533,089*
CENTRALLY PLANNED ECONOMIES			*615,160*
DEVELOPING COUNTRIES			*1,194,390*
Net Oil Exporters			
Algeria	4,800	Libya	160
Angola	9,664	Malaysia	1,319
Bolivia	18,000	Mexico	20,344
Burma	75,000	Nigeria	1,515
Congo	9,040	Peru	12,500
Ecuador	21,000	Saudi Arabia	900
Egypt	3,800	Tunisia	29
Gabon	17,520	Venezuela	11,644
Indonesia	30,000	Zaire	132,000
Iran	10,196	Subtotal	*379,431*
Net Oil Importers			
Subtotal	*814,959*		
Africa			
Benin	1,792	Morocco	975
Botswana	2,984	Mozambique	11,920
Burundi	..	Niger	9,600
Cameroon	22,960	Reunion	82
Central African Rep.	11,040	Rwanda	..
Chad	3,440	Sao Tomé and Principe	..
Equatorial Guinea	2,400	Senegal	4,400
Ethiopia	9,214	Seychelles	..
Ghana	1,615	Sierra Leone	3,000
Guinea	6,400	Somalia	240
Guinea Bissau	120	Sudan	16,000
Ivory Coast	780	Swaziland	700
Kenya	13,440	Tanzania	20,800
Lesotho	490	Togo	480
Liberia	6,000	Uganda	12,000
Madagascar	64,000	Upper Volta	12,000
Malawi	100	Zambia	3,834
Mali	3,520	Zimbabwe	5,000
Mauritania	2,000	Subtotal	*253,406*
Mauritius	80		
Asia			
Afghanistan	6,000	Papua New Guinea	17,762
Bangladesh	1,307	Philippines	7,504
Fiji	400	Sri Lanka	1,180
India	70,000	Syria	1,000
Republic of Korea	5,514	Thailand	6,242
Laos	..	Turkey	15,200
Lebanon	..	Viet Nam	53,598
Nepal	80,000	Western Samoa	15
New Caledonia	139	Subtotal	*285,861*
Pakistan	20,000		

Table 2.11. World Hydroelectric Potential[a] (Estimated gross theoretical capacity, in megawatts). (*Continued*)

Latin America and Caribbean			
Argentina	48,120	Honduras	4,800
Brazil	90,240	Nicaragua	3,600
Chile	15,780	Panama	2,400
Colombia	50,000	Paraguay	6,000
Costa Rica	4,326	Surinam	260
El Salvador	900	Uruguay	2,512
Guatemala	1,176	Caribbean (total)	2,400
Guyana	12,000	Subtotal	*244,514*
Europe			
Portugal	6,188	Yugoslavia	16,957
Romania	8,033	Subtotal	*31,178*
WORLD TOTAL	*2,342,639*		

[a]Includes all installed and installable capacity, assuming average flows.
. . Not available.
Source: World Bank, *Energy in the Developing Countries*, p. 86.

mixtures that are recovered from suitable hot-water bearing rock strata through the drilling of wells. Sources of steam are usually much more attractive because they can be used directly to drive a steam turbine, and because steam sources contain fewer contaminants. Steam/hot-water mixtures must be utilized by transferring the inherent heat energy to a low boiling-point fluid which in turn is used to drive a turbine. Costs of energy from geothermal plants may vary widely. Estimates made for the World Energy Conference for a 100 MW representative plant project costs of US$c1.1/kWh.[29] The U.S. Department of Energy study projects USc1.72/kWh for a steam plant, and between USc1.95 to USc8.53/kWh for various types of hot water plants.[30] The actual operating costs of several 55 MW plants in the Philippines are USc5.3/kWh, with steam costs amounting to USc3.1 of the total.[31]

Overall, it can be assumed that in some countries with favorable geothermal potential such powerplants will become a major source of future power generating capacity.

---

[29]World Energy Conference, *Survey of Energy Resources*, p. 293.
[30]Joseph Kestin, *Sourcebook on the Production of Electricity from Geothermal Energy*, p. 699.
[31]T. L. Sankar and G. Schramm, p. 91.

Table 2.12. Existing and Planned Geothermal Plants.

COUNTRY	ELECTRIC POWER IN MW EXISTING IN 1978	1980	PLANNED 1985	1990	ELECTRIC PRODUCTION, MWh 1978	NONELECTRICAL USES IN 1979 (ESTIMATES) IN THERMAL MW SPACE HEATING	AGRICUL-TURE	INDUS-TRY
USA	608	1000	2000		3×10^6	75	5	5
Italy	421	455	480		2.5×10^6	50	5	20
New Zealand	192	203	272		1.2×10^6	50	10	150
Japan	168	218	1000	6050	0.7×10^6	10	30	5
Mexico	75	153	400		0.6×10^6			
Iceland		64				680	40	50
El Salvador		60	100					
Philippines		58	440					
USSR	3	5	25	200	16×10^3	120	5100	
Republic of China	1	3	50	200				100
China, PR		1						
Turkey		0.5	15	15				
Hungary						300	370	
France						20		
Costa Rica			40	80				
Chile			30	110				
Nicaragua			200					
Indonesia			125					
Totals	1,468	2,220	5,177	6,655		1,305	5,560	330

Source: World Energy Conference, *Survey of World Energy Resources*, p. 296.

Nuclear Energy

Nuclear energy is based on the utilization of nuclear fission chain reactions which, under suitably controlled conditions, produce high-grade heat, which in turn is utilized to produce superheated steam to drive steam-turbine generator sets. Hence, a nuclear reactor is equivalent to the boiler of a fossil-fuel thermal power station. Present reactor designs are based on the controlled fission of uranium 235, which utilizes only a very small proportion of the inherent energy contained in the fuel. Because known uranium deposits are limited, widespread use of these thermal reactors would quickly lead to an exhaustion of known deposits so that within a few decades the expansion of nuclear energy program would have to come to a halt. An alternative to thermal reactors are the so-called breeder reactors. Breeder reactors use fast neutrons to produce more fissionable materials in the process of generating heat. This fissile fuel can be retrieved when the reactor fuel is reprocessed. In other words, breeder reactors produce to a large extent their own fuel. With breeder reactors, presently known uranium resources would last for several thousand years. However, many technical, economic, and safety problems have to be solved before breeder reactors may become a commercial reality.[32]

In 1978 nuclear plants supplied some 7% of the world's electricity from stations mainly located in North America, Europe, Japan, and the USSR. In the U.S.A., they supplied some 13% of the electric energy produced. In several of these countries, expansion of nuclear plants has slowed down drastically, mainly because of unresolved safety and environmental concerns, but also because of very substantial cost increases. While nuclear supporters argue that the former are more psychologically based than the result of factual circumstances, it is difficult to say if and when nuclear reactors will come into favor again.[33] In other countries, such as France, the USSR, Japan, South Korea, Thailand, and Brazil strong nuclear expansion plans are underway. Nuclear plants are available only in relatively large capacities of 600 megawatt per unit or more, a fact which limits the number of countries that could utilize nuclear power in a balanced system. The World Bank projects that in the developing countries about 12% of total new generating plant additions between 1981 and 1990 will be nuclear units.[34] In the United States, nuclear energy is likely to contribute about 25% of total generation by 1990, taking into account only the number of plants operating or presently under construction.[35] In West Germany, where opposition to nuclear power has been strong, only two nuclear plants are in operation at present producing about 12% of total electricity needs, 11 more are under construction. In contrast, France expects to have 50 nuclear stations in operation by 1985 and it will eventually rely on nuclear plants for more than 70% of its power supplies.[36]

Barring serious nuclear accidents with consequences much more serious than those of the Three Mile Island plant, it is likely that nuclear energy will expand in many of the areas of the world in which alternative energy resources are costly and in short supply.

---

[32]For a short concise description of nuclear fission reactors see David Crabbe and Richard McBride, p. 125 ff. A much more detailed, but highly readable account can be found in: Hans Thirring, *Energy for Man,* Chapter 15.

[33]For an analysis of the psychological problems underlying the nuclear controversies see Miller B. Spangler, "Risk and Psychic Costs of Alternative Energy Sources for Generating Electricity," *The Energy Journal,* Vol. II, No. 2, January 1981, p. 37–59.

[34]World Bank, *Energy in the Developing Countries,* p. 48.

[35]Miller B. Spangler.

[36]*From: Fortune,* June 15, 1981, p. 36.

Fossil-Fuel Generating Plants

Fossil fuels provide the major energy source for electric power generation worldwide, and are being utilized in three distinct types of plants: steam, gas turbine, and diesel plants. Of these three, steam generating plants are by far the most important category. This is so because steam plants, although more costly in terms of capital investments, have lower heat rates than gas turbines.[37] Capital costs for diesel engines lie between those of steam and gas turbine plants, while their fuel costs are greater than for steam plants but less than for gas turbines.

Steam plants can be designed for use with a wide variety of fuels. The fuel is burned in a suitably designed boiler to produce superheated steam which drives a steam turbine that is coupled to a generator.

The most important fuels utilized are coal, lignite, fuel oil, and natural gas. A few small plants around the world also operate on wood, mainly in saw mills or pulp and paper plants where the fuel is readily available.[38] Many sugar mills traditionally utilize bagasse as basic fuel for their combined steam-processing and electric power plant requirements. In a few locations peat and oil shale are also used (the former in Ireland and the latter in the USSR). Both in Europe and the United States a few municipal plants have been built that use garbage as their primary fuel. In the United States coal supplies about 69%, oil around 15%, and gas about 17% of the primary energy to electric utility steam plants (see Table 2.13). In the developing countries in 1980 oil supplied about 30.5%, natural gas 6.2%, and coal or lignite some 17.2% of total thermal power generations.[39]

In the industrialized countries, diesel as well as gas turbine generating plants are generally used for peaking purpose only. This is so because they face high operating fuel costs on the one hand but low capital investment costs on the

---

[37]The heat rate, expressed in btu or kcal per kWh produced measures the overall technical efficiency of a generating plant. The less the required heat input per Kwh the more efficient the plant. Well-designed steam plants use about a third less fossil fuels than gas turbines.
[38]The idea of using wood-fired steam power plants instead of costly fuel-oil-fired ones in heavily forested regions has become quite popular lately. However, even with fast-growing species planted in wood plantations around the power plant, generating plant sizes are limited to 25 to 75 megawatt because of the disproportionate rise in the transport costs of the fuel-wood. According to World Bank estimates, transportation costs might rise from about $1.50/ton for a 10 Mw plant to $5/ton for a 100 Mw one. Probably the largest plant presently in operation is at the Jari forest products complex in the Amazon area of Brazil which operats a 55 Mw plant. *From:* World Bank, *Renewable Energy,* Energy Dept. Note No. 53, Washington, D.C. Feb 19, 1980, p. 17.
[39]It should be noted that these percentages include the consumption of diesel and gas turbine power plants which are more common in developing countries than in the U.S., where they are largely used as peaking power plants. *From:* World Bank, *Energy in the Developing Countries,* p. 46.

Table 2.13. Fossil-Fuel Deliveries to U.S. Steam-
Electric Plants in January 1981.

	QUANTITY	PRICE PER U.S.$ MILLION kcal	% OF TOTAL btu DELIVERED
Coal	50.7 million tons	5.65	68.4
Oil	38.1 million bbls	21.52	15.1
Gas	253.8 billion cf	10.08	16.6

Source: U.S. Dept. of Energy, Cost and Quality of Fuels for Electric
Utility Plants, Washington D.C., January 1981, Table 2.

other. In the United States, for example, gas turbine and diesel peaking plants accounted for only 1.1% of the total fossil fuel consumption of all electric utility companies.[40] In developing countries, diesel plants are frequently used as base-load power plants for small isolated regions or as independent sources of electricity for industrial installations. Unit sizes range from a few dozen to several thousand kilowatts. Gas turbines similarly are used for shoulder or more rarely base-load requirements. The reason for the utilization of these less energy-efficient or high fuel-cost plants is that they are more adapted to smaller loads than steam-electric power plants. The latter are subject to very substantial economies of scale so that plants with less than 100 megawatts usually have very high unit costs, while gas turbines, and to a lesser degree diesel plants, have low initial capital costs.[41] Furthermore, the installation time for a diesel or gas turbine set is generally less than two years, while a steam power plant may require 4 to 5 years, and a hydro plant from 8 to 15 years until completion.

Pricing policies that keep the cost of diesel and fuel oils low relative to other petroleum prices also have contributed to a proliferation of the installation of high fuel-cost plants. In Indonesia, for example, in 1979, some 47% of the total installed capacity consisted of privately-owned diesel or gas turbine generators that were not connected to the public utility system.[42] In some countries or regions however, in which natural gas is in plentiful supply and where no alternative markets for that gas are likely to be available in the foreseeable future, the installation of gas turbines as the lowest-cost source of power generation may well be optimal.[43]

---

[40]Department of Energy, Cost and Quality of Fuels, p. 2.
[41]See also the Sri Lanka case study.
[42]T. L. Sankar, Indonesia, Energy Sector Study, Asian Development Bank, September 1980, mimeo, p. 32.
[43]This, for example, was the case in the Cook Inlet area around Anchorage in Alaska in the 1960s and early 1970s. See G. Schramm, The Role of Low-Cost Power in Economic Development.

2.5. TRADITIONAL FUELS

Traditional fuels, consisting mainly of firewood, charcoal, crop residues, and dung, dominated world energy consumption as recently as a century ago (see also Figure 1.2). Today, in terms of total energy consumed, these fuels play a relatively modest role. While accurate data are lacking, estimates range from 4 to 6% of total world energy consumption for cooking purposes to a high of 15% if all types of traditional fuel uses, for heating as well as industrial and commercial purposes, are taken into account.[44]

In terms of their importance to human well-being, however, these relative consumption figures are grossly misleading. It has been estimated that about two billion people out of a world population of 4½ billion depend on these traditional fuels for cooking. But fuel for cooking represents the most crucially important energy use for human survival. The availability of cooking fuel is just as important as the availability of food itself, because most of the latter is indigestible if not cooked.

The overwhelming majority of these 2 billion people have no other alternative source of fuel available to them. They simply do not have the income to buy commercial fuels in the market, but depend, instead, on their own efforts to gather enough combustible materials to cook their daily meal.

In the past, such biomass-based fuels had generally been in ample supply relative to existing local demands, although regional scarcities have been common in the more arid lands for centuries. The unprecedented growth in world population in recent decades has brought about a drastic change in their availability relative to needs. In many regions of the world, encroachment on available supply due to the expansion of agricultural and grazing lands, and overcutting of remaining resources relative to regenerative growth, have become the rule rather than the exception. Table 2.14 provides an overview of potential fuel resources on a per capita basis by country.

The picture that emerges from available data indicates that there are two major problem areas in terms of traditional fuel availability. First, wood is already scarce in many areas so that other fuels, such as dung and crop residues must be used. The use of the last two fuels, however, reduces their availability

---

[44]Hughart estimated total fuel consumption for cooking purposes as 13,516 million gigajoules, while Fritz estimated 980 million tons of firewood, which is equivalent to 21,764 million GJ. Wendorff, as quoted in Fritz, estimated total consumption for all uses of biomass fuels at 3,200 million cubic meters, equivalent to 51,764 million GJ. In comparison, the world's total commercial energy consumption was estimated by the World Bank at 137.8 million bls/day of oil equivalent, or 2.96×10^{11} GJ in 1980. *From:* Markus Fritz, *Future Energy Consumption of the Third World,* Pergamon Press, N.Y., 1981, p. 95; David Hughart, *Prospects for Traditional and Non-Conventional Energy Sources in Developing Countries,* World Bank, Washington, July 1979, p. 38,

Table 2.14. Estimated Organic Resources, 1990 (Giga joules per capita per year)

	SUSTAINABLE FOREST YIELD	DUNG[1]	CROP RESIDUES
Ethiopia	3	6 SH	3
Kenya	1	7 CS	4
Tanzania	54	11 CS	3
Ghana	8	2 CS	2
Nigeria	8	3 CS	2
Sudan	148	18 CS	5
South Africa	2	8 CS	13
Algeria	3	3 SC	1
Egypt	0	1 CH	6
Morocco	6	5 CS	5
Iran	22	6 SC	7
Iraq	1	5 CS	2
Vietnam	12	1 CP	6
Afghanistan	6	8 CS	7
Nepal	15	11 CS	6
Pakistan	1	6 CS	7
India	6	5 CS	6
Sri Lanka	11	2 CS	4
Bangladesh	2	4 CS	4
Burma	82	4 CP	6
Indonesia	63	1 CS	6
Rep. of China	n.a.	n.a.	n.a.
Rep. of Korea	9	1 CP	7
Malaysia	114	2 CP	7
Philippines	12	3 CP	7
Thailand	22	3 CP	9
Chile	71	7 CS	4
Colombia	180	16 CH	4
Peru	245	7 CS	2
Mexico	39	9 CH	9
Brazil	229	15 CP	9
Argentina	104	45 CS	33
Venezuela	211	11 CH	4
PR China	11	3 CP	8
DPR Korea	24	1 CP	13

[1]Primary sources indicated by letter codes:
    C = cattle, buffalo, camel
    S = sheep, goats
    H = horses, mules, asses
    P = pigs
Source: David Hughart, Prospects for Traditional and Non-Conventional Energy Sources in Developing Counties op. cit., p. 42.

as fertilizers, reducing agricultural yields, in the case of the former, and animal feed availability, which reduces animal draught power, in the case of the latter. The main areas facing this problem are the drier regions of Africa, much of South Asia, the People's Republic of China, and some areas of South America. Moreover, country-wide statistics of availability do not truly indicate the real scope of the problem. In most countries remaining wood resources are unevenly distributed. In Bangladesh, for example, with only 9% of the land area under forest cover, more than two thirds of the remaining stands are in mountainous border areas or coastal swamp land far from centers of population.[45] The high transport costs of firewood relative to its energy content make it impossible to harvest such remote stands effectively. The actual available wood resources, on a per capita basis, therefore, are much smaller. As studies in India have shown, for example, the farthest distance that people can travel to gather firewood for their daily needs is 15 miles. In the foothills of the Himalayas, for example, the job of collecting firewood has grown from a task requiring one hour per day to one needing a full day. In Niamey, Niger, a manual laborer has to spend one quarter of his income in order to purchase enough firewood for cooking.[46]

Another consequence of the overcutting of remaining forest stands is the deleterious effect on soil stability, rates of erosion, and downstream flooding. Overcutting in the Himalayas, for example, has led to ever increasing flood losses in the Ganges River flood plains in India and Bangladesh, with losses claimed to be in the billions of dollars.[47]

Given the high costs of alternative, commercial fuels compared to the income levels of most of the people depending on these traditional fuels, it is unrealistic to expect that they will be able to switch-over to the use of the former. Instead, major efforts are needed to maintain and increase the supply of indigenous fuel resources and to teach the people depending on them how to use them more efficiently so that per capita consumption requirements can be reduced (see also section 6.7).

2.5 SUMMARY:

The overall picture that emerges from this discussion is that the world is not about to run out of energy resources, particularly in view of the upward trend in energy costs during the last decade, and the demand response to higher energy prices. Even if no new energy technologies were to be developed, existing and known resources of oil, natural gas, coal, hydro, geothermal. and ura-

---

[45]James J. Douglas, UNDP/FAO/Planning Commission of Bangladesh, *Consumption and Supply of Wood and Bamboo in Bangladesh,* Dacca, June 1981.
[46]Markus Fritz, p. 94;
[47]*Der Spiegel,* Sept. 24, 1978, p. 187.

nium would be sufficiently large to accommodate foreseeable human needs for several hundred years to come.

The major difference in outlook relative to the earlier part of this century is that future energy costs are no longer going to fall in real terms, but more likely will keep rising instead, at least for the next few decades.

This is the result of the gradual decline, in relative terms, in the availability of low-cost, versatile petroleum products. These will have to be replaced at accelerating rates by higher-cost and often less convenient fuels. However, we will not run out of liquid petroleum products. They will be available, from conventional sources, from oil sands, shale, or heavy oil deposits, or from the liquifaction of natural gas or coal. But their real costs of supply will continue to rise, thus making it more and more attractive to switch to alternative sources of energy, or to more efficient techniques of energy utilization.

The major consequence will be that more of other resources like capital and labor will be needed to provide each additional unit of energy. This shift from cheap petroleum to more capital-intensive alternatives is massive indeed, requiring significant changes in hisoric savings/consumption ratios, and overall structures of national economies.

Another consequence of this necessary shift away from petroleum products is that the specific energy supply source mixes and demand management strategies of different countries and even regions within countries will become much more diverse than they have been in the past. Some nations will put more emphasis on coal-derived technologies, others on hydro, and yet others on atomic power or solar energy. Energy-intensive production processes will shift more and more to those locations in which low production-cost but high transport cost energy resources such as hydro and poor quality lignite are still plentiful. At the same time, international trade in energy resources will rapidly expand from the almost exclusive movement of petroleum and coking coal to a much broader spectrum of fuels including natural gas through extralong pipelines, liquified petroleum gas, thermal coal, methanol, electricity supply through long-distance, high-voltage transmission lines, and enriched uranium for atomic powerplants. Energy will still be available, but only at much higher costs than in the past.

3
Energy Markets, Planning and the Role of Demand Management

3.1 INTRODUCTION

The critical and pervasive role of energy in the modern world has been amply demonstrated in the two previous chapters. In this chapter we will discuss the importance of energy planning and management for the maximization of net benefits from energy use and for the achievement of other societal objectives. Normally, at least in western-style market economies, such maximization is expected to come about by the interplay of free market forces. However, in the areas of energy supply and demand, many factors, such as monopolistic practices, externalities, lock-in effects and noncompatibility of societal goals and objectives, interfere with this ideal, so that market failures become the rule, rather than the exception. It is because of these many market imperfections that the need for establishing and maintaining proper planning, evaluation, and management policies arises. In fact, such policies are being used and applied in practically all countries of the world in some form or another. However, they often are either undertaken on an ad hoc basis, or are applied to particular energy subsectors for specific policy purposes only, with little or no attention given to the effects on other sectors or objectives. Such oversights have proved to be rather costly in many cases. Proper energy planning and management, which take account of these wider effects of specific policies and that recommend corrective actions if required, are needed precisely in order to obtain and maintain the basic goal of free market economies: the maximization of net benefits to society from the use of energy.

We first briefly review the basic principles of economic optimization in a free market context. We then discuss the various factors that interfere with these ideal results and make it necessary to take corrective action. We then present an overview of the many direct and indirect policies and actions by government that effect and influence energy markets. After this review, we turn to the basic concern of this book, a discussion of the various aspects of demand management. Demand management is one of the major elements of energy planning. However, because the different components of energy planning are closely interrelated, we first summarize the process of overall energy planning, including demand management. Other aspects of energy planning, such as supply management, investment planning and the role of the institutional framework are discussed in this and subsequent chapters to the extent that they interact with or illuminate the role of demand management.

3.2   FREE MARKET OPTIMIZATION

Theoretically, in a free, competitive market economy the supply, demand, and pricing of energy resources is self-regulating and results in the maximization of net economic benefits from the use of energy without the need for overall planning and management. The basic conditions for such efficiency solutions to come about are the existence of freely competitive markets, of perfect knowledge, of smooth transferability of resources from one activity to another, and the absence of any nonpriced externality. As is well-known from any economic textbook[1], to bring about welfare maximization, optimization must occur simultaneously in three types of market interactions: among consumers, among producers of energy and of all other goods and services, and between consumers and producers. Given static criteria, the basic conditions are that on the supply side prices must equal marginal costs and that on the demand side the marginal rate of substitution of energy for all other goods and services must be equal to the ratio of their prices (and, hence, their marginal costs).[2]

Market Response to Nonmarginal Price Changes

It is clear that in today's world market for energy, and for hydrocarbon products in particular, prices are not determined on the basis of free or competitive markets largely due to the actions of major oil producing countries and oil companies. Nevertheless, given such externally determined prices, market responses will tend to lead to an optimal allocation of resources between the use of energy and all other goods and services, at least on the consumer side of the market. This has been illustrated by the diagrams in Figure 3.1.

The consumer's initial equilibrium position, given his income Y, the price of energy P_{E1} and the price of all other goods and services P_O; is established at A_1. I_1 represents the consumer's optimal indifference curve which reflects his willingness to exchange, at the margin, additional units of energy for sacrifices in all other goods and services (or vice versa). The shape of the indifference

---

[1]See, for example: Edwin Mansfield, *Principles of Microeconomics,* W. W. Norton & Company, N.Y., 1980, Ch. 4 and 7–10.
[2]For a more detailed exposition of the economic efficiency of the basic marginal cost pricing rule, see Chapter 4.

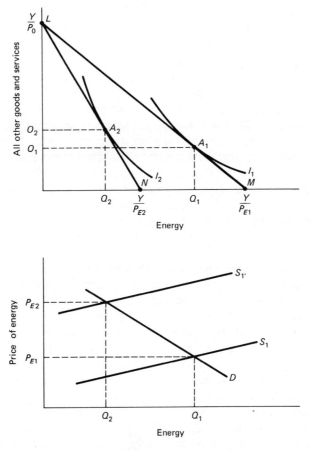

Figure 3.1.   Consumer response to rising energy prices; static case.

curve is convex to the origin on the assumption that the marginal value of additional units of energy declines, just as the marginal value of all other goods and services increases as more and more of them have to be given up.

Assume a relative price increase of energy from P_{E1} to P_{E2}, resulting in a shift of the supply curve S_1 to $S_{1'}$, with income and the price of all other goods and services remaining constant. As a result the consumer's budget (i.e., income) line is shifted from LM to LN. While he still can acquire the same amount of all other goods and services with his given income Y, fewer units of the now higher-priced energy can be bought. This relative price increase reduces consumer welfare and forces the consumer onto an inferior indifference curve such as I_2. A new equilibrium position will be reached at A_2 where once again the marginal equality between the ratio of prices and the marginal rate of substitution is reestablished, albeit at a lower level of overall consumer welfare and energy consumption Q_2.

In the real world of changing incomes, expanding economies, lock-in effects, changing technology, changing expectations, and dynamic responses on both the supply and demand side, the adjustment process is likely to be more complicated than depicted in Figure 3.1. Such a dynamic response path has been shown in Figure 3.2. The initial energy supply curve is given by S_1 and demand by curve D_1. The resulting equilibrium price in period 1 is P_1. In the absence of any disturbance, the equilibrium quantities and prices would grow to Q_2 and P_2 in period 2, and Q_3 and P_3 in period 3 respectively. Given an energy price shock that displaces the market supply curve to $S_{1'}$, price in period 1 increases to P_1' while quantity consumed decreases to Q_1'. This reduction is apt to be relatively small however, largely because of the lock-in effects of existing appliances and equipment, modes of behavior, and locational factors that determine transportation needs.

This observation is supported by empirical evidence. As various studies have shown, the short-run own price elasticity of residential energy demand is low, ranging from $-.1$ to $-.5$,[3] with industrial elasticities somewhat, but not very much, higher. Another factor that effects consumption in the intermediate term is expectation. For example, after the initial shock of the first oil price increase in 1973 wore off, American car buyers returned en masse to the gas-guzzling, full-sized cars they had used in the past, expecting that gas prices would eventually return to their previous, much lower levels. However, once they had purchased such cars, they were in fact "locked-in" regardless of the costs of gasoline. Because of this interaction of lock-in effects and slow changes in perceptions, the so-called "short run" may in fact stretch over a considerable period of time—sometimes a period of several years. In South and East Asia, for example, petroleum consumption between 1973 and 1978 not only increased in absolute terms (by some 55%) but also in relative terms, from 74% of total commercial energy consumption in 1973 to 76% in 1978.[4] This, in part, reflects the long lead time in the design and construction of energy using industrial and other equipment. An oil-fired powerplant, once committed and under construction, may require several years to completion and will have a useful, technical life of over 25 years; it may not be easily replaceable by a nonoil using facility and such replacement could require many more years to come on line.

[3]Robert S. Pindyck, *The Structure of World Energy Demand,* M.I.T. Press, Cambridge, MA, 1979, Table 4.12. According to the World Bank, the average short-run price elasticity of industrial countries is -0.2 and the long-run elasticity -0.4 (ranging between -0.2 and -0.6). For developing countries the long-run elasticity is estimated at -0.3 (-0.1 to -0.6). *From:* World Bank, *World Development Report 1981,* Washington, DC, August 1981, Table 4.3.
[4]T. L. Sankar and G. Schramm, *Asian Energy Problems,* Praeger, New York, 1982, Annex 2.

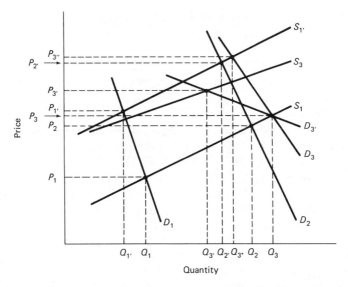

Figure 3.2.   Short-run versus long-run consumer response to changing energy prices.

In the meantime, oil consumption increases, regardless of price. These long lags in the adjustment of energy consumption to higher energy prices have been indicated in Figure 3.2 by the shift of the demand curve from D_1 in period 1 to D_2 in period 2; demand is still highly inelastic with respect to price and relative consumption falls only from Q_2 to $Q_{2'}$ while total consumption has increased as a result of economic growth and increases in income.

Ultimately, however, these constraints to changes in energy consumption are gradually removed. Old, high-energy consuming equipment and appliances can be replaced by more efficient ones. Energy equipment manufacturers, in turn, develop more energy efficient engines and appliances, responding to the changed relative costs of energy, capital, and labor. As a result, energy consumption relative to total output and consumption of all goods and services declines. Energy demand, in the long run, becomes apparently much more elastic, although there are obvious limits to the adjustment process (at least as long as the new, higher energy price remains constant relative to the costs of labor and capital).

Another important development affecting the ultimate equilibrium price[5] is

_____

[5]It should be noted, however, that there really is no such thing as an "ultimate" equilibrium price. We live in a world with constant changes in income, tastes, technology, population, and locations of economic activities. As a result, the demands for and supply of energy change as well, and so does its price.

that the higher market price for energy attracts new entrants on the supply side; energy resources that appeared submarginal in economic terms at the previous price become attractive (e.g., petroleum from tar sands and shale, oil and gas from the Arctic, the North Sea, or Siberia) and are being developed. As these new resources enter the market (which, however, may take years or even decades), they start to compete with established supplies. This, in turn, exerts pressure on prices, particularly if the established, higher oligopolistic price leaves a high and profitable margin for the new suppliers. Ultimately, therefore, market prices will decline to the costs of supply at the margin, although these costs may be substantially higher than the incremental costs of the intramarginal suppliers who might reduce supplies in order to maintain prices.

The overall results are likely to be that quantities consumed as well as prices will be lower than they would be in the absence of these long-term adjustments on both the demand and supply sides. Without them, price would be $P_{3''}$ and quantity consumed $Q_{3''}$. With the adjustments, prices are reduced to $P_{3'}$ and quantities consumed to $Q_{3'}$.

What is important to note, in addition, is that these shifts in demand and supply are long-term in nature. A structure, once insulated, will require less energy input for heating or cooling regardless of the price of energy. Once these technical adjustments on the demand and supply side have been made they will continue to operate at least until the equipment or operations that they support have to be fully replaced. Even then, the new technologies and habits, developed in response to high energy costs, will remain operative. Hence the shift to lower energy consumption and use will be permanent even if prices fall to lower levels once again.

What the preceding discussion shows is that hp.t, given time, market forces tend to bring about the necessary adjustments towards an efficient equilibrium between supply and demand, without overt interference by nonmarket forces. However, market forces do not account for equity issues, nor do they account for externalities caused by energy production and utilization patterns. These issues are discussed in the following sections.

3.3   EQUITY ISSUES, MARKET FAILURES AND NONMARKET OBJECTIVES

Exogeneous increases in world market prices raise serious equity and efficiency issues for domestic energy pricing. If free market forces are allowed to prevail, prices will increase to the levels dictated by the dominant suppliers (at least in the short run). This will not only bring about a massive income shift from domestic energy users to foreign suppliers, but also to domestic producers. If, on the other hand, domestic producers are constrained through price controls to sell at their current marginal costs, domestic supplies will be lower than they should in terms of efficiency criteria.

This has been illustrated by Figure 3.3. Initially, prices are assumed to reflect world-wide marginal costs as given by MC_W. Domestic marginal costs MC_D are depicted by the line ADH. This results in domestic consumption of Q_0 at prices P_0. Of the total, domestic producers supply Q_{D1}, while the balance is imported. Domestic producers enjoy a producer surplus equal to area ABCD, while foreign producers have one equal to area KCJ. Consumer surplus is bounded by BJI and the intersection of the demand curve D_M with the price axis.[6]

After world market prices have been raised by concerted action to P_W, domestic prices, if uncontrolled, would rise to the same level. This would result in an immediate windfall profit gain to domestic producers equal to area BFHC. It is the sheer magnitude of these potential windfall gains (which represent a direct income transfer from domestic consumers to domestic producers) that led most national governments, including those of the United States and Canada, to freeze domestic producer prices at levels close to prevailing domestic marginal costs (i.e., P_0 in Figure 3.3), following the first oil price shock in 1973. Such price controls require complex, administered pricing averaging schemes (between low-priced domestic and high-priced imported fuels). These are not only cumbersome and costly, but also invite cheating and fraud.

Another important consequence of such price controls is that domestic output remains stagnant at levels determined by the controlled price[7] which means that imports, as represented by the difference between Q_M and Q_{D1}, are larger than they should be if efficiency criteria were to be applied. If, instead, domestic prices were allowed to rise to the world market level, this would increase domestic supplies from Q_{D1} to Q_{D2} and result in a net efficiency gain to the domestic economy equal to area CGH. Imports would be reduced at least by the same amount[8] and pressures on foreign exchange balances would be

---

[6]Producer surplus is defined as the total difference between market price and the marginal cost schedule of suppliers. It is called a "surplus" because producers would normally be willing to supply the same quantity if they would be paid no more than their marginal costs. Similarly, consumer surplus measures the intramarginal difference between the market price and the willingness of consumers' to pay, as shown by the demand schedule.

[7]In fact, it may fall if producers expect price controls to be lifted later so that it would be worthwhile for them to forego current income from price-controlled sales for higher income later that compensates them for the costs of waiting (given by the net reduction in current incomes times the opportunity costs of funds).

[8]The reduction would generally be greater if domestic market prices were maintained at levels below world market prices, because this would result in higher domestic consumption rates than those given by Q_M.

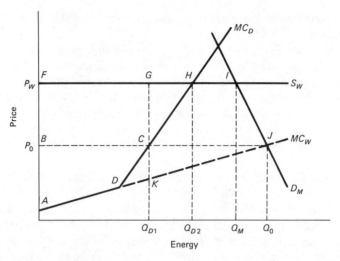

Figure 3.3.   Equity and efficiency issues in domestic energy pricing.

reduced. It would also result in less dependence on foreign supplies, an important security consideration.[9]

Inspite of these potential efficiency gains, most national governments with significant domestic hydrocarbon production have been unwilling initially to allow prices to rise to world market levels. Ultimately, most of them have done so, but only after instituting various measures of taxing away at least some of the windfall profits to domestic producers.[10] It is interesting to note that this equity issue was much less important in those countries that depend almost entirely on imports. For them the income transfer did not go to domestic producers, but to foreign suppliers over which they had no control. This eliminated the political ramification of massive income transfers from one domestic group to another.

---

[9]A recent study has estimated that deregulation of natural gas prices in the United States would result in an increase of domestic gas production equivalent to some 3 million bbls/day of oil equivalent, leading to a reduction of some \$40 billion/year in payments to foreign suppliers (in 1980 prices). *From:* Glenn C. Loury, *An Analysis of the Efficiency and Inflationary Impact of the Decontrol of Natural Gas,* Natural Gas Supply Association, April 1981.
[10]The USA did so by imposing a crude oil windfall profit tax which became effective February 29, 1980. This tax has been described officially as a "temporary excise, or severance, tax applying to taxable crude oil produced in the United States." *From:* Stephen L. McDonald, "The Incidence and Effects of the Crude Oil Windfall Profits Tax", *Natural Resources Journal,* Vol. 21, No. 2, April 1981, p. 331.

Decreasing Costs, Monopolies, Oligopolies

Several of the most important energy supply systems are subject to decreasing average costs. This means that a single firm supplying the whole available market can always do so at lower prices than any combination of two or more competitive, smaller firms. Such situations lead to the formation of so-called "natural" monopolies, i.e., the emergence of a single firm as the sole supplier in a given marketing region. Because it potentially results in lower costs of supply to everyone, the establishment of such single suppliers is in the economic interest of everybody. Electric utility companies and natural gas supply systems are the most prominent examples, although petroleum refineries or coal mines may also become natural monopolies in limited, regional markets with high transport costs from outside suppliers.

While the formation of natural monopolies should be encouraged where appropriate, their pricing and output decisions must be regulated. This is so because profit maximization principles would dictate to the management to expand output only to the point at which marginal revenue is reduced to marginal costs. From society's point of view, however, the optimal output level would be reached at the point at which (long-run) marginal costs are equal to willingness to pay at the margin, i.e., to demand. This has been illustrated in Figure 3.4. For the firm the optimal output is given by Q_p. If output is

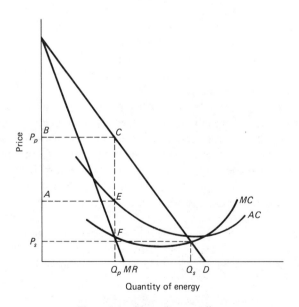

Figure 3.4.   Natural monopolies.

expanded beyond this level, net income declines. From society's point of view, however, optimal output levels are given by Q_s, the point of equality of marginal costs and willingness to pay (see also Chapter 4).

A second, important issue calling for the regulation of natural monopolies (or any other form of monopoly, cartel, or oligopoly arrangement that would give control over market prices to suppliers), is the fact that there is a potential for high, extraordinary profits, particularly if the demand for the output is highly inelastic. This, as we have seen, is almost always the case in the various areas of energy supply. In Figure 3.4, this potential for extraordinary profits from market control has been indicated by area ABCE, the excess of gross revenues over total costs at output level Q_p. Equity as well as efficiency considerations, therefore, call for outside interference in such firm's pricing and output decisions.

This raises complex regulatory and pricing issues. These are discussed in greater detail in Chapter 5. Here it suffices to say that important sectors of our overall energy supply systems[11] must be regulated, if optimal allocation decisions are to be forthcoming.

Externalities

Even if all energy markets were purely competitive, efficiency in production or consumption might not be achievable because of the incidental creation of so-called externalities, or side effects. These lead to differences between private and social costs and benefits. Externalities have been defined as consequences stemming from an action for which the party responsible cannot be charged (if it is a cost) or for which revenues cannot be collected (if it is a benefit). Therefore, it will normally not take these consequences into account in its production or consumption decisions.

The best known and widely discussed types of externalities are pollution and/or health effects. Sulphur or particle emissions by refineries or power-plants, thermal pollution from heat discharges into waterways, or the potential for radiation leaks from atomic powerplants are prominent examples of negative externalities from energy production. Smog creation from automobile exhaust, chemical cleaning establishments, and gas filling stations are examples of negative externalities resulting from energy consumption decisions. In developing countries, an important source of externalities results from the over-cutting of forests that lead to erosion and increased flooding of downstream regions. Traffic congestion, traffic fatalities, or noise pollution from aircraft are other examples. Positive externalities are less readily found. Examples are the

---

[11] In the USA, for example, over 40% of all energy is supplied by natural gas or electricity, the two forms of energy that need natural monopolies in distribution in order to keep unit costs down.

construction of access facilities to energy facilities in remote areas (e.g., oil or gas fields in the Arctic, Siberia, or in various developing countries) that then reduce transportation costs for everyone.

In situations in which the costs (or benefits) of external effects can be quantified and converted into economic magnitudes, efficient resource allocation decisions may be possible. This has been illustrated in Figure 3.5 which indicates the respective optimal private and social production or consumption decisions in the presence of negative production or consumption externalities. If these externalities are not taken into account, output will expand to Q_p. Externalities may be created in the production of energy (e.g., pollution from powerplants) resulting in combined private and social marginal costs of MSC. Optimally, output should be reduced to Q_s. In the case of a consumption externality (e.g., pollution from cars or congestion) the socially desirable level of demand would be measured by D_s, the private willingness to pay as given by D_p minus the social costs of consumption imposed on others. Equivalently, these costs could be added to the marginal private costs. In either case, consumption should be reduced from Q_p to Q_s to bring about efficiency in resource use.

Unfortunately, in most cases the costs or benefits from external effects cannot be measured very easily. Furthermore, because their distributional impact on different groups of society are different and compensation payments are usually impossible to negotiate, assessments of the benefits and costs are extremely difficult and almost always highly controversial. This, however, does

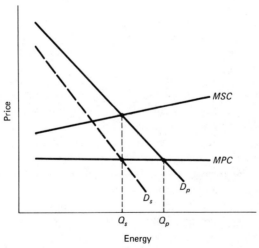

Figure 3.5.   Private versus social output/consumption decisions in the presence of negative production or consumption externalities.

not abstract from the necessity of taking them into account in energy output and consumption decisions. This calls for intervention in the free interplay of market forces that are shaped by private decision criteria only.

Public Goods

Public goods are defined as goods whose provision to one person or party makes them automatically available to all at zero additional costs. Common examples cited in the economic literature are national defense, police protection, or the provision and maintenance of clean air. In many cases, public goods are simply some special form of positive externality. In the case of energy, a typical example is the creation of downstream benefits as a result of upstream reforestation, undertaken primarily for the production of firewood. Another example is the incidental recreational or fishery benefits created from the construction of reservoirs for power projects. In the USA, such recreational benefits have been found to be a major contributor to the total identified benefits from multipurpose water resources developments, while in some developing countries the reservoir fisheries benefits created by power dams are comparable to the value of the electric energy produced.[12] However, the agencies responsible for energy production usually have no means to collect for the incidental benefits created. Therefore, in their design and management decisions, they are likely to disregard them. Reservoirs might be operated to maximize power benefits, with the result that water levels might fluctuate drastically, destroying potential recreational or fisheries benefits. In the interest of overall social benefits, therefore, public interference in these investment and management decisions are called for.

Macroeconomic Effects and National Security Considerations

The massive macroeconomic effects of the oil price changes in the 1970s have already been discussed in Chapter 1. Coordinated fiscal, monetary, and regulatory policies are needed to reduce their negative impacts on real income, inflation, the balance of payments, and levels of foreign indebtedness. Market intervention may be necessary to accelerate the switch-over from imported hydrocarbon sources to substitutes, be they domestic energy resources or replacements in the form of more capital and/or labor intensive, but more energy efficient, structures and appliances.

National security considerations, both in terms of economic and political/ military security requirements, call for market intervention as well. For exam-

---

[12]This, for example, was the unexpected result of the construction of a major power dam in Thailand.

ple, it will rarely be the case that levels of emergency oil storage or diversification of supply sources will coincide with those dictated by free market considerations only.

3.4   THE ROLE OF GOVERNMENT IN ENERGY PRICING AND DEMAND MANAGEMENT

In practically all countries of the world, including those that fully subscribe to a free market philosophy, government plays a pervasive role in the allocation and pricing of energy resources. In order to show how important and multifaceted these governmental policies are, we briefly review the many policy instruments that affect prices.

A government's influence on price can be direct through such means as ownership of supply sources and price controls, or indirect, through taxes, import duties, subsidies, market quotas, taxes on energy-using equipment, and government-guided investments in energy resources. In many countries overlapping policies, regulations, taxes, subsidies, and ownership patterns, often imposed by several layers of government, create a thick tangle of cost and price related consequences that are not easily separated from each other. These problems are exacerbated by the pervasive joint product-nature of most energy resources, such as the multiproduct output of a refinery complex of the multiplant supply network of a typical electric utility operation. Given the multiple, and oftentimes mutually inconsistent goals of government policies (see Section 3.6) that directly or indirectly affect energy use and prices, it is not surprising that in many cases some of these policy effects counteract each other. Another consequence of governmental pricing policies may be that short-run objectives create long-run problems, for example, current price controls or heavy taxation on crude-oil production that lead to a reduction in exploration activities and ultimate supply shortages.[13]

Government's role appears to be most obvious in cases in which the respective energy supply system is owned by the government itself. Such government ownership is common in electric utility systems. It is less so in petroleum and natural gas production, refining, and distribution, where, however, joint ownership is frequent, or where government-owned systems operate side by side with privately-owned ones. Coal mining and distribution more often than not is privately-owned. But a central government's direct influence on prices or supply patterns of publicly-owned energy supply systems may not be strong if these operations are under the management control of an independent board, or are owned by a lower level of government. In the USA, for example, the

---

[13]This in recent years, has been an important problem in many countries such as Argentina, Colombia, Peru, and Ecuador, to mention but a few.

Federally-owned Tennessee Valley Authority in the Southeast and the Bonneville Power Authority in the Northwest are highly independent organizations that control their own pricing policies, while in Colombia the major electric utility companies are owned by the respective municipalities who set prices according to their own internal priorities.

Another major pricing policy instrument, direct price controls, has become the rule, rather than the exception in most countries in recent years. Such price controls may be imposed at the production level, if the respective energy resources are domestically produced, or they may be imposed at the refinery, transportation, wholesale, or retail level. Frequently, multiple price controls are applied at several levels simultaneously. In Thailand, for example, ex-refinery prices for petroleum products are controlled by a complex, product by product formula that takes account of world market prices for imported crude or petroleum products. At the same time, retail prices at the pump are also fully controlled. Both in the USA and Canada price controls are imposed at the production level[14] with cross-subsidies paid to those regions or producers that have to rely on higher-priced imports.[15] In addition, in the U.S. gasoline retail prices were subject to ceiling prices for several years.

Price controls may be applied to privately as well as government-owned energy supply systems. In Mexico, for example, the Federal government sets product prices for PEMEX, the government-owned petroleum and natural gas production and distribution monopoly. It also imposes excise taxes of 12% on PEMEX-supplied petrochemicals and 16% on all other products, and a 50% tax on all crude oil exports.[16]

In the majority of countries with price controls, prices are usually set to reflect specific developmental, as well as socioeconomic and income distributional goals. Some products, such as diesel fuels for transport and kerosene for lighting and cooking may be heavily subsidized (i.e., sold at less than full supply costs), while others, such as gasoline, might be heavily taxed. In a number of countries, such subsidy policies have led to substantial financial losses of the respective energy supply system. Such losses then have to be made up through governmental subsidies (as in the case of Argentina's utility companies) or result in a lack of maintenance, deteriorating service levels, and inability to expand output in order to meet growing demand.

Just as frequent, but at least financially more responsible, are government-

---

[14]In the USA, price controls for oil were lifted in February 1981, but most domestic production remains subject to a windfall profits tax.

[15]For an analysis of the rather complex U.S. price control and cross-subsidy program see Paul MacAvoy, *Federal Energy Administration Regulation,* American Enterprise Institute for Public Policy Research, Washington, D.C., 1977.

[16]U.S. Department of Energy, Office of International Affairs, *The Role of Foreign Governments in the Energy Industries,* Washington, D.C., Oct. 1977, p. 59.

imposed cross-subsidies from one user group to another. In Thailand, for example, rural domestic electricity rates in 1978 were an average 3% lower than rates charged to urban customers in the Bangkok Metropolitan area, while marginal supply costs were some 36% higher for the former.[17]

Another common form of mandated cross-subsidies are uniform tariffs or prices for specific energy sources to specific customer groups on a countrywide basis regardless of delivery costs. Such national tariffs for electricity are used in Thailand, Mexico, and Bangladesh, for example.[18] In Pakistan, they are applied to retail prices of gasoline and diesel fuel, with the result that oftentimes neither is available in outlying regions (at least at official prices).

In almost all countries, delivered energy prices are substantially affected by import and export duties, and by excise and sales taxes that are imposed at various levels in the production and distribution stages. Discriminatory rates that differentiate among user groups and products are common. Quite often, import duties, excise taxes and sales taxes are levied successively at several stages of the import production-distribution-final sales cycle, and often these various tax levies are imposed by different levels of government (central, state, and local). Excise tax rates, in some countries, may reach several hundred percent of the original product price for some products, but may be zero for others. Export taxes are common in countries that are exporters of energy products.[19]

Less direct, but nevertheless highly significant in their effects on final energy product prices are a whole host of government levies and fees. Property taxes, franchise fees, water rights and user charges, royalties, and charges for exploration rights (land bonuses) are the more important ones among them.

Even more indirect, but obviously quite significant are the various schemes for partial government ownership or profit sharing that have become popular in recent years, particularly in the granting of exploration and development rights for oil and gas resources. Such arrangements vary from country to country and usually from one concession to the next. The specific terms and conditions depend obviously on the respective bargaining power of the government vis a vis the prospective concessionaires and the riskiness of the undertaking to the latter both in economic and political terms.[20]

---

[17]See the case study on Thailand's power sector.

[18]In Bangladesh, uniform prices are charged countrywide, even though the fuel costs of the isolated western grid system which depends on oil are some 15 times higher than those of the natural gas-based eastern grid. One of the consequence is that the government-owned utility company incurs huge losses.

[19]In Colombia a rather special form of export tax was applied to crude oil exports until 1977. Export earnings from such sales were converted to local currency at a special, discriminatory lower exchange rate.

[20]For producer's of Norway's rich share of North Sea oil and gas deposits, for example, governmental levies amount to between 85 to 90% of gross revenues. *From: Financial Post,* Toronto, Oct. 18, 1980, p. 25.

Energy prices may also be affected to a significant degree by indirect allocation policies. Import quotas for certain types of energy resources may favor the use of higher-priced, less attractive domestic substitutes. Mandatory, interest-free deposit requirements well ahead of the date of import may represent another form of import duty, particularly in countries with high rates of inflation. Energy prices may also be affected by conservation rules (i.e., the maximum extraction rates for oil and gas or cutting rates for firewood) or by hydropower reservoir operating rules that may give priority to other uses, such as irrigation. Another source of governmental influence on prices may result from environmental rules and regulations. These, in the case of developed countries, have become quite prominent in recent years.[21]

In many countries, direct or indirect subsidies have had significant effects on energy prices. Mandated lower tariffs or user prices for specific user groups or products have already been mentioned above. Other subsidies may consist of special tax concessions (income tax holidays or reduced rates, accelerated depreciation rules, rebates of import duties on imported supplies or equipment), export bonuses, governmental loans or financial guarantees, outright grants, or direct subsidies for job creation in depressed areas are some of the policies and means that are used to entice and promote energy developments. All of them, of course, eventually affect energy supply prices.

More indirect are the effects of government investment or regulatory policies. For example, strong efforts in reforestation may ultimately reduce firewood and charcoal prices. High taxes on automobiles may reduce ownership and, hence, gasoline consumption—combined with price-controlled domestic petroleum production, this may reduce the costs of such products to other users.

This list of governmental actions and policies that could and do affect energy prices could be readily expanded. Suffice it to say that governmental influence is pervasive and can be expressed through many policy instruments. The major difficulty is that the true, net effects of any single policy or policy measure can rarely be defined with precision in the absence of an overall evaluation framework, since all of them are imposed within the dynamics of complex, interacting economic forces.

3.5   THE NEED FOR INTEGRATED ENERGY PLANNING AND DEMAND MANAGEMENT

Because of the many interactions and nonmarket forces that shape and affect the energy sectors of every economy, decisionmakers in an increasing number

---

[21]Requirements to install sulfur-dioxide removal equipment in coal-fired powerplants in the USA, for example, increase plant investment costs by 20 to 30% and reduce net energy output by 4–6%. *From:* National Academy of Sciences, *Implications of Environmental Regulations For Energy Production and Consumption,* Washington, D.C., 1977, p. 103 ff.

of countries have realized that energy sector investment planning, pricing and management should be carried out on an integrated basis, e.g., within a national planning framework which helps analyse energy policy options ranging from short-run supply-demand management to long-run planning.[22] However, in practice, most policies are still carried out on an ad-hoc and, at best, regional, partial, or subsectoral basis. Thus, typically, electricity and oil subsector planning have traditionally been carried out independently of each other as well as of other energy subsectors. Environmental planning focused on pollution effects of energy systems, but gave little attention to resulting consequences in terms of alternative choices of energy resources and the overall costs of these choices to the economy. As long as energy was relatively cheap such partial approaches and the resulting economic losses were acceptable, but lately, with rising energy costs (especially of oil), drastic changes in relative fuel prices, and increasing substitution possibilities, the advantages of more integrated energy policies have become evident.

Coordinated energy planning and pricing require detailed analyses of the interrelationships between the various economic sectors and their potential energy requirements[23] on the one hand, and of the capabilities and advantages and disadvantages of the various energy subsectors such as electric power, petroleum, natural gas, coal, and traditional fuels (e.g., firewood, crop residues, and dung) to satisfy these requirements on the other. Nonconventional sources, whenever they turn out to present viable alternatives, must also be fitted into this framework. The discussion applies both to the industrial and the developing world. In the former, the complex and intricate relationships between the various economic sectors and the prevalence of private market decisions both on the energy demand and supply sides make analysis and forecasting of policy consequences a difficult task. In the latter, substantial levels of market distortions, shortages of foreign exchange and human and financial resources for development, larger numbers of poor households whose basic needs somehow have to be met, greater reliance on traditional fuels, and relative paucity of energy as well as other data, add to the complicated problems faced by energy planners everywhere.

The broad rationale underlying energy planning is the need for a flexible and continuously updated energy assessment or master plan (EMP) that will promote the best use of energy resources in order to further overall socioeconomic development and improve the welfare and quality of life of a country's citizens.

---

[22]See, for example: U.S. Department of Energy, *National Energy Plan II*, Washington, D.C., May 1979. Government of Canada, *The National Energy Program*, Ottawa, Oct. 1980.

[23]The word "requirement" should not be interpreted as absolute "needs." As will be seen from the discussions in this book, the role of energy demand management is to identify and utilize to best advantage the many trade-offs and substitution possibilities between energy consumption, income, capital, labor, and human welfare.

Energy planning, therefore, is an essential part of national economic planning, and should be carried out and implemented in close coordination with the latter. However, the word "planning", whether applied to the national economy or the energy sector in particular, does not imply some rigid framework along the lines of centralized and fully planned economies. Planning, whether by design or deliberate default, takes place in every economy, even in those where so-called free market forces reign supreme. In energy planning, the principal emphasis is on the detailed and disaggregated analysis of the energy sector, its interactions with the rest of the economy, and the main interactions within the various energy subsectors themselves.

Supply and demand management make it easier for the energy policymaker to forecast and achieve energy supply-demand balances, thus preventing major economic disruptions and consequent reductions in national welfare. Supply management includes identification and optimal exploitation of all energy resources, investment planning, transformation, refining and distribution of energy, and so on.

Demand management includes all means of influencing the magnitudes and patterns of energy consumption. As discussed later the so-called hard tools of demand management such as physical controls and rationing, mandatory regulations relating to the pattern of energy production and use, and technological options such as energy saving retrofits, are most effective in the shorter term. The "soft" tools of demand management such as pricing, taxation, financial incentives and subsidies, and education and propaganda, are more useful in the medium and long run.

3.6 SCOPE AND OBJECTIVES OF PLANNING, POLICY TOOLS, AND CONSTRAINTS

The scope of integrated national energy planning (INEP) may be clarified by examining the hierarchical framework depicted in Figure 3.6. At the highest and most aggregate level, it must be clearly recognized that the energy sector is a part of the whole economy. Therefore, energy planning requires analysis of the links between the energy sector and the rest of the economy. Such links include the input requirements of the energy sector such as capital, labor, raw materials, and environmental resources such as clean air, water, or space, as well as energy outputs such as electricity, petroleum products, woodfuel, and so on, and the impact on the economy of policies concerning availability, prices, and taxes in relation to national objectives.

While some of these relationships are at the macrolevel—such as foreign exchange requirements for energy imports, or investment capital requirements for the energy sector—others are more directly linked with and limited to specific activity levels. For example, policies affecting the transport sector such as

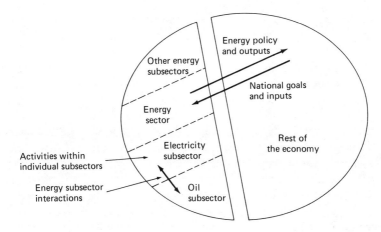

Figure 3.6.   Hierarchy of interactions in Integrated National Energy Planning (INEP).

subsidies to public transport, construction or nonconstruction of superhighways or airports, the levels of license fees for vehicles or excise taxes on diesel versus gasoline vehicles, tax credits for energy conservation, pollution control legislation or specific end-use planning policies may have as profound an impact on energy demands as more overall broad-based energy pricing, allocation, or supply management policies.

The second level of INEP treats the energy sector as a separate entity composed of subsectors such as electricity, petroleum products, and so on. This permits detailed analysis of each subsector with special emphasis on interactions among them, substitution possibilities, and the resolution of any resulting policy conflicts such as competition between natural gas, bunker oil, or coal for electricity production, diesel or gasoline use in transport, kerosene and electricity for lighting, or woodfuel and kerosene for cooking.

The third and most disaggregate level pertains to planning within each of the energy subsectors. Thus, for example, the electricity subsector (or regionalized subsectors) must determine its own demand forecasts and long-term investment programs; the petroleum subsector, its supply source, refinery outputs, distribution networks, and likely demands for oil products; the woodfuel subsector, its consumption projections and detailed plans for rotation or reforestation, harvesting of timber, and so on.

In practice, the three levels of INEP merge and overlap considerably. For example, a class of issues that affects both macro and micro aspects of energy planning are those related to energy substitutions or energy conservation. (See also Chapter 6.) Within certain limits many energy resources are substitutes for each other although convenience in use and overall systems cost may vary widely. Hence appropriate supply and pricing policies may bring about significant shifts in energy demand for specific energy resources, at least in the long run.

Similarly, individual actions or deliberate policies aimed at bringing about energy conservation, i.e., reductions in energy usage relative to levels that would prevail in their absence, may significantly affect energy consumption.[24] Such conservation may simply be achieved at the expense of some loss in personal comfort or convenience (like reducing thermostat settings, driving within mandated speed limits, or switching off lights in unoccupied rooms). Other means may consist of the substitution of energy by capital or labor, for example, the replacement of pilot lights by electronic switches, the reduction in the curb weight of automobiles, recirculation of process heat in industrial plants through better engineering or lighter materials, or the installation of insulating materials in buildings.[25]

Objectives

Integrated energy planning should result in the development of a coherent set of policies which can meet the needs of many interrelated and often conflicting national objectives. For most of them, energy in its many varied forms will be needed. Energy planning, therefore, must be developed to meet the overall national objectives as efficiently as possible. The specific tasks of energy planning include: (a) the determination of the detailed energy needs of the economy to achieve growth and development targets, (b) the choosing of the mix of energy sources to meet future energy requirements at lowest costs, (c) the minimizing of unemployment, (d) the conservation of energy resources and elimination of wasteful consumption, (e) the diversification of supply and reduction of dependence on foreign sources, (f) the meeting of national security and defense requirements, (g) the supply of basic energy needs to the poor, (h) the saving of scarce foreign exchange, (i) the contribution of specific energy demand/supply measures to contribute to possible priority development of special regions or sectors of the economy, (j) the raising of sufficient revenues from energy sales to finance energy sector development, (k) price stability, (l) the preservation of the environment, etc.

---

[24]The significant reduction in petroleum consumption in the United States from an average of 18.5 million bpd in 1979 to 17.1 million bpd in 1980 is largely attributed to oil energy conservation measures.
[25]For example, estimated savings in specific vehicle consumption of petroleum products between 1979 and 1985 in the USA are in the 30% to 50% range. *From:* Marc H. Ross and Robert H. Williams, *Our Energy Regaining Control*, McGraw-Hill, New York, 1981, Figure 9.4.

Policy Tools and Constraints

To achieve the desired objectives, the policy tools available to a government for optimal supply-demand planning and management include: (a) physical controls; (b) technical methods (including research and development); (c) direct investments or investment-inducing policies, tax policies; (d) education and propaganda; and (e) pricing. Since these tools are interrelated, their use should be closely coordinated for maximum effect.

Physical controls are most useful in the short run when there are unforseen shortages of energy. All methods of limiting consumption by physical means such as load shedding, or rotating power cuts in the electricity subsector, reducing or rationing the supply of gasoline or banning the use of cars during specified periods, are included in this category. However, physical controls can also be used as long-run policy tools. Examples are the U.S. federal legislation that prohibits the construction of additional oil or gas-fired powerplants, and the prohibition of exports of Alaskan crude oil.

Technical means applied to the supply of energy include the determination of the most efficient means of producing a given form of energy; choice of the least-cost or cheapest mix of fuels; research and development of substitute fuels such as oil from shale, coal, or natural gas; the substitution of alcohol for gasoline; and so on. Technology may be used also to influence energy demand, for example, by promoting the introduction of higher efficiency energy conversion devices such as more fuel-efficient automobiles, better stoves for woodfuel, research into solar heating devices, etc.

Investment policies may have a major effect on energy consumption patterns in the long run. The extension of natural gas distribution networks, the building of new power plants based on more readily available fuels such as coal, and the development of public urban transport networks are just some of these policies. It should be noted that many of them may well be undertaken by sectors other than energy—examples are investments in transportation facilities or the systematic installation and/or electrification of deep-well irrigation pumps. Close cooperation between the energy administration and the planning authorities of these other sectors are obviously called for.

Taxation is a major policy tool that can profoundly affect energy consumption patterns in the long run. For example, the long-standing difference in automobile engine sizes and fuel consumption between North American and European cars (which antedates the 1973 oil crisis) is basically the result of high European taxes on fuels and engine displacements, with low taxes on the former and practically no taxes on the latter in the U.S. and Canada.

The policy tool of education and propaganda can help to improve the energy supply situation through efforts to make citizens aware of cost-effective ways to reduce energy consumption, of the energy use implications of specific appliances or vehicles, and of the potential for substitution of energy by capital (e.g., proper insulation).

Pricing is a most effective means of demand management especially in the medium and long run, as discussed further in Chapters 4 and 5. Controlling the use of energy through coordinated use of these policy tools is the principal goal of energy demand management. Practical aspects of such policy coordination are discussed in Chapters 6 and 8.

In the context of developing countries we generally face additional constraints on energy policies. There may be severe market distortions due to taxes, import duties, subsidies, or externalities which cause market (or financial) prices to diverge substantially from the true economic opportunity costs, or shadow prices. Therefore, on the grounds of economic efficiency alone we may have to use (second-best) shadow pricing decisions (see Chapter 4). However, these again may have to be modified in anticipation of energy user reactions that will be based on market prices rather than underlying economic cost considerations. Furthermore, there often are severe income disparities and social considerations which call for subsidized energy prices or rationing to meet the basic energy needs of poor consumers (see Chapter 5). Finally, there are usually many additional considerations that affect policy decisions, such as considerations of future investment requirements, financial viability and autonomy of the energy sector, and regional development needs, as well as sociopolitical, legal, and other constraints.

3.7 ENERGY PLANNING PROCEDURES

As explained earlier, INEP is the process which identifies the goals and objectives of an energy master plan (EMP). Since both the means and the ends are closely interlinked, describing the planning procedure provides better understanding of the plan itself, and vice versa. In the previous section the three hierarchical levels which define the scope of INEP were described, i.e., analysis of: (a) interactions between the energy sectors and the rest of the economy. (b) interactions among different subsectors within the energy sector, and (c) activities within each energy subsector. Next we will discuss the hierarchy of time horizons—short, medium, and long term—that must be considered in planning, and also outline the problems of uncertainty, before describing the INEP procedure.

Planning Horizons and Uncertainty

In the short-run, within a time scale of about one year, national energy planning is mostly concerned with supply-demand management decisions to meet unforseen problems. Even if past planning decisions and policies had been completely successful, given the information available or the time, unforeseen (and forseeable) problems are bound to arise from time to time. When such sudden difficulties occur (e.g., unavailability of hydroelectric power following a drought or shortages of petroleum-based fuels because of supply disruptions in oil producing areas) contingency plans including physical rationing, price surcharges or others must be used to minimize adverse effects on the economy and the various national objectives.

Energy planning for the medium term is more flexible because there is sufficient time to make significant policy changes or bring about some structural changes within the approximately 1 to 10-year time horizon involved. The most important decisions include the planning, evaluation, and implementation of energy projects, e.g., the building of new power stations or oil or gas pipelines, pricing, interfuel substitution, and conservation policies.[26]

Long-run energy planning horizons generally extend at least 10–20 years. They involve the least restricted strategies of all. Typically, a variety of alternative scenarios can be examined to determine the choice of less or more energy intensive patterns of economic development, the gradual changeover from dependence on some energy sources to others, optimum energy supply development programs (usually consisting of a series of individual but frequently sequential projects), such as long-range electric power system expansion, search for and development of new oil and gas deposits, the building of new major pipelines as those from Siberia to Central Europe or the Alaska gas pipeline,[27] reforestation programs, and so on.

While the distinction between long-, medium-, and short-run planning is convenient for conceptual and analytical purposes, it should be noted that in practice there are no sharp dividing lines among the three categories. There is a hierarchical relationship involved in which the short-run decisions and medium-run policies should merge as smoothly and consistently as possible with the long-run strategy. However, just as the short-run policies are mainly responses to unforeseen events, so must the medium- and long-run policies be modified to take account of changing circumstances. Any form of integrated energy planning and any basic energy master plan must be subject to continuous changes and adjustment to adapt to new circumstances including the emergence of new national goals and objectives.

Any type of planning must deal with the problem of uncertainty. However, uncertainty raises greater difficulties in energy planning because of this sector's widespread interaction with the rest of the economy, and its present vulnera-

---

[26]As will be shown later any of the latter policies are essentially medium- to long-run policies in terms of their ultimate effectiveness in changing energy consumption patterns.
[27]See also the case study on the Alaska gas pipeline.

bility to international events. Specific problems of uncertainty are caused by:
(a) long planning horizons for energy investment decisions, high capital intensity, long lead times required for energy resource development, the high capital
intensity of energy supply projects, and the danger of resultant "lock-in"
effects; (b) incomplete knowledge of the national energy resource base, and the
possibility of finding oil, gas, coal, etc., in the future; (c) uncertainties about
the reliability and costs of energy supplies from foreign sources; (d) changes in
the patterns of energy use; and (d) changes in the technology of energy supply
and end-uses. Therefore, it is important to consider various assumptions
regarding the future and to develop several alternative planning scenarious
based on these assumptions.

Socioeconomic Background

As shown in Figure 3.7 the INEP procedure itself may be broken down into
the following steps: (a) identification of the socioeconomic background and
national objectives; (b) energy demand analysis; (c) energy supply analysis; (d)
energy balance; (e) identification of possible resource constraints, impact analysis, and consistency checks; and (f) policy and energy plan formulation. These
divisions are conceptual ones; in practice there will be considerable overlap
among them.

Steps (a), (e), and (f) examine broad relationships with the rest of the economy, corresponding to the highest hierarchical level discussed earlier (see Figure 3.6), although the identification of resource constraints under (e) may indicate considerable microanalysis. Thus in step (a), national priorities and
objectives, as well as the overall development of the economy, say over the next
10 to 30 years, must be examined. Typical questions that should be addressed
concern economic targets like the desired rates of sectoral GNP growth, bal-

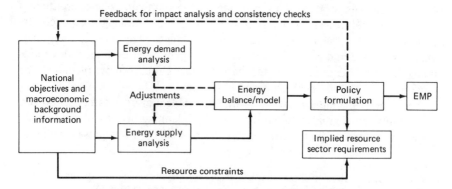

Figure 3.7. Basic steps in implementing INEP.

ance of payments, and overall savings/investment targets, and equity issues such as the minimum levels of energy for disadvantaged groups, or special requirements for the development of depressed regions or rural areas. Information is also required on the energy-use implications of overall or specific sectoral socioeconomic development plans, e.g., whether rapid industrialization will occur, where and of what type, whether projected agricultural growth will depend on significant use of energy intensive activities such as farm mechanization, pumped irrigation, heavy fertilizer use, etc. A related set of questions concern technically and socially feasible capital, labor, or interfuel substitution possibilities and policies, such as encouraging or discouraging households to switch from traditional/noncommercial fuels like woodfuel to commercial fuels like kerosene or gas, electrification of railways versus greater reliance on diesels or road transport, promotion of public transport and/or discouragement of the use of private automobiles for commuting purposes, and so on. Finally, projections are needed of future domestic and import availability and prices of fuels like crude oil or coal; the financial resource requirements for energy imports or the development of domestic energy sources; the development of appropriate new technologies that are optimal for the country given its underlying energy needs and available domestic resource endowments (of energy, labor and capital), including the development of nonconventional energy sources (e.g. solar, biogas, minihydro, coal or natural gas liquifaction, etc.); and the impact of specific energy supply and demand policies on the environment.

Supply and Demand Analyses

Steps (b), (c), and (d) relate mainly to the two lower levels in the INEP hierarchy shown in Figure 3.6, i.e., energy sector and subsector planning. It should be noted that these stages in INEP will be strongly influenced by the information and assumptions from the preceding step (a). While these steps deal with forecasts spanning the full long-run horizon of INEP, medium- and short-run projections should be developed within the same consistent framework. The principal objective of step (b), energy demand analysis, is to determine future energy requirements by type of fuel and by consumer category (or type of usage).[28] Past and present energy usage patterns must be analyzed in detail, relative to other factors such as prices, incomes, levels, and types of economic activity, supply constraints, stocks of energy using equipment, and so on. The results of this analysis form the basis for forecasting the structure of national energy demands. In general, the energy requirements of productive sectors of the economy, e.g., industry, may be analyzed on the basis of technological rela-

---

[28]For a more detailed discussion of energy demand forecasting see Chapter 7.

tionships and production functions underlying these activities. However, a thorough analysis of household energy demand, particularly for traditional fuels (which often constitutes the bulk of the fuel supply in developing nations) usually requires detailed surveys and economic studies.[29]

Energy supply analysis or step (c), involves the systematic determination of all possible future energy supply options, disaggregated by energy subsector. First, past and present data on energy resources availability and production, imports and exports, generation, storage, refining, transportation, distribution and retailing, financial and manpower requirements, costs, and output prices must be examined. Then this information together with projections of future domestic resource development projects and programs, energy subsector output capabilities, capacity to import or export, application of new technologies and nonconventional sources, financial, manpower, and organizational resources, etc., must be combined to provide forecasts of the availability and likely costs of the different forms of energy.

Energy Balance

Step (d), supply-demand balancing, consists basically of matching specific energy sources with corresponding uses. A simple but typical matrix for doing this is shown in Figure 3.8 where the rows indicate energy uses (including losses in processing, refining, transportation, retailing, and so on). Examination of past and present energy balances allows the energy analyst to determine the past evolution of supply and demand within a comprehensive framework, the bottlenecks that exist, and how supply and demand have to adjust to meet identifiable constraints. Next, supply-demand balances must be developed for future years. Projected energy shortages and surpluses by fuel type and usage category must be reconciled, for example, by increasing or decreasing energy imports or exports, or interfuel substitution (where technically, economically, and socially feasible), by augmenting domestic, conventional and nonconventional energy sources, or by reducing demand through pricing, rationing, or other physical controls.

A major issue that must be confronted is that of identifying likely resource constraints (step e).[30] Depending on their severity, they may be dealt with by

---

[29]See, for example, M. Munasinghe and C. J. Warren, "Rural Electrification, Energy Economics and National Policy in the Developing Countries," *Future Energy Concepts*, Publ. No. 176, Institution of Electrical Engineers, London, 1979.

[30]Another important set of constraints may be political or constitutional. In Canada, for example, the ownership and legal rights to resources, including oil and gas, rest with the provinces, not the federal government. As a result, any national energy planning must evolve in the context of federal-provincial agreements, rather than coercion. See John F. Helliwell, "The national energy compact," *Canadian Public Policy*, Vol VII, No. 1, Winter 1981, pp. 15–23. Similar constraints govern federal-state relationships in India.

IN THOUSANDS OF METRIC TONS OF OIL EQUIVALENT /A

	1 PRIMARY SOLID FUELS	2 SECONDARY SOLID FUELS	3 CRUDE & NGL	4 PETROLEUM PRODUCTS	5 GAS	6 NUCLEAR POWER	7 HYDRO & GEOTHM	8 ELECTRICITY	9 TOTAL COLUMNS 1 TO 8	10 NON-COMMERCIAL ENERGY	11 TOTAL COLUMNS 9 & 10
1 INDIGENOUS PROD.	62971		8893		975	999/B	10689/B		84527	31206	115733
2 IMPORTS +	-312		14411	2517				-2	16929		16929
3 EXPORTS -				-82					-396		-396
4 MARINE BUNKERS -				-631					-631		-631
5 STOCK CHANGE +/-	-2312	23	73	-405					-2621		-2621
6 TOTAL ENERGY REQUIREMENTS	60347	23	23377	1399	975	999	10689	-1	97808	31206	129014
7 STAT. DIFFERENCE	1	1			1			-1	2		2
8 TRANSFERS											
9 ELEC. GENERATION	-15292			-1552	-147	-999	-10689	7605	-21074		-21074
10 GAS MANUFACTURE	-44				28				-16		-16
11 REFINERIES			-23377	21600					-1777		-1777
12 OTHER TRANSFORMATIONS	-9151	6994							-2157		-2157
13 E. SECTOR USE & LOSS				-197	-326			-655	-1178		-1178
14 TOTAL FINAL CONSUMPTION	35861	7018		21250	531			6948	71608	31206	102814
15 INDUSTRY	24537	6964		2994	503			3728	38726	1661	40387
16 IRON AND STEEL	6020	5266		472				318	12076		12076
17 CHEMICAL & P CHEM	760	1010		762	503				3055		3055
18 OTHER INDUSTRY	17736			1317					19053		19053
19 TRANSPORTATION	8336			8725				175	17236		17236
20 ROAD				7068					7068		7068
21 RAIL	8322			819				175	9316		9316
22 AIR				616					616		616
23 INLAND WATERWAY	14			223					237		237
24 OTHER SECTORS	2816	54		5678	28			1803	10379	29545	39924
25 AGRICULTURE				140				791	931		931
26 COMMERCIAL											
27 PUBLIC SERVICE								199	199		199
28 RESIDENTIAL	2816	54		4618					7488	29545	37033
29 NON-ENERGY USES				3853					3853		3853
30 NOT INCLUDED ELSEWHERE	172							1242	1414		1414

NOTE A: ONE METRIC TON OF OIL EQUIVALENT IS DEFINED AS TEN MILLION KCAL
NOTE B: CALCULATED EQUIVALENT OF POWER PLANT INPUT ASSUMING 28% EFFICIENCY.

Source: International Energy Agency, Workshop on Energy Data of Developing Countries—Vol. II, OECD, Paris, Dec. 1978, p. 203.

Figure 3.8. Basic energy balance. (In Thousands of Metric Tons of Oil Equivalent /A)

measures such as pricing, rationing, or additional borrowing from abroad. However, some of these remedies may directly affect the overall socioeconomic goals of the country (i.e., step (a) of the planning process); this may mean that some of these goals may have to be adjusted to meet the limitations of available resources for derived energy developments. Thus, some of these equilibrating measures may require going back to steps (a), (b), and (c), to readjust the supply and demand analyses and forecasts.

Additional considerations that enter into the analysis are those related to environmental effects. This is true not only in developed countries. For example, air pollution loads in cities such as Mexico, Santiago, Sao Paulo, or Bangkok are such that local demands for abatement and change can no longer be ignored.

Policy Formulation and EMP

The series of steps described above is designed to yield a set of energy policies for the management of supply and demand which will ensure that future national energy requirements can be satisfied. Since there are likely to be many different policy combinations, there will also be several alternative packages or sets of policies that will yield somewhat different, but nevertheless feasible, results, depending on given policy objectives and perceived constraints. The final step (e) consists of formulating these alternative packages and then testing them for their probable impact on the rest of the economy.

In this process, certain consistency checks must be made. Thus, the consequences of some energy policies may imply violation of other national objectives or assumptions regarding the evolution of the economy made in step (a). For example, a certain highly growth oriented industrialization strategy may entail unacceptably high balance of payments deficits for direct energy or energy-related capital goods imports. This may drain financial resources from sectors like agriculture and/or drastically curtail the energy supply to those sectors, thus infringing on other priorities. Another set of conflicts may arise with environmental goals. For example, in both West Germany and the United States the expansion of nuclear power has virtually come to a halt in recent years because of strong popular resistance. If some of the basic objectives must be changed as a result of such identified constraints then the INEP process of steps (a) to (e) must be repeated using the new assumptions. Thus, the INEP process summarized in Figure 3.7 must be designed to allow for dynamic consistency checks through iterative feedbacks.

Finally, the set of energy policies that are consistent with broad national goals and best satisfy future energy needs may be selected by the decision-makers. This policy package and the associated set of supply and demand forecasts and balances constitute the initial Energy Master Plan (EMP). This plan must be continuously updated and its respective components changed and adjusted in the light of changing circumstances that were not foreseen at the time of initial plan formulation. This is an ongoing and never-ending process. A capable team of professionals with free and immediate access to data should undertake the task of analyzing and formulating plans and energy development programs in both the public and private sectors.

3.8   PROBLEMS OF IMPLEMENTATION

In this section we discuss problems associated with the practical implementation of energy planning. These difficulties may be examined in two convenient categories: (a) data collection and analysis, (b) institutional structure and manpower needs.

Data Collection and Analysis

Energy planning can be carried out at different levels of sophistication, depending on data availability and the capability to analyse this information (i.e., computer facilities, skilled manpower, etc.). In countries where these constraints are severe, and especially when there has been little prior experience in energy planning even in individual subsectors, INEP may have to be implemented progressively.

For example, the first phase might consist of energy planning at a relatively uncomplicated level. A basic socioeconomic accounting matrix (or small input-output table) might be used to generate information regarding the economic background. Similarly, simple time trend projections could be used for energy supply and demand forecasts (with judicious assumptions where data is unavailable, particularly with respect to traditional fuels). The energy balance might consist of a basic table like Figure 3.8, with reliance on direct policies like increasing oil imports or shedding electricity load (power cuts) to make equilibrating adjustments. Thus the initial version of the EMP would consist of simple supply and demand projections and a straightforward set of policies, with little scope for impact analysis or iteration. In brief, such a first attempt at INEP would rely principally on physically based data, extrapolation of past trends in energy supply and demand (i.e., assuming, for lack of better information, that energy resources and technology, consumer behavior, external factors, and so on would essentially continue unchanged), very basic consistency checks in the energy balance, and relatively uncomplicated policy analysis.

At this stage it is worthwhile stressing the importance of data collection and analysis in the INEP procedure. Building a good energy data base is an important requirement of planning. In particular, the data should as far as possible be easy to gather, accurate, convenient to manipulate and analyze, internally consistent, relevant to policy work, compatible with the work of other sectors in the economy, and consistent with internationally accepted standards. A simple data base might be a set of tables, while a more sophisticated version would involve a computerized data bank. While basic data (by energy subsector) may be in diverse physical units (e.g., kilowatt-hours (kWh) of electricity, barrels of oil, tons of coal, cubic meters of wood etc.), alternative presentations should be made in terms of a common energy unit for comparison and preparation of an overall energy balance.[31] It should also be made clear whether the energy is in gross or net terms, e.g., kWh of electricity delivered to the consumer, or liters of fuel oil burned at a thermal power station before accounting for efficiency of conversion, losses, and so on.[32]

Even the first version of an EMP based on a simple planning procedure can often be used effectively. At the very least, it provides a consistent and comprehensive approach to analysing national energy problems. This is superior to the traditional uncoordinated planning by subsector. Furthermore, by focusing on data needs and analysis, it makes it apparent where information is poor, what type of data collection and organization must be improved, what kind of manpower and analytical skills are needed, and so on. In addition, it forces energy planners to relate the process to explicit objectives and policies. More generally, INEP and the EMP facilitate the recognition of key energy issues, and help to analyse these problems in a consistent way. The identification of such issues and the formulation of policies to solve them (or establishing guidelines to study them further before deciding on appropriate policies), is one of the most important consequences of INEP. Typical energy related issues might include: (a) data problems (as discussed earlier), (b) manpower and organizational needs (to be discussed next), (c) conservation, (d) import/export and dependence on foreign sources, (e) environmental degradation, (f) pricing, (g) investment and financing, (h) shortages and rationing, etc.

Once experience has been gained and skills built up by working with a simplified planning procedure, more sophisticated approaches may be developed. Thus, a multisector macroeconomic model could be used to establish the socio-economic background, including analysis of several alternative scenarios which take uncertainty into account, e.g., high versus low economic growth rates, more versus less energy intensive growth, high versus low costs of energy imports/exports, and so on. The supply and demand projections could also be more complex, using multiple correlation techniques to include the effects of

---

[31]Commonly used basic energy units include: (a) Joule = 0.239 kilocalories (UN; (b) ton of oil equivalent or TOE = 10^7 kilocalories, net calorific value (OECD); and (c) ton of coal equivalent or TCE = 7×10^3 kilocalories.

[32]However, it must be remembered that energy balances expressed in physical units do not identify energy-use systems costs which may vary substantially between different energy resources.

economic variables like price and income, and to capture the impact of policy induced shifts in energy usage and supply patterns, e.g., government programs to promote natural gas or LPG or to discourage the use of kerosene.

The energy balance could be expanded to include more subcategories of supply and demand, and details of losses in extraction, refining, conversion, transportation, etc. Ultimately a sophisticated computerized energy model might be developed, which would incorporate the effects of many items of physical, economic, and behavioral information, and government policies.[33] These models generally fall into two broad categories: (a) optimization models, in which some form of objective function is optimized subject to physical, economic, and policy constraints, e.g, minimization of the costs of meeting total energy needs given the unavailability of domestic oil resources; and (b) simulation models, in which alternative scenarios are explored in a self-consistent framework, subject to constraints. Type (a) models usually yield a well defined optimal EMP, but may suffer from uncertain data inputs and methodical constraints. With type (b), the policymaker has to choose from among the alternative simulated scenarios.[34] Type (b) models allow more latitude with respect to judgement and incorporation of considerations which cannot be quantified in the model—a situation which may be more relevant in developing countries.

The distinction between these two broad categories of models may be summarized rather simply, using the following notation. Let S_i be the initial state of the economy (including, particularly the energy sector), P be the set of policy options available to the policymaker, and T be the transformation mechanism (represented in the model by mathematical expressions of technical, economic, and behavioral relationships existing in the economy). The final state of the economy, S_f, is derived by applying T to S_i and P, represented symbolically by the expression:

$$T(S_i, P) \rightarrow S_f$$

Often S_i is the present state and S_f is some future state.

In optimization models, S_i and T are given and the model yields an optimal final state S_f^* (which optimizes the chosen objective function). The optimal policy options P^* which yield S_f^* are also usually determined. With simulation models, S_i and T are given, but the planner selects different policy options and investigates the different outcomes or final states that result.

Optimization models are usually of the mathematical programming type,

---

[33]See, for example: National Academy of Science, *Energy Modeling for an Uncertain Future,* Washington D.C., 1978; and Raphael Amit and Mordecai Avriel (eds.), *Perspectives on Resource Policy Modeling: Energy and Minerals,* Ballinger Publ. Co., Cambridge, MA, 1982.
[34]For a discussion of the problems affecting energy demand models see Chapter 7.

with linear as well as quadratic objective functions and constraints. Optimal control models seek to provide a continuous path over time of the control (or policy) variables that will yield the optimal outcome. The emphasis is not so much on the static final state as on an optimal path for the economy from the present to some final future date.

Simulation models are often econometric in nature where technical, economic, and behavioral equations represent relationships among different variables in the economy. The values of coefficients and parameters in these econometric equations are estimated and the models "tuned" on the basis of past data. The emphasis here is on consistency. Whatever the scenario selected, the evolution towards the final state is constrained by all the variables having to grow in a mutually consistent manner, according to prespecified relationships generally established on the basis of historical data. We note that optimization models may also be based on econometric equations.

In input-output (I/O) models, energy-economic relationships are captured by a matrix of coefficients representing the quantities of different primary factors and other inputs required to produce one unit of output in various sectors. I/O models may be used in both a simulation or optimization type framework. They are particularly useful in modeling the complex linkages between the energy sector and the rest of the economy. Another family of models, the systems dynamics models, are especially formulated to incorporate the effects of feedback, lagged variables, and dynamic changes in rates of change.

We note that all these models must represent both the demand and supply relationships for energy and other commodities.[35] Either the supply or the demand side may receive greater attention in any given model. Thus econometric process analysis models based mainly on technical relationships are used for detailed analysis of energy flows on the production or supply side. Similarly, econometric market models that focus on the economic-behavioral reactions of consumers to prices, incomes, and other variables are commonly used for modeling the demand side.

Policy impact analysis and feedback can be facilitated by using a computerized energy model, which interfaces with existing macroeconomic models. A more complex set of national objectives could be formulated, and explicit trade-offs among them explored. In summary, as data and analytical capability improve, the INEP procedure can be progressively upgraded with greater pay-offs in improved energy sector planning and supply-demand management.

---

[35]For an example of detailed modeling in a single energy subsector, see: Mohan Munasinghe, "Optimal investment planning and pricing policy modeling in the energy sector: electric power," in: R. Amit and M. Avriel (eds.), Chap. 6.

Institutional Structure and Manpower Needs

The major organizational problems in many countries is that energy-related planning and decisionmaking is scattered among many public and private sector institutions such as electricity authorities or companies, petroleum authorities, or private oil or gas supply companies, forestry departments, and others. Often all of them pursue their own policies with very little coordination and consideration of the effects of their policies on each other and the country as a whole. Ideally, a single energy planning authority should determine overall energy policy, and coordinate it with overall economic and social policies.

A stepwise approach in organizing the energy sector may often be useful. For example, an energy council or board may be created that brings together representatives of all energy suppliers and major users. This body could have a secretariat which would form the nucleus of the network for data collection and analysis, to begin establishing planning procedures.

Once the advantages of coordinating energy policies became evident, a central energy agency or ministry of energy might be set up. The degree of control that such an institution might exercise would range from the establishment of broad policy guidelines and reliance on the market mechanism, to very detailed planning and direct control of the energy sector, depending on the economic philosophy of the government and the organization of the energy sector.

The topic of manpower requirements for carrying out INEP is particularly relevant in developing countries where skilled staff are scarce. The necessary expertise should be drawn from as many government departments and other public and private institutions as possible, to represent a wide variety of viewpoints and skills. This would help develop a balanced approach to INEP and also maintain links with other sectors and institutions, without which planning could become an abstract theoretical exercise. In particular, emphasis on the commercial energy sub-sectors should not be allowed to lead to a neglect of the traditional fuels sub-sector.

The balanced development of energy planning skills both at the management/leadership and technical levels is important. The underlying theme should be self-reliance in energy planning, because the final responsibility for INEP should rest on local staff and policymakers. Although in many cases it may be necessary to use foreign experts or consultants to initiate the process, the training of local counterparts and the goal of eventual transition to completely national staffing should have a high priority. At the same time, because both technical and economic knowledge in the energy area tends to change rapidly, energy planners should have good, up-to-date library and documentation facilities, as well as ready access to international conferences, training courses and so on.

4
Integrated Framework for Energy Pricing: I

4.1 OBJECTIVES OF PRICING

Before developing a comprehensive and practical framework for energy pricing, the goals of pricing must be identified. The objectives of energy pricing, which is one of the most effective long-run tools for demand management, are closely related to the goals of energy planning, but are more specialized.

First, the economic growth objective requires that pricing policy should promote economically efficient allocation of resources both within the energy sector and between it and the rest of the economy. In general terms, this implies that future energy use would be at optimal levels, with the price (or willingness-to-pay of the consumer) for the marginal unit of energy used reflecting the incremental resource cost of supply to the national economy. This incremental resource cost should normally reflect the highest of one of the following three basic measures: the real economic costs of supply, the foregone opportunity cost of the energy in alternative uses, and the foregone future value of the energy if it is a depleting asset. Relative fuel prices should also influence the pattern of consumption in the direction of the optimal or least-cost mix of energy sources required to meet future demand. Distortions and constraints in the economy necessitate the use of shadow prices and economic second-best adjustments, as described later.

Second, the social objective recognizes the basic right of all citizens to be supplied with certain minimum energy needs. Given the existence of significant numbers of poor consumers and also wide disparities of income in many societies, this may imply subsidized prices, at least for low income consumers.

Third, the government would be concerned with financial objectives relating to the viability and autonomy of the energy sector. This would usually be reflected by pricing policies which permit institutions (whether government or privately owned) in the different energy subsectors to earn a fair rate of return on assets and to self-finance an acceptable portion of the investments required to develop future energy resources, without, at the same time, fostering inefficiency and higher costs because of lack of competition.

Fourth, energy conservation is also an objective of pricing policy. While prevention of unnecessary waste is an important goal, there often are additional

100

reasons underlying the desire to conserve certain fuels. These include the desire for greater independence from foreign sources (e.g., oil imports), the goal of minimizing environmental degradation, the need to reduce the consumption of woodfuel due to deforestation and erosion problems, etc.

Fifth, we recognize a number of additional objectives, such as the need for price stability to prevent shocks to energy users and consumers from large price fluctuations, the need for simplicity in energy pricing structures to avoid confusing the public and to simplify metering and billing, and so on.

Finally, there are other specific objectives such as promoting regional development (e.g., local mining activities or rural electrification) or specific sectors (e.g., export-oriented industries), as well as the consideration of other sociopolitical, legal, and environmental constraints.

Because the objectives mentioned above are often not mutually consistent, a realistic integrated energy pricing structure must be flexible enough to permit trade-offs among them. To achieve this, the formulation of energy pricing policy must be carried out in two stages. In the first stage, a set of ideal or first-best prices, which strictly meets the economic efficiency objectives, is determined, based on a consistent and rigorous framework.[1] In cases where significant market distortions exist, the strict efficiency pricing of energy will require the use of shadow prices instead of market prices; the resulting modified energy prices are called second-best prices because they are efficient in the context of economic distortions. The second stage of pricing consists of adjusting these efficient prices (established in the first step) to meet all the other objectives and constraints. The latter procedure is more ad-hoc, with the extent of the adjustments being determined by the relative importance attached to the different objectives. In the rest of this chapter, we develop the economic framework which permits the efficient pricing of energy. The second stage adjustments due to noneconomic factors are discussed in the next chapter.

The viewpoint adopted is that of a national government which attempts to maximize its various objectives over time. As the discussion of the preceding chapter has shown, there are likely to be many different objectives, not all of which can be readily reconciled with each other. Here it will be assumed that the basic, underlying goal is the maximization of economic efficiency, i.e., the maximization of national income over time. This does not mean that income maximization should be the overriding goal of a comprehensive energy pricing policy. It only means that whatever other objectives interfere with this basic one, the consequences of following them can be assessed in terms of the

---

[1]Starting with the economic efficiency objective does not imply that this national goal is more important than all the others. It merely serves as a convenient starting point for the quantitative analysis of pricing policy.

national income gains that must be sacrificed for them.[2] National income gains or losses can usually be measured reasonably unambiguously in monetary terms. The same is not true of such potentially important, but in terms of valuation far more elusive, goals as, for example, improved income distribution, regional balance, social cohesiveness, national defense, or environmental protection. However, by having measured foregone national income gains first, these can then be used as a yardstick to estimate the costs, although not the benefits, of satisfying the other goals and objectives.[3] In the analysis that follows, the various objectives of a comprehensive pricing policy will be systematically developed, starting with the simple principles of economic efficiency and welfare maximization in a static framework. Following this base case, more complex issues such as dynamic considerations, peak period pricing, market distortions, pricing of depletable energy resources, and others will be considered and analyzed in turn.

4.2 BASIC PRICING PRINCIPLES

The fundamental principle underlying rational economic pricing policies for any good or service is the well known proposition that the costs of providing the last unit consumed should be just equal to the willingness of somebody to pay for it. Provided that both the cost and demand functions are well-behaved,[4] and other considerations (described later) do not interfere, this results in a maximization of economic welfare, as illustrated in Figure 4.1. In particular, we will assume for the time being that market prices provide an accurate indication of the true economic value of scarce resources. This condition is relaxed later when the use of shadows prices is discussed to compensate for distortions in market prices.

---

[2]For a formal development of this approach see Gunter Schramm, "Accounting for Non-Economic Goals in Benefit-Cost Analysis," *Journal of Environmental Management,* Vol. I, No. 2, 1973, pp. 129–150.

[3]The only means of establishing a single, "most preferred" outcome is to apply distributional weights to the nonnational income goals so that they can be converted into national-income equivalents. However, weighting inevitably requires prior value judgments. The above approach does not. For a formal approach to the application of weighting see Lyn Squire and Herman C. Van der Tak, *Economic Analysis of Projects,* Johns Hopkins Press, Baltimore, 1975, Chapters VI and VII. For application of this approach in the energy sector, see Mohan Munasinghe, *The Economics of Power System Reliability and Planning,* Johns Hopkins University Press, Baltimore, 1979, Chapter 9.

[4]Meaning that marginal costs (MC) are usually increasing or constant as output rises while the willingness to pay, or demand, is declining; or, if MC are declining, they must fall at a lesser rate than demand in the relevant price range.

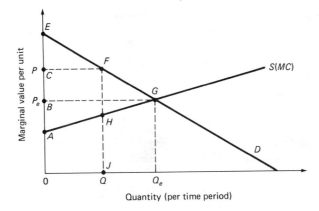

Figure 4.1. The basic pricing principle.

D is the consumer's (downward sloping) demand curve which indicates the amount that he is willing to purchase at any given price. The upward sloping supply schedule, S, is the marginal cost (MC) of supply, or the cost of producing one more unit of energy at any given level of output. We note that MC will, more generally, represent the economic opportunity cost of the marginal unit of energy supplied (as elaborated in Chapter 5), and may include a user cost or depletion premium component in the case of nonrenewable energy sources (see Section 4.8 for details). At the price P, the consumption level is Q. The total benefit of consumption is given by the area OEFJ, under the demand curve.[5] Correspondingly, the cost of supply is represented by the area OAHJ, under the supply curve. Therefore, the net benefit of consumption or total benefit minus cost is the area AEFH, which is the sum of the consumer surplus CEF and the producer surplus (or profit) ACFH.

Clearly, net benefits AEG are maximized at the price and output combination (P_E, Q_E). The result holds for all slopes of D and S curves, provided the slope of the supply curve exceeds that of the demand schedule. In other words, when the price P is higher than P_E, the marginal benefit or willingness-to-pay FH exceeds the marginal cost HJ, and net benefits could be increased by decreasing the price and increasing demand. Conversely, if the price is below marginal cost and the full demand (which will exceed Q_E) is being supplied,

---

[5]Strictly speaking, the relevant demand curve is the compensated demand curve, or schedule of consumer responses to price changes, while keeping consumer welfare (or utility) constant. See, for example, Ezra J. Mishan, *Cost Benefit Analysis,* Praeger, New York, 1976, Chapter 6.3.

the marginal cost exceeds the willingness-to-pay. Therefore, net consumption benefits will rise if the price is increased toward P_E and consumption falls to Q_E.

A simple mathematical expression for net benefits is given by:

$$\text{NB} = \int_0^Q p(q) \cdot dq - \int_0^Q \text{MC}(q) \cdot dq \qquad (4.1.)$$

where $p(q)$ and $\text{MC}(q)$ are the equations for the demand and supply curves, respectively. The first order (necessary) condition for maximizing NB is:

$$\frac{d\,\text{NB}}{d\,Q} = p - \text{MC} = 0 \qquad (4.2.)$$

Thus, net benefits are maximized when price equals marginal cost at the market clearing point G.

What is frequently overlooked, however, is that even under these most simplifying assumptions there is not just a single price as depicted by P_E, which will maximize national welfare, but a whole range of possible prices.

If discriminatory prices could and would be charged at the margin, price schedules could be set above P_E for the initial units consumed, decreasing to P_E for the marginal unit and the same amount of supplies would be forthcoming and utilized. Implicitly, the fact is well known and widely used in energy pricing. For example, most electricity or natural gas pricing schedules of utilities throughout the world apply discriminatory pricing principles which oftentimes bear little or no relationships to underlying sectoral marginal costs.[6] The important condition that must be met to apply such discriminatory prices is the ability to separate the various customer groups, and to prevent intersectoral resales from one group to another; otherwise the costs of implementing the discriminatory pricing policy, relative to uniform pricing, may become excessive and outweigh gains in revenue to the producer.

However, while aggregate welfare, or economic efficiency, does not change as a result of well-designed discriminatory pricing practices, what changes profoundly is the distribution of income between different groups. If marginal prices above P_E are charged, potential consumer surplus (i.e., income and well-being) is reduced; if marginal prices below P_E are enforced, producer surplus is cut. Whether real aggregate welfare changes as a result of such discriminatory pricing depends on the respective weights that must be attached to the

---

[6]In Bogota, Colombia, for example, the municipally-owned electric utility company charges higher prices to industrial than to domestic users, even though marginal costs of supply are much lower to the former.

income changes of the affected groups, and to the benefits that may be produced if some of the difference is extracted by the government in the form of taxes which in turn are used to finance other activities. However, apart from the distributional issues, the important point to note is that such discriminatory pricing may offer an important tool for generating governmental revenues without affecting economic efficiency objectives.

4.3 GROWING DEMAND AND DYNAMIC EFFECTS

The above analysis has been conducted in comparative static terms, abstracting from any potential growth in demand. However, economic growth is one of the basic objectives of any government, particularly so in the developing world. Growth in the demand for energy resources will result simply from increases in population or output even if everything else remains equal. Furthermore, increases in real income per capita will generally result in increased demand per capita for most energy resources as well, except for those that are considered to be inferior, such as traditional fuels. This increased demand per capita will result in an upward displacement of the demand schedule (i.e., a greater willingness to pay for any given quantity). Shifts in demand schedules may also result from changes in productive input mixes (e.g., changes from oxen to tractors in agriculture) or from changes in relative prices of substitute energy resources (i.e., scarcity). Whatever the cause of these increases in demand, they all lead to shifts in the demand curve from DD_1' to DD_2' to DD_3' in Figure 4.2. Generally, there will be corresponding changes in the amounts consumed and in optimum prices or price schedules. If marginal costs remain constant over time (supply curve SS' in Figure 4.2) the single, optimal price P_0 also remains constant while consumption rises from Q_1' to Q_2' to Q_3'. However, if marginal costs rise (supply curve SS'') the optimal price also must be raised from P_1 in period 1 to P_2 and P_3 in periods 2 and 3, respectively, while consumption increases from Q_1 to Q_2 to Q_3. The same applies to the levels of discriminatory price schedules. Hence, if demand grows and marginal costs increase prices must be increased as well to maintain economic efficiency. Annual or, in the case of rapidly rising marginal costs, more frequent price revisions are needed to reflect these changing conditions.

We now consider another type of dynamic effect due to the growth of demand from year 0 to year 1, which leads to an outward shift in the market demand curve from PD_0 to PD_1 as shown in Figure 4.3. Assuming that the correct market clearing price P_0 was prevailing in year 0, excess demand equal to GK will occur in year 1. Ideally, the supply should be increased to Q_1 and the new optimum market clearing price established at P_1. However, the available information concerning the demand curve PD_1 may be incomplete, making it difficult to locate the point L.

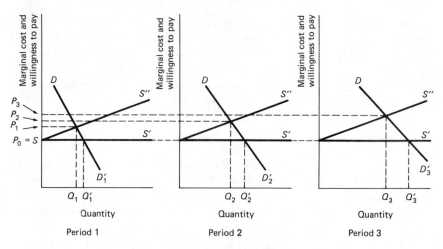

Figure 4.2.   Demand-induced changes in price.

Fortunately, the technical-economic relationships underlying the production function or known international prices usually permit the marginal opportunity cost curve to be determined more accurately. Therefore, as a first step, the supply may be increased to an intermediate level Q', at the price P'. Observation of the excess demand MN indicates that both the supply and, if necessary, also the marginal cost price should be further increased. Conversely, if we overshoot L and end up in a situation of excess supply, then it may be necessary to wait until the growth of demand catches up with the oversupply. In this iterative manner, it is possible to move along the MOC curve towards the optimum market clearing point. It should be noted that, as we approach it, the optimum is also shifting with demand growth, and therefore we may never hit this moving target. However, the basic guideline of pegging the price to the marginal opportunity cost of supply and expanding output until the market clears is still valid.

Next, we examine the practical complications raised by price feedback effects. Typically, a long-range demand forecast is made assuming some given future evolution of prices, a least-cost investment program is determined to meet this demand, and optimal prices are computed on the basis of the latter. However, if the estimated optimal price which is to be imposed on consumers is significantly different from the original assumption regarding the evolution of prices, then the first round optimal price estimates must be fed back into the model to revise the demand forecast and repeat the calculation. (See Chapter 8 for details.)

In theory, this iterative procedure could be repeated until future demand, prices, and MC estimates become mutually self-consistent. In practice, uncertainties in price elasticities of demand and other data may dictate a more pragmatic approach in which the MOC would be used to devise prices after only one iteration. The behavior of demand is then observed over some time period and the first round prices are revised to move closer to the optimum, which may itself have shifted as described earlier.

4.4 SHORT-RUN VERSUS LONG-RUN MARGINAL COST PRICING

Marginal costs are defined as the net change in supply costs resulting from an incremental change in output. This means that in the short-run only variable costs (i.e., the costs of those inputs that vary with changes in output) form part of the marginal cost accounting framework. The fixed costs of existing plant (e.g., capital equipment, buildings, etc.) remain constant by definition. Therefore, when output changes and marginal costs are calculated, fixed costs are netted out. As a consequence, prices that are set to reflect marginal costs ignore the fixed costs of past investments. Such a pricing policy is appropriate from the point of view of economic efficiency, because prices that reflect marginal costs are equal to the net opportunity costs of resources at the margin needed to bring forth the additional supply.

However, the strict application of such prices is appropriate—or feasible—only in a static world in which there is no change, in which demand remains

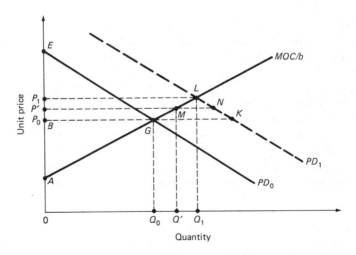

Figure 4.3. Dynamic effects due to demand curve shifts.

constant or declines, in which no lumpy investment is ever needed to increase capacity or to replace worn-out equipment or depleted resource deposits.[7]

Practical difficulties with this pricing approach are encountered when new investments are needed, whether they are just for replacement purposes or because of needed net additions to supply capacity. New investments usually are lumpy and require large amounts of resources that must be committed first before any additional output can be produced. Since investments prior to their irrevocable commitment are variable, their costs have to be included in the calculation of overall marginal costs. However, as soon as they have been made their costs become "sunk" costs so that they no longer affect decisions at the margin. As a consequence, marginal costs again fall to the incremental level of operating (i.e., variable) costs, and investment costs once again are ignored.

This situation has been depicted in Figure 4.4. Short-run marginal costs are represented by the line ABCFGHIJ. To simplify the analysis, they were assumed to remain constant, i.e., equal to price P_0, except in those ranges in which preexisting capacity limits are approached (section BC and GH) or new capacity needs to be added (Sections CF and HI).

Initially, existing demand is represented by demand schedule D_0 and the quantity demanded at price P_0 = SRMC (short-run marginal cost) is equal to Q_0. As demand grows, existing capacity levels are encroached upon and short-run marginal costs start to rise (section BC). At output Q_1 and SRMC-based price P_1, full capacity is reached and no additional supplies are forthcoming in response to increases in the willingness to pay. In order to maintain its market clearing function, however, price must increase (section CF). Within this section, the SRMC-based price acts as a rationing device, eliminating lower-valued energy uses in favor of higher-valued ones. Once price P_3 (point F) is reached new additions to capacity become feasible. Once this new capacity is installed price falls to P_0. This process repeats itself as demand keeps growing and the newly established capacity limits are reached (section GHIJ).

The amplitude of these price fluctuations in typical energy supply systems would be rather substantial, if the costs of the additional, required capital investments are to be borne by the consumers of the energy. This is clearly a prime requirement of any economically efficient pricing scheme. But price fluctuations of such magnitude would be unacceptable for any economy. They would certainly be highly disruptive to any energy-cost sensitive activity such as cement, pulp and paper, or steel production, or transportation. They would

---

[7]This, in fact, was the world in which marginal cost pricing rules were first developed—the world of the Great Depression in the 1930s. For the first presentation of marginal cost pricing principles see Harold Hotelling, "The General Welfare in Relation to Problems of Taxation and of Railway and Utility Rates," *Econometrica*, 1938.

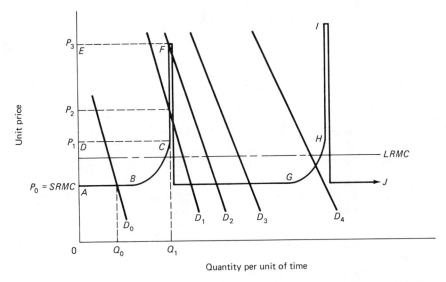

Figure 4.4.   The effects of capital indivisibilities on price under strict marginal cost pricing rules.

also be unacceptable to domestic consumers. Economic as well as political considerations would rule out the adoption of such pricing patterns.

This means that modifications of the basic marginal cost pricing are needed. These modifications should meet three criteria: First, they should maintain the basic integrity and advantages of marginal cost pricing, aiming at the equivalence of willingness to pay and incremental cost of supply at the margin. Second, they should assure that all supply related costs are borne by the respective consumers. And third, they should maintain reasonable long-term price stability or price predictability to facilitate forward planning and energy-use related investment patterns.

Two possible alternative approaches offer themselves. The first is to utilize some form of two-part tariff, which would consist of a fixed periodic charge (or one-time connecting charge) reflecting capital costs and another reflecting the SRMC of the energy supplied. Such tariffs have been particulary recommended for situations in which peakload capacities are needed. These have been discussed in greater detail below. Two-part tariffs can be utilized only for those energy resources that depend on fixed connections with metering devices (e.g., electricity and natural gas). They could be used also in those specific cases in which a particular large energy user agrees to participate directly in the financing of a new energy supply source. However, they would be impractical for all energy resources which could easily be resold outside formal market channels. This applies to all petroleum products, natural gas liquids, coal, charcoal, and wood.

Two-part tariffs would be increasingly more inefficient the higher average capital costs are relative to operating costs. This is so because higher average capital costs would lead to high fixed charges and low energy costs. A potential energy user would either have to pay the fixed charges, or do entirely without this source of energy. Once he agrees to pay, the fixed charges no longer affect his consumption pattern. Only the energy unit costs are relevant to his decisions. If the latter are low, wasteful usage is likely to result.[8] This waste,[9] in turn, would result in higher growth rates, which would require larger and more frequent additions to capacity. But capacity costs, once again, would not affect energy use, creating a vicious cycle of rapidly rising, wasteful energy use patterns. Such tariffs would tend to exclude the poor from obtaining the benefits from that particular energy source since they could not afford to pay the high connection charges. The share of capital costs on total costs for most energy supply systems are high, except for those that depend largely on high-cost imported fuels. Hence, two-part tariffs for the purpose of financing all capital costs do not appear to be useful for an economically efficient management of energy supply except in cases in which short-run marginal costs are a substantial proportion of long run ones.

The other alternative for dealing with indivisibilities would be to utilize a forward looking averaging approach.[10] The costs of forthcoming investments, i.e., the marginal investment costs, would be spread over an appropriate period, usually the life expectancy of the asset or, sometimes, its financing period.[11] These leveled-out capital costs, annuitized at the appropriate interest rate, would be divided by the energy units supplied per year and added to the marginal operating costs. The total unit charge, then, would reflect long-run marginal costs, or LRMC, in contrast to the short-run marginal costs, or SRMC, defined above (see Figure 4.4).

Including the annuitized capital cost charge in the marginal cost price struc-

---

[8]Evidence of this type of behavior is well known from studies of the relationship of water tariffs and water consumption. Whenever charges were fixed, e.g., based on size of building or land area, regardless of consumption levels, consumption was higher than in comparative situations in which water charges were tied to quantities consumed. See, for example: Gunter Schramm and Fernando V. Gonzalez, "Pricing Irrigation Water in Mexico; Efficiency, Equity and Revenue Considerations," *The Annals of Regional Science,* Vol. XI, No. 1, March 1977, pp. 15–35.
[9]Waste is defined relative to full economic costs of making these units of energy available. Full costs include capital costs.
[10]It should be noted that in the examples presented in Table 4.1 below, capital costs were already made subject to an averaging process, since they were assumed to be distributed over at least one year. Without this assumption, marginal costs would have been astronomical.
[11]The latter may be appropriate if it is shorter than the life expectancy of the asset and if it is uncertain that refinancing would be available at reasonable terms.

ture actually is a vitally important signal to the energy consumer of the real costs of his consumption. With growing demand, each additional unit consumed encroaches upon existing capacity and raises the specter of additional future investment costs. The levelized capital costs and charges, therefore, are nothing but a measure of these future costs.[12] It is assumed here that if consumers are willing to pay the annuitized cost of a long-lived investment in its first year of operation, in a situation where demand is expected to grow steadily, this implies willingness to bear those costs throughout the lifetime of the asset.

4.5 PEAK-LOAD AND SEASONAL PRICING

The demand for energy is not uniform over time but fluctuates, between seasons, between days of the week, or between hours of the day.[13] This is usually not a serious problem for those energy resources that can be stored at relatively low costs—examples are most petroleum products, coal, wood, and charcoal—although for some of them storage costs can be significant under special circumstances.[14] Costs are usually considerably higher for natural gas and electricity supply systems. Both may face substantial seasonal fluctuations in demand, particularly in more temperate regions with large variations in heating or cooling requirements. This calls for the installation of adequate storage or supplemental seasonal supply capacity. Daily or weekly load fluctuations particularly affect electric utilities since electric energy cannot be stored in significant quantities, except indirectly in the form of water in a storage reservoir.

There are various ways to deal with fluctuating demand. From the supplier's point of view the simplest or least costly one is to limit supply to a fixed amount and to selectively or uniformly cut-off specific user groups during those periods in which demand exceeds this available supply. This is sometimes arranged on a contractual basis between utility companies and certain users that agree to purchase interruptible energy at special, lower tariffs. Examples are cold storage facilities which are disconnected during peak-load hours, or off-peak nighttime rates for electric heating customers that install heating systems with special thermal storage facilities that are activated for daytime use.[15] Also quite

---

[12]Alternatively, we might think of the capital cost charge as a unit rental charge for the scarce resource (i.e., the capital) provided.

[13]For some typical annual and daily electricity load curves in a tropical country see the case studies of Sri Lanka and Thailand (Chapters 9 and 10).

[14]Such as the coal storage costs during the winter season for U.S.-Canadian powerplants around the Great Lakes that depend on waterborne shipments, or the need for large gasoline storage facilities to accommodate continuous, year-round United States refinery production on the one hand and peak summer gasoline demands on the other.

[15]Such installations have become popular in various European countries in recent years.

common in many nations are forced cuts in supply, first by reducing voltage or pressure (in the case of gas supplies), then, if needed, by systematically shutting-off increasing numbers of users until the remaining demand and available supply balances.[16] While the effects of such a strategy in utility revenues may be slight (or even beneficial in terms of net income) the economic costs to the utility's customers are usually very high, amounting to a large multiple of the value of energy not supplied.[17] Because of these high user costs, mandatory rationing is a rather undesirable and economically inefficient tool for managing peak-load demands.

The other alternatives to meet seasonal or shorter-duration peak-load demands are to install either energy storage facilities or special peak-load supply capacity.[18] Various alternatives are available. In the case of natural gas, storage could be provided in pressurized tanks or underground sealed caverns.[19] Another alternative could be to maintain shut-in wells with enough spare capacity to serve those peak-load demands. A third is to install special peak-load gasification plants based on the conversion of alternative liquid hydrocarbon fuels. With the skyrocketing costs of the latter, this alternative is no longer economically attractive. Electric supply systems can be augmented by additional gravity or pumped-storage hydro capacity, or by designated peakload units such as older, inefficient thermal plants or gas turbines.

We can demonstrate mathematically the implications for economically efficient pricing of peak demand by recognizing that during such periods demand

---

[16]Haiti's major utility company provided a typical example in the late 1960s. Its available capacity was some 12 percent lower than its market's daily peak-load demand. To manage this imbalance, the company used a system of rotating shut-offs by subdistricts, limiting their duration in each district normally to no more than half-hour periods. The overall reduction in total energy sales was a surprisingly low 1% of total sales (even assuming that no temporal substitution took place). However, the costs to consumers was high as can be seen indirectly by the fact that the total capacity of privately-owned standby generator sets amounted to some 57% of the utility company's installed capacity. From: Gunter Schramm, "Le Secteur de L'Energie Electrique," *App. I of Organ. of American States, Mission D'Assistance Technique Integree,* Washington, D.C., 1972.

[17]For example, industrial outage costs to industry range from US$1 to 5 per kWh not supplied in Brazil (expressed in 1976 US dollars). See Mohan Munasinghe, *The Economics of Power System Reliability and Planning,* Johns Hopkins University Press, Baltimore, 1979, Appendix E.

[18]The question of how much capacity should be installed to meet stochastic (i.e., uncertain) future demand is an intricate one that requires the comparison of the potential net losses from nonsupply with additional costs of higher supply security. This issue has been analyzed in considerable detail in Mohan Munasinghe, *The Economics of Power System Reliability and Planning; Theory and Case Study,* Johns Hopkins Press, Baltimore, 1979.

[19]This is a frequently used type of storage in the USA. It depends, obviously, on the existence and availability of such caverns.

will infringe on capacity constraints. Thus equation (4.1) may be reformulated as:

$$\text{Maximize NB} = \int_0^Q [p(q) - \text{MC}(q)] \, dq \qquad (4.3)$$

subject to $Q < \overline{Q}$, where $\overline{Q}$ is the capacity limit.

The above problem is formally equivalent to unconstrained maximization of the Lagrangian:

$$L = \text{NB} - m(Q - \overline{Q}) \qquad (4.4)$$

The first order (necessary) condition is:

$$\partial L / \partial Q = p - \text{MC} - m = 0 \qquad (4.5)$$

In equation (4.5), MC may be interpreted as the short-run marginal cost of supplying more output with capacity fixed, while m has the usual meaning of marginal cost of adding to capacity. Therefore, we have the following results. During off-peak periods, $Q < \overline{Q}$ and $m = 0$ yielding $p = \text{MC}$. During peak periods, $Q = \overline{Q}$ and $m > 0$ yielding $p = \text{MC} + m$.[20]

Hence per unit costs of energy supplied during peak periods exceed those of off-peak periods, sometimes by a substantial margin. Table 4.1 provides some typical examples based on electricity utility data.

These higher marginal costs of supply during daily and seasonal peak demand periods should be reflected in tariff structures so that energy users are confronted with them during peak demand periods. If they are not, consumption during peak periods would be larger and during slack periods somewhat lower. This has been illustrated in Figure 4.5, where the demand during the peak period D_{pk} is greater than off-peak demand D_{op}. If a uniform, nondifferentiated tariff p_{av} was used that reflects an average of the baseload and peak period marginal supply costs, the quantity consumed during the peak period would be equal to Q_1. However, if true peak period costs p_{pk} were charged consumption would be reduced to Q_3 instead. As a result, overall capacity needs would be lower.

During the off-peak period, consumption would tend to be correspondingly greater (i.e., increase from Q_2 to Q_4) because of the lower price p_{op} relative to the averaged-out price. Another reason for increased consumption during that period would be the partial, intertemporal substitution from the higher-priced

---

[20]See the mathematical Annex I to this chapter for a more rigorous dynamic version of this result.

Table 4.1. Typical Marginal Costs of
Supply at Different Periods of the Day
(1980 US¢ per kWh).[1]

	FUEL COST	CAPACITY COST
Peak Period[2]	6–10	2–3
Off-peak Period[3]	1–4	

[1]Based on an all thermal system.
[2]Gas turbines. All capacity costs including associated transmission and distribution costs have been allocated to the peak period.
[3]Nuclear, coal-fired steam, or diesel.
Source: Authors' estimates.

peakload to the base load period. Hence the consequence of a peak load tariff would be an overall reduction in needed total supply capacity together with a more efficient, i.e., more uniform, utilization of the remaining capacity. This would lead to overall reductions in marginal costs and, hence, prices to energy users (or to higher profits to suppliers or higher government income, as the case may be). Overall economic efficiency would be increased.

Peak load pricing requires time-of-use metering, however, and such metering

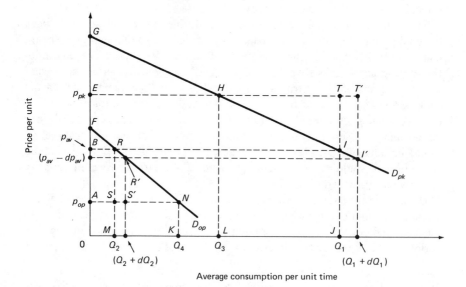

Figure 4.5. Uniform versus peak load pricing and the metering decision.

is usually more costly than simple volume metering only. One of the questions to be addressed, therefore, is whether the additional costs of metering are justified by the resulting savings and increase in the efficient utilization of energy producing equipment.

This is not an issue if the problem is one of adjusting prices to reflect seasonal demand variations. Seasonal tariffs could be instituted and metered by standard volumetric metering equipment without significant additional costs.[21] Seasonal tariffs can be justified on equity as well as efficiency grounds, because consumers would be confronted by the true seasonal supply costs of their consumption. Seasonal price differentials may not lead to major substitutions from peak to off-peak periods uses because postponing energy consumption for several months would be rather impractical. The main effects on consumption would come from a genuine reduction in demand in response to the higher seasonal price (i.e., a movement along a given demand curve).

Metering of residential demand by time-of-day is a more difficult problem because the costs of an average sized electric household meter would increase several fold. Thus the use of time-of-day metering devices would not be economically justified for residential users in low income countries, if the net benefits realized through savings in energy used were less than the increased costs of the more complex metering. The costs of time-of-day metering are much less of a problem for larger users, such as large commercial establishments or industrial plants, because the added costs of metering are small relative to the amount of energy consumed. Most peak-load pricing schemes, therefore, have concentrated on such users, although several developed countries have attempted to apply such tariffs in the residential sector as well.

The degree of sophistication of metering—for example, by time of day— should be determined by the net benefit of metering, the practical problems of installation and maintenance, ease of billing, and so on. The economic rationale for making the metering decision depends on the net benefit of metering based on a cost-benefit analysis that compares the lower supply costs of reduced consumption with the cost of metering and billing plus the decrease in net consumption benefits. This procedure is illustrated in the following discussion, using a two-period tariff.[22]

Figure 4.5 depicts the analysis underlying the meter decision to implement a two-period, time-of-day tariff for electricity supply. D_{pk} and D_{op} are the demand curves for an average hour during the peak and off-peak periods of the

---

[21]Such seasonal and peak load tariffs have been used for several decades in Europe and are becoming popular in the United States. For details, see: Mohan Munasinghe, "Principles of modern electricity pricing," *Proc. IEEE,* March 1981, pp. 332–48.

[22]For an economic analysis of the merits of a multiperiod tariff involving several generation technologies, see: Edward F. Renshaw, "Expected welfare gains from peak-load electricity charges," *Energy Economics,* January 1980, pp. 37–45.

day, each lasting H hours and $(24 - H)$ hours, respectively. If a simple uniform price p_{av} is levied throughout the day, the daily consumption will be $Q_1 \cdot H$ and $Q_2 \cdot (24 - H)$ kilowatt-hours respectively during the peak and off-peak periods. Instead, if a two-period tariff (p_{pk}, p_{op}) is charged during the peak and off-peak hours, the corresponding levels of consumption will be $Q_3 \cdot H$ and $Q_4 \cdot (24 - H)$. Suppose that the LRMC of supplying a kilowat-hour during the peak and off-peak periods are MC_{pk} and MC_{op}, respectively, where the former includes both capacity and energy costs and the latter includes only the energy cost.

The net consumption benefit of the uniform tariff is given by:

$$NB_I = [B_I] - [C_I]$$
$$= [(\text{area OGIJ}) \cdot H + (\text{area OFRM}) \cdot (24 - H)]$$
$$- [Q_1 \cdot H \cdot MC_{pk} + Q_2 \cdot (24 - H) \cdot MC_{op}]$$

Similarly, the net consumption benefit of the two-period tariff is given by:

$$NB_{II} = [B_{II}] - [C_{II}]$$
$$= [(\text{area OGHL}) \cdot H + (\text{area OFNK}) \cdot (24 - H)]$$
$$- [Q_3 \cdot H \cdot MC_{pk} + Q_4 \cdot (24 - H)MC_{op}]$$

The change in net benefit therefore is:

$$\Delta NB = NB_{II} - NB_I$$
$$= [(\text{area MRNK}) \cdot (24 - H) -$$
$$(\text{area LHIJ}) \cdot H] + [(Q_2 - Q_4)$$
$$\cdot (24 - H) \cdot MC + (Q_1 - Q_3) \cdot H \cdot MC_{pk}]$$

A further simplification is possible if it is assumed that the two-period tariff reflects strict LRMC; that is, $p_{pk} = MC_{pk}$ and $p_{op} = MC_{op}$. Then, $(Q_4 - Q_2) \cdot MC = \text{area MSNK}$, and $(Q_1 - Q_3) \cdot MC_{pk} = \text{area LHTJ}$. Therefore, the following result is obtained.

$$\Delta NB = (\text{area HTI}) \cdot H + (\text{area SRN}) \cdot (24 - H)$$

Assuming no change in this average daily cycle during the first year, the annual increase in net consumption benefits is given by:

$$\Delta ANB_1 = 365 \cdot \Delta NB$$

Let the difference between the two streams of metering and billing costs (installation, maintenance, administration, etc.) associated with the two-period and uniform tariffs be $\Delta C_1, \Delta C_2, \ldots, \Delta C_T$, over the lifetime T of the meter.

Suppose that the corresponding stream of net benefits can also be estimated as $\Delta ANB_i, \ldots, \Delta ANB_T$. If the condition

$$\sum_{t=1}^{T} (\Delta ANB_t - \Delta C_t)/(1 + r)^t$$

is satisfied, implementation of the two-period price structure would be economically justified.

Even if the decision to install more complex metering is negative, further analysis is helpful. Thus, it is clear that applying the uniform price p_{av} instead of the strict LRMC prices $p_{pk} = MC_{pk}$ and $p_{op} = MC_{op}$ will result in a loss of net efficiency benefits:

$$NB^1 = (\text{area HTI}) \cdot H + (\text{area SRN}) \cdot (24 - H)$$

The level of p_{av} may be adjusted, however, to minimize this loss of net benefits.

Suppose that a reduction in price from p_{av} to $(p_{av} - dp_{av})$ results in increased consumption given by $(Q_1 + dQ_1)$ and $(Q_2 + dQ_2)$ in periods 1 2 respectively. The new net benefits may be written:

$$NB^2 = (\text{area HT'I'}) \cdot H + (\text{area S'R'N}) \cdot (24 - H)$$

The change in net efficiency benefits is:

$$
\begin{aligned}
d(\Delta NB) &= (NB^2 - NB^1) \\
&= (MC_{pk} - p_{av}) \cdot dQ_1 \cdot H - (p_{av} - MC_{op}) \cdot dQ_2 \cdot (24 - H)
\end{aligned}
$$

Using the familiar first-order condition $d(\Delta NB)/dp_{av} = 0$, yields the optimal uniform price that will minimize the loss of net benefits:

$$p^*_{av} = \frac{(H \cdot dQ_1/dp_{av}) \cdot MC_{pk} + [(24 - H) \cdot dQ_2/dp_{av}] \cdot MC_{op}}{(H \cdot dQ_1/dp_{av}) + [(24 - H) \cdot dQ_2/dp_{av}]}$$

This result may be generalized. Thus, suppose the tariff analyst wishes to consolidate n pricing periods of the day, where H_i is the duration, Q_i is the average hourly consumption, and MC_i is the LRMC during the ith period. The optimal uniform price would be:

$$p^*_{av} = \left[\sum_{i=1}^{n} (H_i \cdot dQ_i/dp_{av}) \cdot MC_i \right] \Big/ \left[\sum_{i=1}^{n} (H_i \cdot dQ_i/dp_{av}) \right]$$

Note that derivatives of the form $H_i \cdot (dQ_i/dp_{av})$ refer to a change in consumption during a given pricing period i due to a uniform change in price across all the pricing periods that need to be consolidated under the single price regime. Generally, in most countries information on demand would be inadequate to determine the magnitude of such derivatives accurately. In this case, a practical approximation might be:

$$p^*_{av} = \left[\sum_{i=1}^{n} (H_i \cdot Q_i) \cdot MC_i \right] \Big/ \left[\sum_{i=1}^{n} (H_i \cdot Q_i) \right]$$

Here the uniform price is set equal to the weighted average LRMC using the consumption levels in the different pricing periods as the weights.

While such a partial approach to peak-load pricing is preferable to one which disregards the issue completely, it also raises a number of problems. Ideally, the added costs of supplying energy during peakload periods should be fully charged to the consumers using it during that time. Unfortunately, residential demand fluctuations are usually substantially higher than industrial and commercial ones. Hence the residential sector is a major contributor to the extra peak period costs.

However, as pointed out above, peakload metering often cannot be used in the household sector. This means that peakload pricing schemes can only be partially applied. On equity as well as efficiency grounds the full utility system marginal cost differential between base- and peak-load periods should not be charged against those users that can be metered. To achieve economically efficient consumption patterns among them, they should be confronted with no more than their proportional share of the total added costs of peak-load supplies. For example, if metered users account for, say, 40% of total peak period consumption, their peak period tariffs should be raised to cover 40%, but no more, of the added costs of peak period supply. The other 60% should be added into the averaged-out LRMC-based uniform tariffs of the remaining groups. It should be noted that the three prices charged to the various user groups—base-load and peak-load prices to users with time-of-day meters, average prices to all others—should be based on the respective long-run marginal costs, i.e., the marginal costs of supply including the annuitized costs of new investments. The reasons for this requirement have been detailed in section 4.4 above.

It should also be noted that both per unit marginal capital and energy costs will differ between peak-load and base-load periods. Because of their higher utilization rates, base-load supply installations tend to be more capital intensive but have relatively lower energy operating costs compared to peak-load plants. The opposite is true for peak-load plants that are sensitive to capital costs, but for which per unit running costs are less significant because of their shorter operating hours.[23]

---

[23]The classical example is capital-intensive but energy efficient atomic or thermal powerplants whose capital costs may range between $500 to $1,500/kW with heat rates between 9,000 to

In energy supply systems that have both short-term (i.e., daily and weekly) and seasonal demand variations both seasonal and time-of-day peak-load tariff structures should be combined. If seasonal peaks affect base-load as well as peak-load demands,[24] separate seasonal tariff adjustments should apply to both to reflect the additional costs of base-load and peak-load supply capacity, respectively. If only peak-load demands increase, adjustments should be limited to the peak-load tariffs only.

Two part tariffs, consisting of a fixed capacity charge and volumetric energy charges are widely used in order to reflect the higher costs of peak-load supplies. The fixed portion of such tariffs are usually based on the installed capacity of equipment (size of motors or processing equipment) or the maximum capacity of the supply connection (size of supply pipe, capacity of wiring, etc.). Another alternative is to install metering devices that register the maximum demand in a given meter reading interval. Such meters are less costly than time-of-day metering equipment. Typical examples of such tariff structures are shown below in Table 4.2. While such tariffs are relatively easy to implement and administer and do not require the high time-of-day metering costs they do not really solve the peak-load problem. This is so because the capacity charge is invariant with respect to time. In other words, the charge is the same regardless of the period of time in which the full installed capacity is utilized. No incentive exists to users to reduce consumption during peakload periods. Hence, while overall installed user capacity might be lower in response to prevailing capacity charges, consumption patterns of installed capacity would not be affected. Time-of-day metering, therefore, is a more preferable demand management and pricing tool.

The discussion in this section has proceeded as if there existed only one basic type of user and marginal cost and two periods of on-peak or off-peak. This is an obvious simplification. Many different user classes exist, and the marginal costs of supply to them will differ as well (for example, by different voltage levels, in the case of electricity supply). Peak demand periods may well be broken down into more discriminating time periods to better reflect variations in the marginal costs of supply.

4.6   SECOND-BEST CONSIDERATION AND THE NEED FOR SHADOW PRICING

The economic arguments underlying the efficient pricing of energy presented earlier in this chapter were derived assuming that market prices in the economy

---

10,000 btu/kWh, compared to gas turbines that use expensive fuels and have heat rates of 13,000 to 16,000 btu/kWh, but cost only $200 to $250 per kW.

[24] A typical example would be electric loads from deepwell irrigation pumps, or gas demands from crop-drying operations.

Table 4.2. Typical Peak Load Electricity Tariff Structures (1980)[1]

		TARIFF		
UTILITY CO. AND COUNTRY	CUSTOMER CATEGORY	DEMAND CHARGE ($/ KW/MONTH)	ENERGY CHARGE (¢/KWH)	FIXED CHARGE ($/MONTH)
1. Northwestern Area Electricity Board (U.K.)	>650 Volts	April–Oct.: 0 Nov.–March: 6.33	0700–2400 daily 5.91 2400–0700 daily 2.78	1.0 per kVA
2. Electricite de France (France)	5–30 kilovolts	3.21	Nov.–Feb.[1] 0700–0900 and 1700–1900: 10–36 (except Sunday) Oct.–March[2] 0600–2200: 6.08 (except Sunday) 2200–0600 and Sunday: 3.06 April–Sept. 0600–2200: 3.87 (Except Sunday) 2200–0600 and Sunday: 2.42	
3. Kenya Power Company (Kenya)	11 and 33 kilovolts >100 Megawatt-hours per month	6.74	0800–2200 Monday to Friday: 3.75 All other times: 2.25	53.93

[1]All prices have been converted to mid-1980 US$ using the appropriate official exchange rates.
[2]Excluding Nov.–Feb. 0700–0900 and 1700–1900

are a good approximation to the true economic opportunity costs of scarce resources. If this is not the case, meeting the economic efficiency objective requires the use of so-called "shadow prices" in the economic analysis instead of market prices. The problem is very relevant to the major energy producing and using sectors of the economy where the rapid shifts in the real costs of energy supply have often not been accurately reflected in the market prices of energy products, as governments have sought to protect consumers or postpone the effects of violent price changes. Therefore, in this section we outline the basic concepts of shadow pricing. A more complete description of shadow pricing for investment planning is provided in Chapter 8, while the use of social weights for income distributional purposes is discussed in Chapter 5.

Shadow pricing theory has been developed mainly for use in the cost-benefit analysis of projects.[25] However, since investment decisions in the energy sector are closely related to the pricing of energy outputs, for consistency the same shadow pricing framework should be used in both instances. Shadow prices are used instead of market prices (or private financial costs) to represent the true economic opportunity costs of resources.

In the idealized world of perfect competition, the interaction of atomistic profit maximizing producers and atomistic utility maximizing consumers yields market prices which reflect the correct economic opportunity costs, and scarce resources including energy will be efficiently allocated. However, in the real world, distortions due to monopoly practices, external economies and diseconomies (which are not internalized in the private market), interventions in the market process through taxes, import duties, subsidies, etc., all result in market prices for goods and services, which may diverge substantially from their shadow prices or true economic opportunity costs. Therefore, shadow prices must be used in investment and output pricing decisions, to ensure economically efficient use of resources. Moreover, if there are large income disparities, we will see later in Chapters 5 and 8 that even these "efficient" shadow prices may be further adjusted, especially to achieve socially equitable energy pricing policies for serving poor households.

It is important to realize that lack of data, time, and manpower resources, particularly in the developing country context, will generally preclude the analysis of a full economy-wide model when making energy-related decisions.[26] Instead, a partial approach may be used, where key linkages and resource flows between the energy sector and the rest of the economy, as well as interactions among different energy subsectors are selectively identified and analyzed, using appropriate shadow prices such as the opportunity cost of capital, shadow wage rate, marginal opportunity cost for different fuels, and so on . In practice, surprisingly valuable results may be obtained from even relatively simple models and assumptions.

To clarify the basic concepts involved in optimal energy pricing, we first ana-

---

[25]For general use of shadow prices in developing countries see: L. Squire and H. Van der Tak, *Economic Analysis of Projects,* Johns Hopkins, Baltimore, 1975; and I. M. D. Little and J. A. Mirrless, *Project Appraisal and Planning for Developing Countries,* Basic Books, New York, 1974. For more specific application to the energy sector, see: Mohan Munasinghe, *The Economics of Power System Reliability and Planning,* Johns Hopkins, Baltimore, 1979, Chapter 9.

[26]This holistic approach or general equilibrium analysis is conceptually important. For example, the efficient shadow price of a given resource may be represented by the change in value of aggregate national consumption or output, due to a small change in the availability of that resource. A more detailed discussion of general versus partial equilibrium in relation to energy sector analysis is given in Mohan Munasinghe, *The Economics of Power System Reliability and Planning,* 1979. For a discussion of economywide energy models, see for example: National Academy of Sciences, *Energy Modeling for an Uncertain Future,* Washington, D.C., 1978.

lyze a relatively simple model. The process of establishing the efficient economic price in a given energy subsector may be conveniently analyzed in two steps. First, the marginal opportunity cost (MOC) or shadow price of supply must be determined. Note that this MOC is the shadow priced equivalent of MC discussed earlier this chapter. The terminology MOC is used here to emphasize the shift from market to shadow prices. Second, this value has to be further adjusted to compensate for demand-side effects arising from distortions in the prices of other goods, including other energy substitutes. From a practical viewpoint, an optimal pricing procedure that begins with MOC is easier to implement, because supply costs are generally well defined (from technological-economic considerations), whereas data on the demand curve are relatively poor.

Suppose that the marginal opportunity cost of supply in a given energy subsector is the curve $MOC(Q)$ shown in Figure 4.6. For a typical nontraded item like electricity, MOC which is generally upward sloping is calculated by first shadow pricing the inputs to the power sector and then estimating both the level and structure of marginal supply costs (MSC), based on a long-run system expansion program.[27] For tradeable items like crude oil and for fuels which are substitutes for tradeables at the margin, the international or border prices of the tradeables (i.e., c.i.f. price of imports or f.o.b. price of exports with adjustments for internal transport and handling cost) are appropriate indicators of MOC.[28] For most developing countries, such import or export MOC curves will generally be flat or perfectly elastic. Other fuels such as coal and natural gas could be treated either way depending on whether they are tradeables or nontraded.[29] The MOC of nonrenewable, nontraded energy sources will generally include a "user cost" or economic rent component, in addition to the marginal costs of production. (For details, see the next section on the pricing of nonrenewable energy).[30] The economic value of traditional fuels are the most difficult to determine because in many cases there is no established market. However, as discussed later, they may be valued indirectly on the basis of

---

[27]For a detailed discussion of the procedures used in the electric power subsector, see: Mohan Munasinghe, "Electric Power Pricing Policy," *Staff Working Paper No. 340,* World Bank, Washington, D.C., June 1979. In this subsector MOC is also called the long-run marginal cost (LRMC).

[28]We note that the use of border prices does not require the assumption of free trade, but implies that the numeraire or unit of value for shadow pricing is essentially uncommitted foreign exchange (but converted into local currency at the official exchange rate). For details, see L. Squire and H. Van der Tak, *op. cit.*

[29]A nontraded item is generally characterized by a domestic supply price that lies above the f.o.b. price of exports, but below the c.i.f. price of imports.

[30]Some interesting dynamic aspects that are relevant in determining MOC for a nonrenewable, nontraded indigenous energy resource are analyzed in: Mohan Munasinghe, "An Integrated Framework for Energy Pricing in Developing Countries," *The Energy Journal,* Vol. 1, July 1980, pp. 1–30.

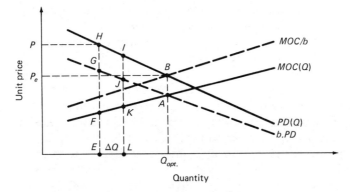

Figure 4.6.   Efficient pricing with shadow prices.

the savings on alternative fuels such as kerosene, the opportunity costs of labor for gathering firewood, and/or the external costs of deforestation and erosion.

Thus, for a nontraded form of energy, MOC is the opportunity cost of inputs used to produce it (including quantifiable externality costs) plus a user cost where relevant, while for a tradeable fuel or a substitute, MOC represents the marginal foreign exchange cost of imports or the marginal export earnings foregone. In each case, MOC measures the shadow priced economic value of alternative output foregone, because of increased consumption of a given form of energy. After identifying the correct supply curve, we next examine demand-side effects, especially second best corrections which capture interactions between different energy subsectors.[31] This second step is just as important as the first one, and therefore it will be examined in some detail.

In Figure 4.6, the market priced demand curve for the form of energy under consideration is given by the curve $PD(Q)$, which is the willingness-to-pay of consumers. Consider a small increment of consumption dQ at the market price level p. The traditional optimal pricing approach attempts to compare the incremental benefit of consumption due to dQ, i.e., the area between the demand curve and x-axis, with the corresponding supply cost, i.e., the area between the supply curve and x-axis. However, since MOC is shadow priced, PD must also be transformed into a shadow priced curve to make the comparison valid. This is done by taking the increment of expenditure $p \cdot dQ$ and asking the question: What is the shadow priced marginal cost of resources used

[31]The general theorem of the second-best shows, for example, that if the price of a given fuel is not set at its MOC, then the efficient price of a close substitute also must diverge from its own MOC. For a detailed discussion of the theory of the second-best in economics, see: D. M. Winch, *Analytical Welfare Economics,* Penguin Books, Harmondsworth, U.K., 1971.

up elsewhere in the economy if the amount $p \cdot dQ$ (in market prices) was devoted to alternative consumption (and/or investment)?

Suppose that the shadow cost of this alternative pattern of expenditure is $b(p \cdot dQ)$, where b is called a conversion factor. Then the transformed PD curve which represents the shadow costs of alternative consumption foregone is given by $b \cdot \text{PD}(q)$; where in Figure 4.6, it is assumed that $b < 1$. Thus at the price p, incremental benefits EGJL exceed incremental costs EFKL. The optimal consumption level is Q_{opt}, where the MOC and $b \cdot$ PD curves cross, or equivalently where a new pseudosupply curve MOC/b and the market demand curve PD intersect. The optimal or efficient selling price to be charged to consumers (because they react only along the market demand curve PD, rather than the shadow priced curve $b \cdot$ PD) will be: $p_e = \text{MOC}/b$, at the actual market clearing point B. At this level of consumption, the shadow costs and benefits of marginal consumption are equal i.e., $\text{MOC} = b \cdot \text{PD}$. This was the same result derived from Figure 4.1, using market prices instead of shadow prices. Since b depends on user specific consumption patterns, different values of the efficient price p_e may be derived for various consumer categories, all based on the same value of MOC. We clarify the foregoing by considering several specific practical examples.

First, suppose that all the expenditure $(p \cdot dQ)$ is used to purchase a substitute fuel, i.e., complete substitution. Then the conversion factor b is the relative distortion or ratio of the shadow price to market price of this other fuel. Therefore $p_e = \text{MOC}/b$, represents a specific second-best adjustment to the MOC of the first fuel, to compensate for the distortion in the price of the substitute fuel.[32] Next, consider a less specific case in which the amount $(p \cdot dQ)$ is used to buy an average basket of goods. If the consumer is a residential one, b would be the ratio of the shadow price to the market price of the household's market basket (here, b is also called the consumption conversion factor). The most general case would be when the consumer was unspecified, or detailed information on consumer categories was unavailable, so that b would be the

---

[32]For example, MOC_{EL} could represent the long run marginal cost of rural electricity (for lighting), and the substitute fuel could be imported kerosene. Suppose that the (subsidized) domestic market price of kerosene is set at one half its import (border) price, for sociopolitical reasons. Then $b = 2$, and the efficient selling price of electricity $p_e = \text{MOC}_{EL}/2$ (ignoring differences in quality of the two fuels, capital costs of conversion equipment such as light bulbs, kerosene lamps, partial substitution effects, etc.—a more refined analysis of substitution possibilities would have to incorporate these additional considerations). It would be misleading, however, to then attempt to justify the subsidized kerosene price on the basis of comparison with the newly calculated low price of electricity. Such circular reasoning is far more likely to occur when pricing policies in different energy subsectors are uncoordinated, rather than in an integrated energy pricing framework. We note that all these energy sector subsidies must be carefully targeted to avoid leakages and abuses, as discussed later.

ratio of the official exchange rate (OER) to the shadow exchange rate (SER), which is also called the standard conversion factor (SCF).[33] This represents a global second-best correction for the divergence between market and shadow prices averaged throughout the economy (see also the Annex to Chapter 8).[34]

4.7   THE PRICING OF DEPLETABLE OR NONRENEWABLE ENERGY RESOURCES

Depletable, exhaustible, or nonrenewable energy resources such as crude oil, natural gas, and coal account for most of the current commercial energy consumption in the world. In the U.S.A., for example, they supply some 94% of total and in the developing countries some 83% of commercial energy consumption.[35] Since exhaustible energy resources are subject to depletion, a critical question arises as to whether they should be used up now, or saved for future use. Thus special consideration must be given to the fact that they have a potential future value or opportunity cost.

Starting from the seminal work by Harold Hotelling, a large literature has developed on this subject.[36] Under certain simplifying assumptions (described later in this section), Hotelling's basic result, subsequently referred to as the Fundamental Principle of the Economics of Exhaustible Resources, indicates that the optimal price of the resource net of extraction costs must rise over time at a rate equal to the rate of interest.[37] This conclusion is valid both in the case of a competitive resource extracting industry seeking to maximize the present discounted value of profits, as well as for an economically efficient extraction path which maximizes the present discounted value of the net benefits of resource consumption.

This general principle has been graphically represented in Figure 4.7. Price at time 0 is equal to marginal costs plus *ab*. It rises at the rate of interest to

---

[33]Note that, with the foreign exchange numeraire, conversion of domestic price values into shadow-price equivalents by application of the SCF to the former is conceptually the inverse of the traditional practice of multiplying foreign currency costs by the SER (instead of the OER) to convert to the domestic price equivalent.

[34]For example, suppose the border price of imported diesel is 4 pesos per liter (i.e., US$0.20 per liter, converted at the OER of 20 pesos per US$). Let the appropriate SER that reflects the average level of import duties and export subsidies be 25 pesos per US$. Therefore SCF = OER/SER = 0.8, $P_e = \frac{4}{0.8} = 5$ pesos per liter.

[35]Commercial energy consumption excludes consumption of so-called traditional fuels such as firewood, dung, and crop residues. From: World Bank, *Energy in the Developing Countries*, Washington, 1980, Table 23. US data from: US Bureau of Mines, Yearbook, Annual.

[36]Harold Hotelling, "The economics of exhaustible resources," *Journal of Political Economy*, Vol. 39, April 1931, pp. 1372–75.

[37]Solow, Robert M., "The economics of resources or the resources of economics," *American Economic Review* 64:1–14, 1974 In its simplest form, Hotelling's original rule stated that the price of an exhaustible should rise at a rate equal to the interest rate, assuming a zero extraction cost.

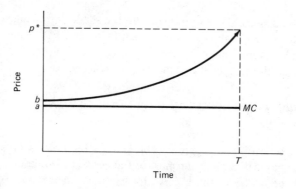

Figure 4.7.   The optimal price path of exhaustible resources over time.

P* when simultaneously the deposit is exhausted and demand for it falls to zero. The usual assumption is that at that point higher-priced energy substitutes will enter into use. The necessary adjustment process, given a stationary economy, would have to come about either by a continuous decline in the output of each deposit, or by a gradual reduction in the number of producers. This postulated reduction in output is required to accommodate the usual assumption of a downward sloping demand curve. As output prices rise over time at the rate of interest, demand necessarily falls and production must decline. A less tortuous and somewhat more realistic assumption made by some analysts is that the demand curve continuously shifts outward (i.e., market demand is growing over time). This, then leads to higher demands at each given price, or unchanged demands at rising prices.

More formally, the validity of this basic proposition can be shown as follows (see Mathematical Annex 2 to this chapter for more detailed analysis). In a competitive market $r_t = P_t q_t - cq_t$, where: $\qquad$ (4.6)

r_t = net profit in period t
P_t = market price in period t
q_t = output in period t
c = marginal cost of extraction (assumed constant)

The producers will attempt to maximize the present value of his discounted profits over the life expectancy of the deposit, subject to the constraint that the total quantity of the resource is fixed. Hence,

$$\sum_{t=0}^{T} q_t = \bar{q} \qquad (4.7)$$

where
T = time horizon of production to exhaustion
$\bar{q}$ = the total quantity of the resource in place.

The maximization problem becomes:

$$\text{Max} \sum_{t=0}^{T} r_t(1 + i)^{-t} = \sum_{t=0}^{T} (P_t q_t - cq_t)(1 + i)^{-t} \qquad (4.8)$$

where
i = appropriate market rate of interest.
Forming the Lagrangian:

$$L = \sum_{t=0}^{T} (P_t q_t - cq_t)(1 + i)^{-t} - \lambda \left(\sum_{t=0}^{T} q_t - \bar{q} \right). \qquad (4.9)$$

The first order (necessary) conditions for maximization of equation (4.9) are:

$$\frac{\partial L}{\partial q_t} = (P_t - c)(1 + i)^{-t} = \lambda \qquad (4.10)$$

Thus,

$$P_t = c + \lambda(1 + i)^t \qquad (4.11)$$

This says that price P_t in period t must be equal to the marginal extraction cost plus an expression $\lambda(1 + i)^t$ which is usually called "user cost" or "depletion premium." Let us define

$$R_t = P_t - c = \lambda(1 + i)^t \qquad (4.12)$$

then

$$\frac{R_{t+1} - R_t}{R_t} = i \qquad (4.13)$$

which is the definition of the "fundamental principle." R_t is also called the resource "rent" or royalty of the depleting asset.

The fundamental principle was initially derived under a rather stringent set of assumptions, including: perfect foresight and complete certainty of future demand, the existence of a known fixed stock of homogeneous resources, an unchanging market organization, constant marginal extraction costs, and no common access effects. We summarize briefly below, subsequent work that has sought to approximate the real world more closely by relaxing some of these assumptions.[38]

The first broad area of study has been market organization, in particular, the effect of monopoly on the price and depletion rate. Hotelling's original analysis suggested that a monopolist would start off with a higher initial resource price than in the competitive case, and that marginal revenue rather than price would grow at the interest rate. He also concluded that the depletion rate would be slower, because the time to resource depletion would tend to infinity for the monopolist, whereas the resource would be exhausted in a finite time under competitive conditions.

Subsequent analysis has tended to confirm the original result that monopoly slows depletion, under a wide range of condtions for the characteristics of demand. Thus, if the price elasticity of demand decreases as the output increases, or if the elasticity increases as the demand curve shifts outward over time, the monopolist would deplete more slowly.[39] Correspondingly, the monopolist would increase output relative to competitive levels in the early period and restrict production later, if the price elasticity decreased as the demand curve shifted over time. The latter case is rather unrealistic, however, because substitute resources are likely to become more available in the future, thus making demand more elastic over time, and accelerated depletion early on could imply prices growing faster than the interest rate, resulting in an unsustainable equilibrium.[40]

Some recent studies have attempted to model an oligopolistic market organization consisting of a few large producers (most often based on OPEC), which is intermediate between the monopolistic and perfectly competitive cases.[41] The results here vary widely depending on the assumptions used.

[38]For a more detailed review of the literature on the economics of exhaustible resources, see: Devarajan, Shantayanan, and Fisher. Anthony C., "Hotelling's 'Economics of Exhaustible Resources': Fifty Years Later," *Journal of Economic Literature* XIX:65–73 (March 1981); and Pindyck, Robert S., "Models of Resource Markets and the Explanation of Resource Price Behaviour," *Energy Economics* 3:130–9 (July 1981).

[39]Weinstein, Milton C., and Zeckhauser, Richard J., "The Optimal Consumption of Depletable Natural Resources," *Quarterly Journal of Economics* 89:371–92 (August 1975); Stiglitz, Joseph E., "Monopoly and the Rate of Extraction of Exhaustible Resources," *American Economic Review* 66:655–61 (September 1976); Lewis, Tracy R., "Monopoly Exploitation of an Exhaustible Resource," *Journal of Economics and Management* 3:198–204 (October 1976); Dasgupta, Partha S., and Heal, Geoffrey M., *Economic Theory and Exhaustible Resources,* Cambridge, UK: Cambridge University Press, 1979.

[40]Dasgupta, Partha S. and Heal, Geoffrey M., 1979.

[41]Schmalensee, Richard, "Resource exploitation theory and the behavior of the oil cartel," *European Economic Review* 7:257–79 (1976); Salant, Stephen W., "Exhaustible Resources and Industrial Structure: A Nash-Cournot Approach to the World Oil Market," *Journal of Political Economy* 84-1079-93 (October 1976); Cramer, Jacques, and Weitzman, Martin L., "OPEC and the Monopoly Price of World Oil," *European Economic Review* 8:155–64 (August 1976); Pin-

The second major area of emphasis in recent studies has been the analysis of the impact of past production on the future prices and depletion rates. Supply-side effects are based on the increase in the costs of extraction as the cumulative stock of output grows. Several studies have analyzed the pattern of optimal depletion of deposits of different qualities, with the highest quality and lowest extraction cost deposits being exploited first.[42] More generally, if the extraction costs are expected to rise at a rate i' (for whatever reason) as cumulative production increases, while i is the interest rate, then the royalty will grow at the rate $(i - i')$, rather than i. In other words, the rate of increase of the depletion premium must equal the rise in the opportunity cost of future use represented by the foregone interest rate i, minus the growth of future extraction costs at the rate i'.[43]

The effects of cumulative production on demand have also been investigated lately.[44] If the stock of a durable resource like gold or silver affects the demand

dyck, Robert S., "Gains to Producers from the Cartelization of Exhaustible Resources," *Review of Economics and Statistics* 60:238–51 (May 1978); Hnyilicza, Esteban, and Pindyck, Robert S., "Pricing Policies for a Two-Part Exhaustible Resource Cartel: The Case of OPEC," *European Economic Review* 8:139–54 (August 1976); Gilbert, Richard J., "Dominant Firm Pricing Policy in a Market for an Exhaustible Resource," *The Bell Journal of Economics* 9:385–95 (Autumn 1978); Ulph, A. M., and Folie, G. M., "Exhaustible Resources and Cartels—An Intertemporal Nash-Cournot Model," *Canadian Journal of Economics* (1980); and Newberry, David M., "Oil Prices, Cartels, and the Problem of Dynamic Inconsistency," *Economic Theory Discussion Paper No. 35*, Department of Applied Economics, Cambridge University, Cambridge, U.K. (1980).

[42]Herfindahl, O. C., "Depletion and economic theory," in: Gaffney, M. (ed.), *Extractive Resources and Taxation*, Madison: University of Wisconsin Press, 1967; Heal, Geoffrey M., "The Relationship Between Price and Extraction Cost for a Resource with a Backstop Technology," *The Bell Journal of Economics* 7:371–78 (Autumn 1976); Solow, Robert M. and Wan, Frederick Y., "Extraction Costs in the Theory of Exhaustible Resources," *The Bell Journal of Economics* 7:359–70 (Autumn 1976); Weitzman, Martin L., "The Optimal Development of Resource Pools," *Journal of Economic Theory* 12:351–64 (June 1976); and Hartwick, John M., "Exploitation of Many Deposits of an Exhaustible Resource," *Econometrica* 46:201–18 (January 1978).

[43]Cummings, Robert G., "Some Extensions of the Theory of Exhaustible Resources," *Western Economic Journal* 7:201–10 (September 1969); Schulze, William D., "The Optimal Use of Non-Renewable Resources: The Theory of Extraction," *Journal of Environmental Economics and Management* 1:53–73 (May 1974); Weinstein, Milton C., and Zeckhauser, Richard J., "The Optimal Consumption of Depletable Natural Resources," *Quarterly Journal of Economics* 89:371–92 (August 1975); Peterson, Frederick M. and Fisher, Anthony C., "The Exploitation of Extractive Resources: A Survey," *Economic Journal* 87:681–721 (December 1977); and Levhari, David and Liviatan, Nissan, "Notes on Hotelling's Economics of Exhaustible Resources," *Canadian Journal of Economics* 10:177–92 (May 1977).

[44]Levhari, David, and Pindyck, Robert S., "The Pricing of Durable Exhaustible Resources," *MIT Energy Laboratory Working Paper No. EL79-053WP*, MIT, Cambridge, MA (1979); Pindyck, Robert S., "Optimal Exploration and Production of Nonrenewable Resources," *Journal of Political Economy* 86-841-61 (October 1978); and Steward, Marion B. "Monopoly and the Intertemporal Production of a Durable Extractable Resource," *Quarterly Journal of Economics* 94-99-111 (February 1980).

for it, but if this stock also depreciates over time, then Hotelling's fundamental principle still holds. However, without depreciation, the growth of the stock may tend to force the price down as time goes on. More generally, with rising extraction costs and depreciating stock, the price will follow a U-shaped path. If the existing stock is augmented by new finds, price again falls initially but rises in the future. The combined effects of cumulative production and different types of market organization such as monopoly, oligopoly, and competition, have not been studied in detail.

The impact of uncertainty on the economics of natural extraction is the third principal area of investigation. Hotelling suggested that uncertainty would cause the exploratory activity of private producers to diverge from socially optimal levels due to two opposing mechanisms. First, certain landowners could benefit from the knowledge that exploration had been successful in neighboring tracts of land. Recent work tends to confirm that exploration activity will be socially suboptimal if potential resource owners wait for others to undertake the high initial risks and costs of exploratory work, thus hoping to get a "free ride."[45] The opposite effect occurs when those who make early finds can file claims and exclude competitors. This can lead to excessive and economically wasteful exploration based on a "gold-rush" mentality, as producers vie to succeed early and block competitors.

Recent work on the uncertainty of supply seems to indicate that for deposits of uniform quality, producers who are uncertain as to the size of their deposits will extract at a slower rate than those who are certain of their resource stocks, in order to avoid running out of resources suddenly.[46] Other studies have also investigated the behavior of producers who are willing to incur additional exploration costs in order to reduce supply uncertainty, under various conditions.[47]

The effects of uncertainty on the demand side have also been examined

---

[45]Joseph E. Stiglitz, "The Efficiency of Market Prices in Long-Run Allocations in the Oil Industry," in: Gerard M. Brannon (ed.), *Studies in Energy Tax Policy*, Ballinger Publ. Co., Cambridge, MA, 1975, pp. 87–94; and Frederick M. Peterson, "Two Externalities in Petroleum Exploration," in: Gerard M. Brannon (ed.).

[46]Kemp, Murray C., "How to Eat a Cake of Unknown Size," in: *Three Topics in the Theory of International Trade*, Amsterdam: North Holland Publ. Co., 1976, pp. 297–308; Gilbert, Richard J., "Optimal Depletion of an Uncertain Stock," *Review of Economic Studies* 46:47–57 (January 1979); Loury, Glenn C., "The Optimal Exploitation of an Unknown Reserve," *Review of Economic Studies* 45:621–36 (October 1978).

[47]Arrow, Kenneth J. and Chang, S., "Optimal Pricing, Use, and Exploration of Uncertain Natural Resource Stocks," *Technical Report No. 31*, Dept. of Economics, Harvard University, Cambridge, MA (1978); Hoel, Michael, "Resource Extraction, Uncertainty, and Learning," *The Bell Journal of Economics* 9:642–45 (Autumn 1978); Pindyck, Robert S., "Uncertainty and the Pricing of Exhaustible Resources," *MIT Energy Laboratory Working Paper No. EL79-021WP*, MIT, Cambridge, MA (1979); and Devarajan, Shantayanan, and Fisher, Anthony C., "Exploration and Scarcity," *CRM Working Paper No. IP-290*, Univ. of Calif., Berkeley, Calif. (1980).

lately.[48] If the uncertainty in future prices increases with time, risk-averse producers will prefer to increase output earlier. However, if the uncertainty is constant (or decreasing), the same resource owner is likely to shift production into the future, since the volume of output at risk will be less as the resource depletes. More obviously, when there is the likelihood of the demand for the resource collapsing at some future date, extraction will be accelerated.

Unfortunately, many of the assumptions used in theoretical models are generally unrealistic in the real world of resource extraction. In fact, rarely, if ever, can we observe price patterns as those described by the fundamental principle. At the macrolevel, prices of depletable resources over time appear either to have fallen in real terms over extended periods of time, or apparently have remained constant.[49] In the area of oil and natural gas, for example, market prices fell steadily in real terms through the post-World War II period until the late 1960s or early 1970s. Similar price declines affected the U.S. coal mining industry. Ensuing price increases appear to have been more the result of ad hoc actions (OPEC, with regard toward oil prices, and environmental and safety regulations for U.S. coal, etc.) than the result of precisely calculated profit-maximizing behavior of depletable resource owners.

The reason for this discrepancy between theoretical conclusions and practical reality is that the analysis abstracts from the necessary constraints of depletable resource production. Extraction rates from any given deposit cannot be varied at will, but must be based on given geological and technological considerations. Given a newly found resource deposit, a prospective producer usually has a certain amount of freedom to choose (a) the data of start-up of production, and (b) the optimal production rate. However, once he has determined those two, production economics and technologies constraints usually force him to stay within relatively narrow limits of the optimum output rate. This has been illustrated in the simple production diagram of Figure 4.8 which illustrates the possible choice between the potential output rates of a given deposit, a, b, and c. In terms of production costs, b is obviously the preferred choice if the producer is a price taker and the quantity of his output does not affect the market price. However, if his production rate varies significantly from output Q_b, his marginal production costs change and rise both at lower or at higher

---

[48]Weinstein, Milton C., and Zeckhauser, Richard J., 1975; Lewis, Tracy R., "Monopoly Exploitation of an Exhaustible Resource," *Journal of Environmental Economics and Management* 3:198–204 (October 1976); Dasgupta, Partha S., and Heal, Geoffrey M., "The Optimal Depletion of Exhaustible Resources," *Review of Economic Studies,* Symposium 1974:3–28 (1974); Long, Ngo Van, "Resource Extraction Under Uncertainty About Possible Nationalization," *Journal of Economic Theory* 10:42–53 (February 1975).

[49]See H. J. Barnett and C. H. Morse, *Scarcity and Growth,* Johns Hopkins Press, Baltimore, 1963. The findings of Barnett and Morse have recently been reevaluated by V. K. Smith, ed., *Scarcity and Growth Reconsidered,* Johns Hopkins Press, Baltimore, 1979.

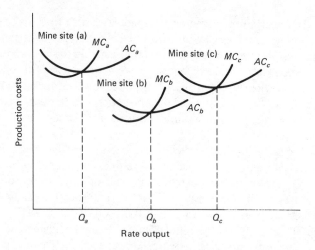

Figure 4.8.   Typical production cost function for mineral resources.

rates of output. What this means is that the resource owner, far from being able to freely choose his rate of output, must carefully evaluate the various potential output rates and design his production plant in such a way as to maximize his net profit over the life expectancy of the deposit. Once the appropriate plant size and production is chosen, deviations from that rate of output become costly and reduce net profits.

Production technological considerations also prevent resource owners from tilting production substantially toward the present, even if the price expectation for the future is for declining or stable, rather than rising, prices (i.e., the "normal expectations" of exhaustible energy resource prices prior to the 1970). This has been illustrated in Figure 4.9.

Let us assume a mining company is facing the decision how quickly to produce a given amount of minerals from a newly discovered deposit. Its discounted, total production costs schedule is given by curve aa′. The present value of revenues, given constant real prices for the output, is represented by bb′. Given the expectation that the value of output will increase by 7% in real terms over time present valued revenues are represented by bb″. Both cost and benefit streams are discounted at 12%. From the mining company's point of view, producing all output at once and selling it now would maximize gross revenues (assuming that market demand is infinitely elastic). This is shown by point b, which represents the highest value of revenues obtainable. However, technical considerations make it impossible to produce all minerals at once. This is indicated by the vertical portion of the production cost curve below point a. As production is stretched out over longer periods of time discounted production costs will fall, mainly because congestion costs are reduced and overinvestment in mining equipment can be avoided. Production costs will be lowest at A where the inherent benefits of economies of scale of the equipment to be used will be balanced by the losses of having to shut down the mine and having to scrap its equipment prematurely because the deposit is exhausted before the mining equipment is really worn out. If even lower production rates are adopted unit costs will rise because of the inefficiency of equipment and methods that then must be utilized at smaller production rates. However, point A will rarely, if ever, represent the most optimal point of production over time. This is so because any optimum depends on the equality, at the margin, between the reduction in the present value of gross revenues and the fall in the present value of production costs. As long as the latter fall at a faster rate than the former, it pays the company to stretch out the rate of production over a longer period of time. Hence, the optimum production rate, which maximizes total net revenues in present values (i.e., the separation between the gross revenue and production cost curves), occurs at the point where the slopes of these curves are equal. In brief, net revenue maximizing production is defined by points E^1 for the discounted gross revenue curve bb', and E^2 for the discounted gross revenue curve bb''. It can easily be seen from the diagram that expected relative price increases brought about by projected future scarcities lead to a stretching out

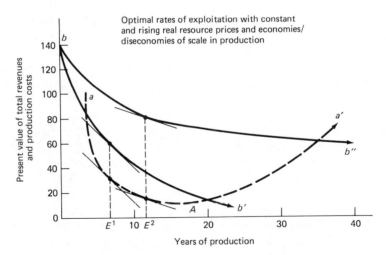

bb'' = present value of gross revenues, 12% discount rate, 7% real annual price rise;
bb' = present value of gross revenues, 12% discount rate, constant real prices;
aa' = present value of production costs, 12% discount rate, constant real prices.

Figure 4.9.   Optimal rates of exploitation with constant and rising real resource prices and economies/diseconomies of scale in producttion.

of the optimum production schedule over time. The mining firm, facing a 12% discount rate and no expected relative price increases for its output will operate at E^1. Faced with an expected relative price rise of 7% per annum, it will choose point E^2 instead. However, once a rate of extraction has been chosen, daily operating conditions and marginal production costs will tend to force the operator to maintain the optimum rate. This optimum rate, however, will likely be at a scale of plant which is greater than would be dictated by a minimizing of production cost approach.

4.8   PRICING NONTRADABLE, DEPLETABLE ENERGY RESOURCES: NORMATIVE ISSUES

The preceding discussion of optimally pricing depleting resources over time was conducted from the point of view of the producer. The objective was to determine the optimum (i.e., economically efficient price) for the supplier. However, the issue of optimum pricing of depletable resources must also be viewed from the demand side, because consumption today creates a potentially positive opportunity cost to the energy user of foregone consumption tomorrow. This issue is addressed in the context of depletable resources that have no outside markets. The pricing of such nontradable, exhaustible energy resources presents special problems because of this lack of outside markets that would determine the foregone opportunity cost of using up the resource domestically.

The question to be analyzed is what the appropriate price should be to consumers, given that the pricing objective is economic efficiency. Two rules appear to bracket the appropriate pricing range: The lower bound is given by the marginal economic extraction costs, plus processing and transportation costs to markets. These costs must be equal to or less than the economic costs of alternative energy resources delivered to the same market under similar conditions. The upper bound appears to be given by the current economic replacement costs of the next best alternative energy resource that could be used instead.[50] As will be shown here, in most cases neither of these two limits adequately represent the true economic value of the depleting resource, even though they determine the appropriate upper and lower bounds. Figure 4.10 indicates these two limits for the case of constant real extraction costs, and the two cases of constant or rising real replacement costs. If the marginal cost pricing rule is applied naively, price would be set equal to $P_C = $ MC and remain constant (under the simplifying assumption of constant real marginal extrac-

---

[50] The current real economic replacement costs may be either constant (if the replacement resource is available in essentially unlimited quantities at constant marginal costs relative to demand), or rising in real terms (if it is subject to increasing scarcity (whether man-made or natural).

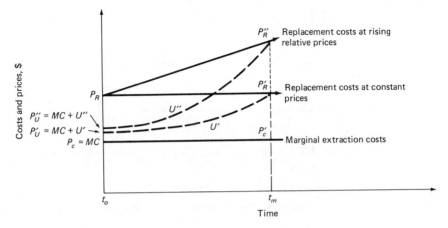

Figure 4.10.   Marginal cost versus replacement cost pricing.

tion costs) until the exhaustion of the deposit at time t_m. It would then rise to P_R' or P_R'', the then prevailing cost of the next best alternative energy resource.[51] However, following the theory of exhaustible resources discussed in the preceding section, prices charged should not only reflect the marginal costs of extraction but also the discounted foregone net present value of the resource, which is determined by its discounted future replacement cost. Hence, the price charged should be equal to MC + U' or MC + U'', where U' or U'' represent the economic rent or user cost, for the two respective cases. The MC + U price path, then, is given by either $P_U P_R'$ or $P_U P_R''$, depending on the level of future replacement costs. $P_R P_R'$ shows the replacement costs of the next best alternative energy resource at constant real prices, while $P_R P_R''$, assumes constantly rising real prices of the next best alternative resource.[52]

While the above rule of MC + U determines the economically efficient minimum pricing path based on present marginal costs plus discounted future opportunity costs, the question remains why today's price should not be raised immediately to the level of the full marginal replacement cost of the best substitute. Such an argument appears quite persuasive, because in an expanding market the sum of the consumer and producer surplus gained in the future

[51]This is the marginal cost of what Nordhaus called the "backstop technology." See W. D. Nordhaus, "The Allocation of Energy Resources," *Brookings Papers on Economic Activity*, No. 3, 1973, pp. 529–76.
[52]In a few cases the potential for substantial economies of scale may actually result in lower real future costs. This would mean that the appropriate price is determined by marginal costs only. Examples of such situations are future, low-unit cost, large-size hydropower sites on the lower Kongo River, the Bramaputra River, and the various Atlantic-watershed hydro sites in Peru.

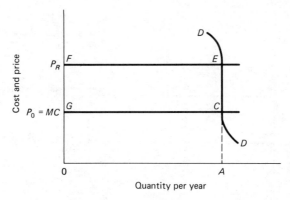

Figure 4.11. Effects of marginal cost versus replacement cost pricing with totally inelastic demand.

from each unit of the low-cost supply is always greater than the net amount of consumer surplus lost from the same unit consumed today, provided demand is not totally inelastic (see also Figure 4.11 below). In order to analyze this question, it is necessary to compare the magnitude of the respective present and future gains and losses of consumer and producer surpluses.[53]

Figure 4.11 depicts the simple case in which demand is perfectly inelastic between the two limiting prices P_C equal to marginal costs and P_R equal to replacement costs. In this situation, from an economic efficiency point of view, it does not matter which price will be charged. In either case demand will be the same and resource exhaustion will occur at the same time.[54] The difference in price times the total quantity available, e.g., FGCE, will be either appropriated as consumer surplus if $P_C = MC$ is charged, or as producer surplus or government income (if the difference is taxed away) if the price is set equal to replacement costs P_R. The effect on national income is zero. In setting the price, no account has to be taken of U, the rent due to depletion, because it would simply change the distribution between consumer and producer surplus.

Figure 4.12 depicts the case in which demand between prices P_C and P_R is not completely inelastic. If P_C equal to marginal costs is charged, (annual) consumption is equal to OB, while demand is reduced to OA per year if P_R, the replacement cost, is charged. As a consequence, annual consumption is

---

[53]Economic efficiency will be used here in its conventional sense of maximizing real consumption benefits over time, given appropriate real interest rates, absence of uncertainty, appropriate efficiency shadow prices, and absence of any weighting of benefits or costs according to their distributional impact (i.e., the objective to be maximized is unweighted real national income).
[54]Consumption levels may change, however, if there are significant income effects as more revenues are extracted from users, and this in turn would affect the time to exhaustion.

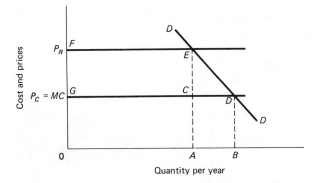

Figure 4.12.   Effects of marginal cost versus replacement cost pricing with elastic demand, annual demand.

reduced by AB, and the life expectancy of the deposit is increased from m years (Figure 4.13) to $m + n$ years, where $n = m$ (AB)/(OB). The change from price P_C to P_R results in a conversion of consumer surplus equal to area FGCE to producer surplus or government revenue. In addition, however, there is a loss of consumer surplus equal to area ECD which is not compensated for by an equivalent increase in producer surplus. Hence ECD represents a true loss in terms of national economic benefits.

This loss, however, is offset by the annual resource savings equal to AB, which translates into an extension of the deposit's life from $m + n$ years (Figure 4.13). This means that the real cost of the energy resource remains equal to P_C for an additional n years. The resulting net gain to the economy is equal to area LNPM.

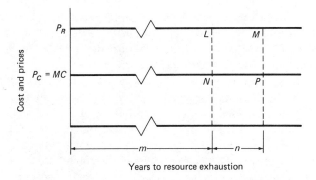

Figure 4.13.   Effects of marginal cost versus replacement cost pricing with elastic demand on life expectancy of deposit.

In order to evaluate the net gains or losses to the economy from charging the price P_R equal to replacement cost instead of price P_C equal to marginal extraction costs, the annual net losses of consumer surplus ECD have to be compared with the net gain from lowered real costs of supplying the resource during period n (i.e., area LNPM).

The net effects can be calculated by the following formula:

$$\sum_{t=1}^{m} c \left(\frac{1}{1+i}\right)^t = \left(\frac{1}{1+i}\right)^m \sum_{t=1}^{n} 2c \left(\frac{1}{1+i}\right)^t \left(\frac{m}{n}\right) \qquad (4.14)$$

where

c = the net loss in annual consumer surplus at price P_R (area ECD)

m = the years to exhaustion of the domestic energy resource if price P_C is charged (i.e., reserves/OB = m)

i = the internal rate of return

n = the years of additional supply from the domestic energy resource deposit resulting from the price-induced reduction in demand (i.e., $n = m$ (AB/OB)).

This formulation is based on the following simplifying assumptions:

1. The demand curve in the relevant price range between P_R and P_C is a straight line.
2. Relative prices of domestic and imported energy resources remain constant.
3. The annual production rate of the deposit, once chosen, remains constant (i.e., equal to either OA or OB).
4. The alternatives are either to charge P_C = marginal extraction costs or P_R = replacement costs.

Equation 4.14 contains four unknowns, c, m, i, and n. Because it cancels out, c can be eliminated. This leaves m, i, and n to be determined. The equation can be solved for i, the internal rate of return that equates the net losses in consumer surplus to overall economic savings resulting from the conservation of the lower-cost domestic resources, given various values for m, the basic life expectancy of the domestic resource in years, and n, the number of years of domestic production added to m as a consequence of the price-induced reduction in consumption. Of course, n is a positive function of the elasticity of demand and of m, the basic life expectancy of the deposit. The more elastic the demand, the greater n becomes because of the greater reduction in demand resulting from the higher elasticity.

If the elasticity of demand is zero, domestic resource conservation is zero as well, and both sides of equation (4.14) vanish since $c = 0$ and $m = n$. All that happens is a conversion of consumer surplus into producer surplus or government revenue. This is an income distribution, but not an efficiency, issue. This situation is illustrated in Figure 4.11.

As the elasticity of demand increases in absolute terms the internal rate of return (associated with raising price from P_C to P_R) decreases. However, the rate of change is relatively modest. An increase in reduction of normal demand from 10 to 30% per year, for example, reduces the internal rate of return from 8 to 7% if the life expectancy of the deposit is 15 years, and from 4 to 3.1% if it is 30 years.

The internal rate of return (IRR) is much more sensitive to the size of the domestic deposit relative to demand (i.e., the deposit's basic life expectancy). If it is 10 years only, rates of return vary between 10.7 and 12.3%. If it is 40 years instead, rates of return are a low 2.5 to 2.9%.

These results are illustrated in Table 4.3 which assumes life expectancies of the deposit from 5 to 40 years, and reductions in annual demand of 10, 20, and 30%, respectively as a result of increases in price.

Given these results two conclusions can be drawn: First, the internal rate of return is not very sensitive to differences in the price elasticity of demand within reasonable ranges.[55] Second, as reserve/production ratios increase the internal rate of return decreases. A reasonable cut-off rate appears to be in the range between 10 to 15 years in which the IRR ranges between 7–12%. If remaining reserves fall below this range, internal rates of return increase very rapidly, making it desirable to increase prices to the replacement cost level. With inventories greater than 10 to 15 years, the IRR becomes unattractive so that it is more reasonable to maintain prices closer to current marginal extraction costs.

For the purpose of (administered) pricing policy, one possible relatively unsophisitcated rule might be to set prices equal to economic extraction costs for as long as the IRR is below a given target rate. Once that target rate is reached due to reductions in remaining resource inventories, prices should be raised regularly to reflect the increased IRR, until at the point of resource exhaustion price equals the long-run marginal replacement costs of the next best alternative resource.[56]

It should be noted that the results of the above analysis depend on a number of specific assumptions. The first is the linearity of the demand function in the relevant price range. If it is strongly convex instead, meaning that a majority of marginal users will be willing to pay no more than the lower price, P_C, the value of lost consumer surplus will be less and the resulting IRR higher (see

---

[55]Given a price increase of 50%, for example, the demand elasticities resulting in a 10% to 30% reduction in demand range from -0.2 to -0.6.
[56]See Mathematical Annex 2 of this chapter for details.

Table 4.3. The Effects of Deposit Life Expectancy and Demand Elasticities on
Potential Economic Benefits

	CONSTANT REAL PRICE CASE	
LIFE EXPECTANCY OF DEPOSIT, YEARS	PERCENT REDUCTION IN DEMAND FROM INCREASED PRICES, %	INTERNAL RATE OF RETURN
5	10	26.0
5	20	24.0
5	30	22.0
10	10	12.3
10	20	11.5
10	30	10.7
15	10	8.0
15	20	7.5
15	30	7.0
30	10	4.0
30	20	3.6
30	30	3.1
40	10	2.9
40	20	2.7
40	30	2.5

Figure 4.14). If it were concave instead, the opposite would be the case (see
Figure 4.15).

Another important assumption is that the real prices of the replacement
energy resources will not increase over time. This may be the case in some
countries such as the U.S., Canada, or Australia, where available substitute

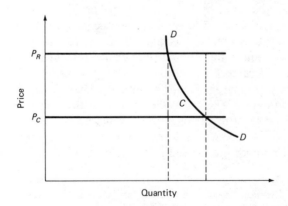

Figure 4.14.   Demand curve convex (to the origin).

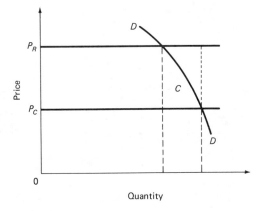

Figure 4.15.   Demand curve concave (to the origin).

resource deposits such as coal are very large. However, if the real costs of the substitutes are likely to rise, the value of the conserved domestic resources will also rise, making it economically more attractive to raise present prices to the prevailing levels of substitutes. This situation has been depicted in Figure 4.16. As can be seen, the future economic benefit from conservation, which was equal to LNPM in the constant price case (Figure 4.13) increases to L'NPM' instead. As a result, the value of future savings increases relative to the sum of the lost consumer surplus over time, and the IRR measuring the former relative to the latter increases. This makes it more desirable to raise current prices.

MATHEMATICAL ANNEX 1 (TO CHAPTER 4): DYNAMIC VERSION OF MARGINAL COST PRICING RULE

In this annex, we present the dynamic version of the basic (static) rule of pricing presented at the beginning of Chapter 4. The results show that setting the price equal to the marginal cost at every instant of time over the period under consideration will maximize the net benefits of energy consumption to society.[1]

The model is formulated in the general context of a peak load pricing case, where capacity constraints could limit the available supply at certain times (see section 4.5). This yields a two-part pricing structure in which the marginal costs of supply are higher during the peak period when energy demand begins to infringe on the capacity constraint.

Consider the case of an energy supply system delivering output to society. We seek to find the set of price (and output) levels over some given period of time that will

---

[1]For a more detailed exposition of this result, see Michael A. Crew and Paul R. Kleindorfer, *Public Utility Economics,* St. Martin's Press, New York, 1979, Chapter 7.

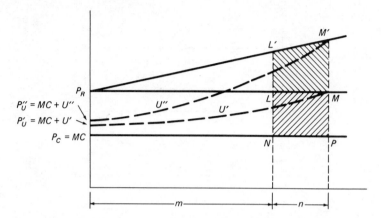

Figure 4.16.   The effects of rising real prices for substitute energy resources on the value of conserved domestic resources.

maximize the present discounted value of net benefits or welfare represented by the area under the demand curve minus the area under the supply curve (see section 4.2). The problem is formulated with time varying continuously to simplify the presentation, but the result also holds for the discrete time case.

The net benefits of consumption at any given time t are represented by the expression:

$$\text{NB} = \int_0^Q p(q,t) \cdot dq - c(q,K,t) - i(t)$$

where
p = consumer's willingness-to-pay or price curve
q = output or production per unit time period
c = production cost function which represents the available technology for energy supply
K = capital stock or available capacity
i = investment rate

The welfare maximization problem may be formulated as the following optimal control problem:

$$\text{Max} \int_0^T \text{NB} \cdot \exp(-rt) \cdot dt$$

subject to

$$K(t) \geq q(t) \geq 0$$
$$I(t) \geq i(t) \geq 0$$
$$\text{and } i(t) = \dot{K}(t) + \delta K(t)$$

where

r = discount rate;

I = upper limit on investment rate

= rate of capital depreciation

and T = time horizon under consideration.

Now we may define the Hamiltonian function:

$$H = NB + \lambda (i - \delta K);$$

and the corresponding Lagrangian function:

$$L = H + m_1 \cdot (K - q) + m_2 \cdot q + m_3 \cdot (I - i) + m_4 \cdot i.$$

By Pontryagin's maximum principle, the solution to the optimal control problem must also maximize the Lagrangian function at each instant of time over the time horizon under consideration, thus giving rise to the following set of conditions:[2]

$$(\partial L/\partial q) = 0$$
$$(\partial L/\partial i) = 0$$
$$\text{and } (\partial \lambda/\partial t) = r - (\partial L/\partial K)$$

Expanding the first of these conditions yields the desired optimal price:

$$p^*(t) = (\partial C/\partial q) - m_1 + m_2$$

We may interpret the above using the complementarity conditions:

$$m_1(K - q) = 0, \ m_2 q = 0; \quad \text{and} \quad m_1 \geq 0, \ m_2 \geq 0$$

Case 1. Peak period—capacity constrained: $K(t) = q(t) > 0.$

$$p_1^* = (\partial C/\partial q) + m_1$$

Thus, during the peak period, the welfare maximizing optimal price is equal to the marginal operating cost (holding capacity fixed), $(\partial C/\partial q)$, plus the marginal cost of adding new capacity, m_1.

Case 2. Off-peak period—no capacity constraint: $K(t) > q(t) > 0.$

$$p_2^* = (\partial C/\partial q)$$

In the off-peak period case, only the marginal operating cost enters into the optimal price.

---

[2]For details, see Takayama, Akiro, *Mathematical Economics,* Holt, Rinehart & Winston, New York, 1974.

These are the well known marginal cost pricing conditions in a relatively simple form, and it is clear that they hold for a dynamic price path over a given time period just like in the static case developed in Chapter 4.

MATHEMATICAL ANNEX 2 (TO CHAPTER 4): THE DYNAMICS OF OPTIMAL PRICING AND EXTRACTION OF AN EXHAUSTIBLE RESOURCE

Notation

$x(t)$ = quantity of exhaustible resource that has been extracted by time t

$\dot{x}(t) = \dfrac{dx}{dt}$ = current extraction rate

z = rate of use of best alternative source

$\dot{x} + z$ = total consumption rate

$u(\dot{x} + z)$ = utility of consumption

c = cost of extraction

i = interest rate

$k(t)$ = unit cost of alternative supply

NB = present value of benefits minus costs

a = available amount of exhaustible resource constraint on exhaustion x
 $\leqq a$

Problem

$$\underset{x,z}{\text{Max}} \int_0^\infty [u(\dot{x} + z) - c\dot{x} - kz]e^{-it}\, dt \qquad (4.1A)$$

subject to

$$x(t) \leq a$$

The Langrangian function for this problem is:

$$\underset{x,z}{\text{Max}} \int_0^\infty \{[u(\dot{x} + z) - c\dot{x} - kz]e^{-it} + \lambda(a - x)e^{-it}]\, dt \qquad (4.2A)$$

Efficiency conditions (Euler-Lagrange) for x:

$$\frac{d}{dt}\{[u' - c]e^{-it}\} = \begin{cases} -\lambda e^{-it} & t \geq T \\ 0 & t = T \end{cases} \qquad (4.3A)$$

where T is the time of exhaustion determined by

$$X(T) = a$$

Efficiency condition for z:

$$z \begin{Bmatrix} = \\ \geq \end{Bmatrix} 0 \leftrightarrow u' - k \begin{Bmatrix} < \\ = \end{Bmatrix} 0 \qquad (4.4A)$$

from which

$$u' \begin{Bmatrix} < \\ = \end{Bmatrix} k(t) \qquad \text{for} \qquad t \begin{Bmatrix} < \\ \geq \end{Bmatrix} T. \qquad (4.5A)$$

Now u' = marginal utility, and p is the efficiency price of one unit of the exhaustible resource at time t. From (4.3A)

$$[u' - c]e^{-it} = \text{constant} = \mu \qquad \text{for} \qquad t < T$$

or

$$u' = c + \mu e^{it}. \qquad (4.6A)$$

To determine μ observe (4.5A) for $t = T$

$$u'(T) = k(T).$$

By continuity, using (4.6A)

$$u'(T) = c + \mu e^{iT} \qquad (4.7A)$$

yielding

$$\mu = [k(T) - c]e^{-iT}$$

Thus

$$p = u'(t) = c + [k(T) - c]e^{i(t-T)} \qquad (4.8A)$$

The price of the exhaustible resource rises continuously from

$$c + [k(T) - c]e^{-iT} \qquad \text{to} \qquad k(T)$$

The premium $[k(T) - c]e^{-iT}$ must be considered a scarcity rent. Its level depends critically on the time T to exhaustion. This time may be determined by equating supply and demand. Notice that from time zero to T only the exhaustible resource is used and that at $t = T$ there is a complete switchover to the foreign resource. (This is a bang-bang solution.)

$$u'(\dot{x}) = p(t)$$

determines consumption $\dot{x}$ as

$$\dot{x} = (u')^{-1}(p(t)),$$

Then the time of exhaustion T is given by

$$a = \int_0^T \dot{x}\, dt = \int_0^T (u')^{-1}(p(t))\, dt$$

As an illustration assume

$$u(\dot{x}) = b \log \dot{x}$$

$$u'(\dot{x}) = \frac{b}{\dot{x}}$$

$$\dot{x} = \frac{b}{p(t)} = \frac{b}{c + [k - c]e^{9(t-T)}}$$

$$a = \frac{b}{c} \int_0^T \frac{dt}{1 + \left[\dfrac{k}{c} - 1\right]e^{i(t-T)}}$$

$$= \frac{b}{c} \int_0^T \frac{dt}{1 + he^{it}} \qquad h = \left(\frac{k}{c} - 1\right)e^{-iT}$$

$$= -\frac{b}{ic} \ln [h + e^{-it}]\ \Big|_0^T$$

$$= \frac{b}{ic} \ln \frac{h}{h + e^{-iT}} = \frac{b}{c} \ln \frac{he^{iT}}{1 + he^{iT}}$$

$$a = \frac{b}{c} \ln \left[1 - \frac{c}{k(T)}\right].$$

Alternatively let

$$u(\dot{x}) = \beta\dot{x} - \frac{b}{2}\dot{x}^2, \qquad k(T) = k$$

$$u'(\dot{x}) = \beta - b\dot{x} = p$$

$$\dot{x} = \frac{\beta}{b} - \frac{p}{b}$$

$$= \frac{\beta}{b} - \frac{1}{b}[c + (k - c)e^{i(t-T)}]$$

$$a = \frac{\beta}{b}T - \frac{1}{b}\int_0^T [c + (k - c)e^{i(t-T)}]\, dt$$

$$a = \left(\frac{\beta - c}{b}\right)T - \frac{k - c}{ib}[1 - e^{-iT}]$$

(4.9A)

The right-hand side is an increasing function of T so that (4.9A) has a unique root. For small values of iT, a Taylor expansion yields

$$1 - e^{-iT} = iT$$

$$a = \left[\frac{\beta - c}{b} - \frac{k - c}{b} \right] T$$

$$T = \frac{ab}{\beta - k}$$

The time to exhaustion increases with the cost k of the alternative source, the available amount a and the rate b at which marginal utility decreases with consumption.

5
Integrated Framework for Energy Pricing: II

In the previous chapter, the many different and conflicting objectives of pricing policy from the national viewpoint were set out. Then the principles underlying the pricing of energy to meet one of these goals—economic efficiency—were described, starting with a simple model to which various complicating features were successively added. The first part of this chapter discusses how these economically efficient prices may be adjusted to meet the other objectives and constraints such as subsidized prices and lifeline rates, financial viability of energy sector institutions, and so on. The terms marginal costs and LRMC are used interchangeably and denote the economic efficiency value. The latter part of Chapter 5 analyzes the economics of some special issues including the case of temporary oversupply, discriminatory pricing, economies of scale, declining markets for energy, and increasing production costs.

5.2 SUBSIDIZED PRICES AND LIFELINE RATES

Sociopolitical or equity arguments are often advanced in favor of subsidized prices or "lifeline" rates for energy, especially where the costs of energy consumption are high relative to incomes of poor households. Economic reasoning based on externality effects may also be used to support subsidies, e.g., cheap kerosene to reduce excessive firewood use and prevent deforestation, erosion, etc. To prevent leakages and abuse of such subsidies, energy suppliers must act as discriminating monopolists. Targeting specific consumer classes (for example, poor households), and limiting the cheap price only to a minimum block of consumption is easiest to achieve, in practice, for metered forms of energy like gas or electricity. Other means of discrimination may also be required such as rationing, licensing, etc.[1] All these complex and interrelated issues require detailed analysis.

The concept of a subsidized "social" block, or "lifeline" rate, for low income consumers has another important welfare-economic rationale, based on the

---

[1] For example, a minimum ration of cheap kerosene for households, or a special license for trucks using subsidized diesel oil and a ban on diesel-driven passenger cars.

income redistribution argument. We clarify this point with the aid of Figure 5.1 which shows the respective demand curves for energy AB and GH of low (I_1) and average (I_2) income domestic users, the social tariff p_s over the minimum consumption block 0 to Q_{min}, and the efficient price level p_e. All tariff levels are in domestic market prices. If the actual price $p = p_e$, then the average household will be consuming at the "optimal" level Q_2, but the poor household will not be able to afford the service.

If increased benefits accruing to the poor have a high social value, then the consumer surplus portion ABF must be multiplied by an appropriate social weight W_c which is greater than unity. Thus, although in nominal market prices the point A lies below p_e, the "socially weighted" distance OA could exceed the marginal supply cost.[2] The adoption of the block tariff shown in Figure 5.1 consisting of the lifeline rate p_s, followed by the full tariff p_e, helps to capture the consumer surplus of the poor user, but does not affect the optimum consumption pattern of the average consumer.[3] In practice, the magnitude Q_{min} has to be carefully determined. to avoid subsidizing relatively well-off consumers. It should be based on acceptable criteria for identifying "low income" groups, and reasonable estimates of their minimum consumption levels (e.g., sufficient to supply basic energy requirements for the household). Q_{min} will vary widely depending on the state of economic development. For example, such minimum levels of electricity consumption in industrialized countries may be several hundred kilowatt-hours per month. In contrast, Q_{min} for most developing countries will rarely exceed 50 kilowatt-hours per month.

The level of p_s relative to the efficient price may be determined on the basis of the poor consumer's income level relative to some critical consumption level.[4] The financial requirements of the energy sector would also be considered in determining p_s and Q_{min}. This approach may be reinforced by an appropriate supply policy (e.g., subsidized house connections for electricity, special supply points for kerosene, etc.).

5.2 FINANCIAL VIABILITY

To introduce more rigor into the analysis of tariff policy and financial performance criteria for public utilities, it is necessary to distinguish between the economic and social objectives of tariff policy, and the minimum level of finan-

---

[2]For a more detailed discussion of social weights and social shadow prices, see Chapter 8.
[3]Ignoring the income effect due to reduced expenditure of the average consumer for the first block of consumption, i.e., up to Q_{min}.
[4]A simple approximate expression might be $p_s = \text{MOC} \cdot (c/\bar{c})$ where c = income/consumption level of poor consumer and $\bar{c}$ is the critical income/consumption level. For an explicit model which derives an expression for p_s in terms of marginal costs and other welfare and economic parameters, see the mathematical annex to this chapter.

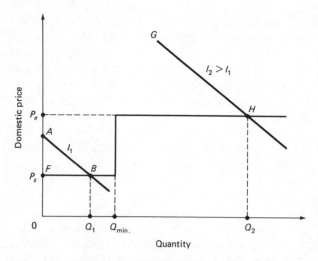

Figure 5.1.   Welfare economic basis for the social or lifeline rate.

cial performance necessary for the continued efficient operation of the enterprise concerned. Such a distinction should not be interpreted as downgrading
the role of financial analysis; it simply reflects the fact that while a minimum
level of financial performance is almost by definition a *necessary* condition, it
is rarely a *sufficient* condition for the achievement of a range of social and
economic objectives that might be assisted by the pricing and investment policies of the enterprise. As described earlier, preeminent among those objectives
that transcend the financial interests of the enterprise itself is the goal of using
marginal cost pricing for influencing consumer behavior in order to improve
resource allocation. Other nonfinancial policy objectives are discussed later in
this chapter.

The principal financial objectives most often encountered relate to the revenue requirements of the sector. They are often embodied in criteria such as
some target financial rate of return on assets or an acceptable rate of contribution towards the future investment program. In principle, for state-owned
energy utilities, the most efficient solution would be to set price equal to marginal cost and rely on government subsidies or taxes to meet the utilities' financial needs. In practice, some measure of financial autonomy and self-sufficiency
is an important goal for the sector. Because of the premium that is placed on
public funds, a marginal cost pricing policy that results in failure to achieve
minimum financial targets for continued operation of the energy sector would
rarely be acceptable. The converse and more typical case, where marginal cost
pricing would result in financial surpluses well in excess of traditional revenue

targets, usually attracts the criticism of consumers. Therefore, in either case changes in revenues have to be achieved by adjusting the strict marginal cost.

A widely used criterion of financial viability for an energy company is its potential to earn an acceptable rate of return on assets, for example, the net operating income after taxes given as a fraction of net fixed assets in operation plus, in some cases, adequate working capital. In the case of private utilities— for example, in the U.S.—the regulatory authorities have traditionally imposed a fair rate of return as an upper limit on earnings (and therefore, on average price per unit sold).[5] Where utilities are government owned, as in most developing countries, the target rate of return is usually considered a minimum requirement to help resist sociopolitical pressures that tend to keep prices too low. If the asset base is defined in revalued terms, then this requirement is more consistent with the forward-looking approach of LRMC. Another future oriented financial criterion that is especially useful when the system expands rapidly, requires the utility to make a reasonable contribution to its future investment program from its own revenues. This self-financing ratio is often expressed by the amount of internally-generated funds available after operating expenses and debt service, as a fraction of capital expenditures.

The application of the financial criteria often raises serious conceptual and practical problems. Thus, if a rate of return test is to be used, then the question of asset revaluation arises. The use of historical costs for working assets, typically original cost less depreciation, would tend to understate their value when capital costs are rising rapidly. If assets are to be revalued, the costs of either (a) exactly reproducing the energy supply system at today's prices; or (b) replacing it with an equivalent system, also at today's prices, might be used after netting out depreciation to allow for the loss of value corresponding to the economic and functional obsolescence of existing equipment. Whichever approach is adapted, significant difficulties of interpretation clearly will occur in its practical application.[6] Furthermore, in certain countries legislation exists that forbids public utilities or corporations to revalue their assets.

Whichever criterion or combination of criteria is used, it is important that the initial tariffs based on strict economic efficiency prices be included in the utility's financial forecast. Then these first round prices may be adjusted through an iterative process until the chosen parameters of financial viability fall within the acceptable range. Although in practice this process is quite ad hoc, some practical guidelines may be effectively used for reconciling marginal

---

[5]See Peter J. Garfield and William F. Lovejoy, *Public Utility Economics,* Prentice Hall, New Jersey, 1974; or James Suelflow, *Public Utility Accounting: Theory and Applications,* Michigan State University, East Lansing, Michigan, 1973

[6]See, for example, G. Robert Faust and H. Gary Larson, "International Valuation Procedures for Electric Utilities," *Public Utilities Fortnightly,* August 14, 1980, pp. 19–25.

costs and the revenue requirement. The relative adjustments to marginal costs between major consumer categories (like residential and industrial), as well as among the different pricing periods (like peak and off-peak) within a given consumer category, will determine the share of the revenue burden to be borne by each user group in a given pricing period.

The simplest practical method of adjustment, which also appears to be the most equitable, is to retain the relative structure of the efficient energy prices and vary the average tariff level by equiproportional changes. However, in general this procedure will not be economically efficient.

The application of the Baumol-Bradford inverse elasticity rule, whereby the greatest (least) divergence from the strict value of marginal costs occurs for the consumer group and pricing period where the price elasticity is lowest (highest), is the most satisfactory adjustment procedure from the viewpoint of economic efficiency.[7] In other words, price adjustments should be smallest for those consumers whose energy usage is most sensitive to price changes, and largest for those who are least sensitive. It is intuitively clear that this guideline would result in the smallest deviations from the "optimal" levels of consumption consistent with the strict application of the marginal cost pricing rule.

In the case of two goods, the following expression applies.

$$(1 - \mathrm{MOC}_1/p_1)/(1 - \mathrm{MOC}_2/p_2) = (1/e_1 + 1/e_{12})/(1/e_2 + 1/e_{21})$$

MOC_i and p_i are the strict marginal cost and price, respectively, of good i; while $e_i = (\partial Q_i/\partial p_i)/(Q_i/p_i)$ and $e_{ij} = (\partial Q_i/\partial p_j)/(Q_i/p_j)$, are the own and cross price elasticities respectively of demand (Q) with respect to price (p). The two goods 1 and 2 may be interpreted as either the electricity consumption of two different consumer groups in the same pricing period or the consumption of the same consumer group in two distinct pricing periods. In practice, a larger number of consumer types and rating periods must be considered, and application of the rule will be limited by lack of data on price elasticities and the need to use subjective estimates. This technique may appear to penalize some customers more than others, thus violating the fairness objective.

Adjustments involving lump sum payments/rebates or changes in customer and connection charges are also consistent with economic efficiency provided consumers energy usage is relatively unaffected by these procedures, i.e., consumption depends mainly on the variable charges. The magnitude of the adjustments that can be made may be insufficient however. Another related approach for reducing revenues is to strictly charge marginal costs only for marginal consumption and reduce the price for an initial block of energy use.

---

[7] See William J. Baumol and David F. Bradford, "Optimal Departures from Marginal Cost Pricing," *American Economic Review*, June 1970, pp. 265–83.

These subsidies on customer charges or on the initial consumption block can also be tailored to satisfy the lifeline rate requirement for poor consumers but such measures tend to complicate the price structure.

In practice, an eclectic approach involving a combination of all these methods is most likely to be successful.

5.3   OTHER OBJECTIVES AND CONSTRAINTS

Aside from "second-best" adjustments and the satisfaction of basic needs of low-income consumers by means of an increasing block rate, several other objectives and constraints face the policymaker in determining an appropriate level and structure of energy prices. In practice, several legitimate economic reasons, quite apart from political or social concerns, could justify a departure from a strictly marginal cost-based tariff policy. Thus, the decision to promote productive energy use in a particularly deprived region in which large external benefits from development might be reaped, may well provide legitimate economic reasons for a departure from strict efficiency pricing.

It is rarely possible to completely disentangle the economic from the noneconomic implications or justification of such policies. Thus, maintaining a viable regional development program or reducing rural to urban migration clearly incorporates notions of income redistribution, general welfare, and economic growth and efficiency considerations. For example, the decision to electrify a remote rural area may require significant subsidies because the beneficiaries are unable to pay a price that fully reflects the relatively high unit costs of rural electrification. Such a decision might be made on grounds that are not completely economic, and the major objective might be, in the simplest terms, to alleviate local political discontent. Whether or not the primary rationale for subsidization lies in economic or social considerations, however, the decision to subsidize should in all cases be made with a clear understanding of the economic efficiency—that is, the marginal costs—that are entailed.

Arguments advanced in favor of subsidized energy prices often may not survive careful scrutiny. Consider the possible use of low energy tariffs to promote industrial development or to attract multinational enterprises. Except in certain specific energy-intensive industries, such as aluminum smelting or chemicals, however, the viability of existing industrial consumers or the decision to set up a new enterprise will depend much less on energy prices than on other factors including the availability of a skilled work force, wage rates, incentives such as tax concessions offered the government, and the provision of an adequate infrastructure. Direct cash subsidies to specific industries from the government are likely to be much more effective than blanket indirect subsidies through energy prices that probably could be taken advantage of by a much broader group of industrial consumers.

In general, decisions to subsidize energy should be approached with care. Subsidization implies either a corresponding drain on public revenues or, if cross subsidization from other energy consumers is the source, at least the frustration of certain legitimate alternative demands for energy. Should the burden of subsidization be placed on public revenues, particular costs are implied in the specific case of a country in which a premium should be placed on public revenues because of a low tax base and inadequate fiscal machinery. In other words, at the margin a rupee or a peso of investment in the public sector is of greater value than a rupee or a peso of private consumption.

Indeed, one might go further and argue that because of this situation, the energy sector could become an appropriate vehicle for the generation of public revenues. Under current cost conditions in which the unit costs of energy are rising fairly dramatically, a pricing policy based on marginal opportunity costs will typically tend to yield considerable financial surpluses to the utility. In this case, efficient resource utilization is fully consistent with the generation of public revenues, and a pricing policy that helps to satisfy both objectives would be advantageous. Moreover, as noted earlier, it is reasonably straightforward to build into the tariff structure an income redistributional element or at least the satisfaction of basic energy needs for low-income groups.

It is particularly difficult to reform pricing policy where low incomes and a tradition of subsidized energy prevail. In practice, price changes have to be gradual, in view of the costs that may be imposed on those who have already incurred expenditures on energy using equipment and made other decisions expecting little or no changes in traditional pricing policies. The efficiency costs of "gradualism" arising from the need to avoid abrupt implementation of marginal cost based tariffs can be seen as an additional implicit shadow value placed on the full social benefits that result from this policy.

Although the more extensive application of peak load pricing based on MOC is clearly justified, the transition period may take many years. First, the truths and myths regarding modern pricing policies must be well understood by customers, producers, and government regulators. Any means of disseminating information will play a key role in this respect. Second, the application of new prices could begin with the larger and better off customers because they are less numerous and may be sensitized more easily. Over the years, more and more consumers would become subject to the new tariffs. Third, the changes in price structure should be initially small, to avoid customer resistance resulting from unfamiliarity or from hardship caused by large increases in their energy bills. Later, prices could be altered more rapidly to approximate marginal costs better. Finally, alternative price structures could be offered simultaneously—that is, both the conventional and new prices—and customers could gradually be won over to the new prices.

Another frequent constraint arises when the government has a policy of uniform national prices for each consumer category throughout the country. This would deviate most from strict efficiency prices when there are many isolated energy supply systems or when disparities between urban and rural supply costs are large, especially for retail distribution. Again, there may be pressing sociopolitical and economic reasons for such a policy, including reducing public discontent, promoting development of remote areas, satisfying basic energy needs, and so on. The benefits of a uniform national tariff policy, however, must be weighed once again against the efficiency costs of deviating from the strict value of marginal costs.

This section concludes with a discussion of an often used macroeconomic argument that appears to be a valid policy constraint, but on closer analysis is not such an important issue. The argument involves the possible inflationary effect of increases in energy prices implied by charging economic efficiency prices. In the case of domestic consumers, moderate energy price increases will not significantly affect the consumer price index since the value of energy they consume usually amounts to a relatively small percentage of total household expenditures. Similarly, for industrial and commercial users, the cost of energy is only a small fraction of total production costs. In general, unless the increase in energy prices was manyfold, its effect on total costs would be quite small, except in the case of a few energy intensive consumers such as metal or chemical plants. Conversely, if demand for energy was unduly stimulated by low prices, then the additional expenditures on supply capacity required to meet this demand may have a much more inflationary impact in the long run; ignoring for the moment the losses in economic efficiency inherent in this situation. For example, in the electricity sector, a consumer's decision to add one kilowatt of demand for cooking or air conditioning at a private cost of about US$100 to 200 will usually imply investments about ten times as great by the utility to meet this demand increment. In certain cases, however, raising energy prices in conjunction with the prices of other government-supplied essential items may have an adverse psychological effect on the inflationary expectations of the public. Also, producers could use the increase as an excuse to raise the prices of their outputs.

5.4   TEMPORARY OVERSUPPLY AND FIXED-TERM MARKETS

One of the consequences of basing energy prices on long-run marginal costs is that excess production capacity may be temporarily created that may remain idle for, sometimes, considerable periods of time. This may occur because of indivisibilities or substantial economies of scale of new plant additions. For example, the economically optimal size (in the sense of lowest average unit costs of output) of a given hydropower site may be narrowly circumscribed by geological and hydraulic conditions, or economies of scale in the design and lay-out of a refinery, a gas or oil pipeline,[8] or a thermal powerplant may be such that the penalty costs of temporary overcapacity may be more than compensated for by the reduction in the investment costs of the installation and the equivalent annuitized per unit costs over its life expectancy.

This situation has been depicted in Figure 5.2. Demand over time, at prices reflecting long-run marginal costs, is assumed to be growing along $D_N D_N$.[9] New capacity must first be added when demand reaches C_B, the existing base capacity of time T_1. The most efficiently-sized plant has a capacity equal to $C_A C_B$ (BC), which means that a gradually declining overcapacity is created initially (i.e., triangle BCE). This overcapacity is finally absorbed at time T_3 when new capacity must be added once again (presumably recreating the temporary overcapacity problem). One policy to reduce the seeming "waste" of underutilized capacity would be to reduce price from long-run to short-run marginal costs. This issue has already been discussed in section 4.4. If this policy were successful,[10], it would shift demand from BE to BD. As a consequence, full capacity utilization equal to C_A would be reached at time T_2 instead of at T_3. This, however, would be inefficient since the additional demand would have been created only as a consequence to price levels that did not cover the full, long-run marginal costs of supply. Furthermore, at those lower prices, if they were uniform, the supply organization would incur long-run deficits and not be able to maintain its financial integrity. For both reasons, such pricing policies would have to be rejected.

A viable alternative, however, would be to charge the lower, short-run marginal cost-based price only for the proportion of total capacity that would remain idle otherwise. This, however, requires price discrimination between different users (see next section). Overall, as long as the price charged for this temporary overcapacity is at least equal to or exceeds short-run marginal costs (SRMC), total average lifetime costs of the plant are reduced because of the net contribution of the additional revenue above SRMC, and economic efficiency is increased because additional units of energy are being supplied at prices above or equal to SRMC but below or equal to the willingness to pay for them. This potential net gain has been illustrated in Figure 5.3 where P_1 and P_2 represent the LRMC- and SRMC-based price levels, respectively. While P_1 is charged to all regular customers, P_2 is charged to temporary (domestic) ones. $D_T, D_{T_2}, \ldots D_{T_n}$ are the respective demand schedules in

---

[8]See, for example, the discussion about the optimum size of the gas pipeline in the Gulf of Thailand in the case study on Thailand (Chapter 10).

[9]Each point on the line $D_N D_N$ represents a market clearing equilibrium point with supply equal to demand and price equal to marginal cost (see Figure 4.3). The line is generated as the demand curve shifts outward with time.

[10]This may not be the case since it would depend on the short-run price elasticity between the levels of long-run and short-run marginal costs.

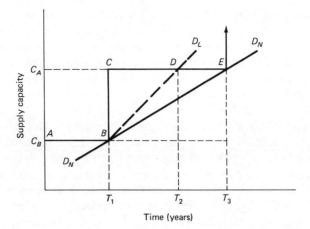

Figure 5.2.   Temporary overcapacity and short-run versus long-run marginal cost pricing.

periods T_1, T_2, ... T_n. As long as total demand at price P_2 is less than total available capacity Q_c, economic efficiency is maximized by selling excess capacity supplies at price P_2 to domestic consumers. This net gain is represented by areas ABC, A'B'C', and A"B"C", respectively. However, if the temporary sales represent exports, the net gains are limited to the excess of the prices charged above P_2 = SRMC times the quantity exported (i.e). areas BDEF, B'D'E'F', and B"D"E"F", respectively).

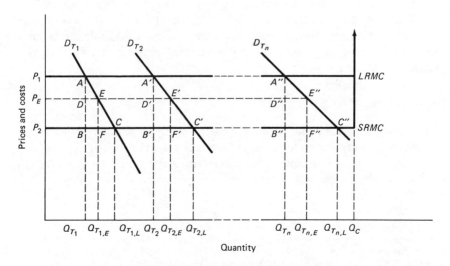

Figure 5.3.   Gains from SRMC pricing.

The question whether such extra sales at lower than LRMC should be made or not is critically dependent on the ability of the supply organization to limit them strictly to the time period in which such excess supplies are actually available. As normal demands at LRMC-based prices grow, SRMC-based sales must be reduced accordingly to stay within total available capacity. What must be avoided under all circumstances is that these temporary, lower-priced supplies lead to the establishment of permanent uses that cannot be cut-off and eliminated whenever it is economically appropriate to do so. Not only must the temporary nature of the supplies be clearly spelled out in the contractual arrangements underlying the sales, but care must be taken that no political lobbying group is established that may press for the continuation of low-priced sales when they are no longer warranted on the basis of temporary excess capacity. For political reasons alone it may be preferable to enter into short-term export contracts rather than to attempt to market the available supplies domestically. (See also section 5.6).

Temporary, nonrepeatable energy supply contracts are not uncommon, of course. Sales of so-called secondary electric energy are widely practiced, particularly by electric generating systems with substantial temporary hydraulic overcapacity.[11] Other examples are oil refineries with partial excess supplies resulting from imbalances between refinery outputs and domestic demands. On occasion the temporary marketing of excess energy supplies is used as a deliberate strategy for efficient energy system development. These issues are discussed in greater detail later in this chapter.

In summary, what can be concluded is that short-run marginal cost pricing has a legitimate place in the pricing strategy of energy supply systems, if the SRMC-based prices are limited to the specific quantities of temporary excess supplies, and if the contracts under which they are made available can be terminated without penalty as soon as normal market demands at LRMC-based prices grow sufficiently.

5.5 DISCRIMINATORY PRICING

Discriminatory pricing can be defined as the charging of different prices to different customer groups for the same goods or services supplied. It is widely practiced whenever it is possible to differentiate between user groups. Typical examples are reduced theater admission to senior citizens or juveniles, or dif-

---

[11]A typical example is provided by the winter-peak, hydro-based Bonneville Power Administration system in the U.S. Pacific Northwest which has large quantities of secondary energy for sale during the spring and early summer run-off period. This oversupply is now being marketed in the U.S. Pacific Southwest as a result of high-capacity transmission interties with these thermal-based, summer-peak systems.

ferential transportation charges for the same quantity, weight, and handling characteristics of different commodities.[12] In the area of energy supplies, discriminatory pricing schedules are common for electricity and natural gas systems. These can easily discriminate among customers because they supply through individually metered connections. Discrimination is less common for other types of energy supplies because of the difficulties to prevent resales. However, within certain limits, price discrimination is also frequently practiced in petroleum product or LPG distribution, for example, when regionally differentiated prices are charged that do not simply reflect the differential in hauling and distribution costs.

Price discrimination can be defined as the charging of prices to selected groups of customers that differ by different margins from the social long-run marginal costs of supplying them.[13] It should be noted from this definition that not all price differentials automatically qualify as "discriminatory" pricing. Differences in the quantity, quality, timing, and location of deliveries will result in differences of marginal costs; these should be appropriately accounted for in the setting of prices.[14]

Furthermore, uniform deviations of prices to all customers from social long-run marginal costs cannot be defined as discriminatory, since no discrimination takes place if everybody is charged uniformly more, or less, at the same rate.[15] On the other hand, the prevalence of uniform prices may well represent price discrimination against some groups. In Pakistan, for example, gasoline and diesel prices at the pump are uniform throughout the country. This represents price discrimination against customers in the Karachi region where the country's oil refineries are located, because fuel delivery costs to this region are much lower than to the rest of the country.

Price discrimination may also be practiced through the differential pricing of close substitutes. Kerosene prices versus LPG, or premium versus regular gasoline are examples. The rationale for such price differentials has been discussed earlier.

There are many reasons why price discrimination is practiced so widely. Income distributional objectives attempt to foster economic development

---

[12]For example, government-approved U.S. railroad tariffs literally contain thousands of different tariff classifications.

[13]"Social" long-run marginal costs in the context of this discussion means simply that all external costs or benefits should be accounted for.

[14]However, differences in social weights on the basis of social merit or income distributional criteria should be excluded; lifeline and similar group preferential rates are properly defined as discriminatory pricing.

[15]Such deviation from marginal costs would almost always be present under the common average cost pricing schedules.

through low energy prices to specific sectors,[16] or simply outright political pressure by powerful groups, have been factors. These issues have been discussed elsewhere in this volume. Here we are not concerned with the underlying rationales of price discrimination but with the question to what extent such pricing schemes are economically efficient or inefficient.

It is clear that price discrimination can only be practiced to the extent that the parties subject to it are prepared to pay the higher prices charged, rather than do without the supplies altogether. Within these limits of the willingness to pay, unweighted aggregate economic welfare, or efficiency, remains unchanged. What takes place is an income transfer. Potential consumer surplus of the energy consumer is being converted into revenue to the energy seller, or producer surplus.

This has been illustrated in Figure 5.4. The demands of two differentiated consumer groups are represented by the two distinct demand curves $D_A D_A$ and $D_B D_B'$ resulting in an overall market demand curve of $D_A S D_M$.[17] If a uniform price P_M = LRMC is charged, total sales amount to Q_3 and both consumer groups reap excess benefits in the form of consumer surplus equal to area $P_M D_A S R$. If the demand schedules were known with certainty and perfect price discrimination possible, the seller would charge differential unit prices following the demand schedule $D_A S R$. All consumer surplus would be appropriated by the seller and converted into producer surplus, but sales and quantities produced and consumed would remain unchanged.

Perfect price discrimination is, of course, impossible. The best that can be achieved with discriminatory pricing schemes is to change different blocks of price to different users. Let us assume prices to consumer group A are increased to P_A, with prices to group B remaining unchanged at P_M. As a consequence, group A's consumption is reduced from Q_A to Q_1, and its consumer surplus from area $P_M D_A T$ to area $P_A D_A K$. Producer surplus equal to area $P_M P_A K N$ is created at the expense of an equal amount of consumer surplus. However, there is a net difference between the amount of producer surplus gained and consumer surplus lost. This difference is equal to area KNT. This is the net loss of aggregate economic welfare, or efficiency, resulting from the discriminatory pricing imposed on group A. It is the area under group A's demand curve above LRMC, but below the discriminatory price P_A. Group B's consumption and consumer surplus remains unchanged of course. Total consumption is reduced from Q_3 to Q_2 (which is equal to group A's reduction in consumption from Q_A to Q_1). What must be concluded then, is that discrimi-

---

[16]Typical are the low rates charged for diesel fuels in countries such as Mexico or Thailand, for example.
[17]The discussion assumes that these are the Hicksian income compensated, not the Marshallian, demand curves.

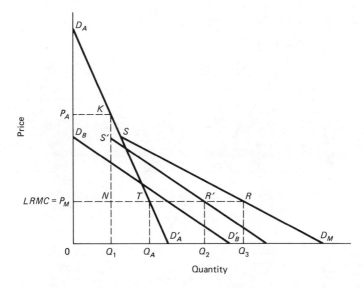

Figure 5.4. Discriminatory pricing and economic efficiency.

natory pricing schemes indeed impose an economic cost on the economy, basically because of the need to limit rates to discrete intervals, and to discrete groups of customers.[18]

5.6 THE EFFICIENCY OF PRICE DISCRIMINATION UNDER ECONOMIES OF SCALE

There are a number of situations in which price discrimination becomes an important tool in bringing about the economically most efficient development of energy resources for a given market. Such conditions arise when new potential additions to supply exhibit substantial indivisibilities and are very large relative to existing markets. Hence, while their average unit costs may be attractively low at full production, market size limitations may be such that the unit costs would be unacceptably high at such lower rates of output. This situation has been depicted in Figure 5.5. As can be seen, at any given single

---

[18]Theoretically it should be possible to impose continuously declining price schedules, provided prices are continuously metered, as in the case of natural gas or electricity. However, apart from the need to know the actual demand schedules (which even the consumers themselves may not know at any given time), such pricing schemes would be an administrative nightmare and, therefore, quite impractical.

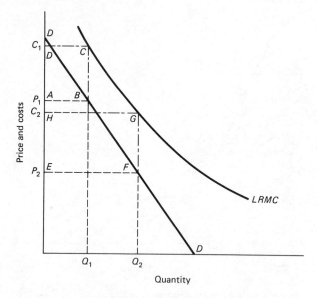

Figure 5.5.   Plant indivisibilities and limited markets.

market price, demand, or willingness to pay, is less than LRMC. At price P_1 quantity Q_1 would be utilized; the LRMC, however, would be equal to C_1 and a net loss equal to area ABCD would be incurred. At price P_2 consumption would be equal to Q_2, LRMC equal to C_2 and the net loss equal to area EFGH. Hence, building of the plant would neither be justified on economic nor on financial grounds.

Situations like this are common, particularly in developing countries in which market size usually is a major constraint. As a result, these countries are saddled with low volume, small-scale energy supply systems whose unit supply costs are usually far higher than those of optimum-scale systems in more advanced countries that serve larger markets.[19]

There are two options—both involving discriminatory pricing—that can sometimes be used to overcome these development limitations imposed by market size. If the relevant domestic market demand is relatively inelastic and the

---

[19]A typical example is provided by Electroperu's system expansion plans in Southern Peru. To serve projected primary demand a 120 MW hydroplant (Charcany V) will be built in preference to another 255 MW plant whose unit costs were estimated to be some 40% lower. Another site, with even lower unit costs—some 75% lower than Charcany—but with a site capacity of some 1,200 MW will probably have to wait well into the 21st century until markets are large enough to absorb its output.

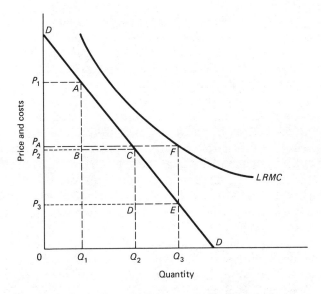

Figure 5.6.   Discriminatory pricing to cover long-run marginal costs.

quantitative difference between market demand and a new project's potential supply is not too large, discriminatory pricing schedules may raise enough total revenue to cover the LRMC. This case has been illustrated in Figure 5.6. As in Figure 5.5, the total market demand schedule lies below the long-run marginal cost schedule at any given single price. However, discriminatory pricing, with P_1 charged to user group 1 which purchases $0Q_1$ at that price, P_2 charged to group 2 purchasing Q_1Q_2, and P_3 charged to group 3 purchasing Q_2Q_3, results in average revenues equal to P_A which are equal to LRMC at F. It should be noted that under such pricing schedules it may well be necessary to sell substantial quantities of output of prices below LRMC—although never at prices below SRMC. If the long-run marginal costs of the project under consideration are lower than those from any competing alternative, aggregate economic welfare will be maximized by choosing this project despite the need for using discriminatory pricing in order to make it economically and financially feasible.[20]

The other potential alternative is to search for additional markets for the

---

[20]Provided, of course, that the necessary price discrimination does not violate existing income distributional goals. However, these could be included in the analysis by attaching the respective social weights to the income effects of the relevant groups affected. For a discussion of social goals and weighting see the Mathematical Annex to this chapter.

project's surplus output, even if it means that this output must be sold at substantially lower prices than those charged in the primary domestic market. Such additional markets can sometimes be found through exports. Examples are the transmountain pipeline in British Columbia, Canada which could not have been built originally without the added expected revenue from export sales to the United States which absorbed about ⅔ of the throughput initially but accounted for less than 40% of total revenue.[21] Other, more recent examples are the Nelson River Power Development of Manitoba Hydro which presently depends for its financial viability on export sales of close to 50% of its output.[22] Other examples are the huge Itaipu hydro project, jointly undertaken by Brazil and Paraguay, with most of the output marketed in the much larger power market of Brazil, or the James Bay Power Development of Quebec Hydro which will deliver substantial quantities of surplus power and energy to U.S. northeastern utilities.

New markets or demands sometimes can also be created by attracting energy-intensive activities to the vicinity of the new project. Examples are Ghana's Volta River power development, which depended on a new exclusively export-oriented aluminum smelter to market most of its power output, or the Aswan Dam in Egypt which sells substantial portions of its electric production to an aluminum smelter and an energy-intensive fertilizer plant.

The rationale for the creation of such low-priced, additional markets has been illustrated in Figure 5.7. Domestic demand is represented by D_D, while export or newly attracted energy-intensive demand is given by D_F. Price P_F equal to D_F is lower than LRMC at any level of output. However, by charging the higher price P_D in the domestic market, average revenue for output $0Q_2$, representing the sum of domestic demand $0Q_1$ plus export demand Q_1Q_2, is sufficient to cover LRMC, and to assure the feasibility of the project.[23]

Before such a development strategy is adopted, a number of important issues must be considered. First, the prices charged to domestic consumers should, in the aggregate, be lower—or at least not higher—than those that would have to be charged under similar conditions for supplies from the next best alternative source.[24] Higher prices (or costs) should be accepted only if they are

---

[21]By now, 25 years later, natural gas prices charged for exports are much higher than the price-controlled domestic ones.

[22]*Financial Post,* Toronto, February 9, 1980. The economic and financial viability of this project was actually rescued by the huge and unanticipated increase in the production costs of competing atomic and thermal powerplants in those export markets. Without these additional markets the Nelson project would have been a very high-cost, inefficient source of electric energy for Manitoba. For an analysis see G. Schramm, *Analyzing Opportunity Costs—The Nelson River Development,* Univ. of Manitoba, Natural Resources Institute, Winnipeg, 1976.

[23]i.e., actual total revenue $P_D \times (0Q_1)$ plus $P_F \times (Q_1Q_2)$ is equal to the revenue $P_A \times (0Q_2)$ under LRMC pricing.

[24]More precisely, the economic costs of domestic supplies from the project must be lower than those of any other alternative.

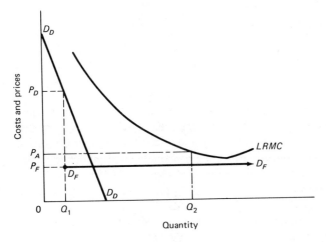

Figure 5.7. Discriminatory pricing to increase total market demand.

temporary, and followed by correspondingly lower prices later on, so that the net present value of the lifetime costs of domestic supplies minus all net benefits from the low-priced outside sales would be lower than the net costs of any other alternative that could be built or utilized at this time. What this would mean in effect is that domestic consumers would be taxed temporarily through higher than necessary current energy prices in order to gain the benefits of lower unit costs resulting from scale economies in the long run. An example might be provided by the alternative choices of a large-size hydro plant whose surplus output could be sold under a term contract at low prices abroad, while financial viability of the project might require temporarily higher tariffs to domestic consumers compared to those from a smaller-sized, alternative plant that could be built instead.[25]

Another possible justification for accepting higher domestic energy costs of such a large-scale project might be provided by the creation of substantial economic benefits from derived activities of such a project. One example could be the creation of energy-intensive industries that would not be attracted otherwise, such as the aluminum smelter complex on the Volta River project in Ghana. Or the project itself may be a multipurpose one that creates substantial benefits in other areas such as irrigation or water supply.

Two other caveats must be kept in mind. The first is related to the question

---

[25]An example of such a situation was provided by the initial sale at low prices of all treaty energy from British Columbia's share of the Columbia River Development between the U.S.A. and Canada, while British Columbia itself was relying for its needs on incremental supplies from the higher-priced Peace River Power Development in Northern British Columbia. For a discussion see John V. Krutilla, *The Columbia River Treaty,* The Economics of an International River Basin Development, Johns Hopkins Press, Baltimore, 1967.

of recapture. As domestic, higher-value demands grow, contractual arrangements should make it possible to reduce and eventually eliminate the low-priced export or industrial sales.[26] The second relates specifically to nonrenewable energy resources. Because these are subject to ultimate exhaustion, export or low-value industrial sales contracts must be related not to the marginal costs, but to marginal costs plus user costs, i.e., the future replacement costs of the exhaustible resources utilized now. Hence, minimum promotional prices should be set not lower than short-run marginal costs plus estimated user costs. Some of these issues have been discussed in greater detail in Chapter 4.

From an economic point of view, the general guideline to be applied in all of these evaluations should be that the sum of the present value of all direct and indirect net benefits of such a project have to be greater than the sum of the combined or separate net benefits of all potential alternatives. The question whether domestic energy users should contribute through temporary or permanently higher energy prices to the feasibility of such a project, however, is a separate one, and must be decided on the basis of the multiple social goals that are the guidelines of the decisionmaking agency.

5.7 PROMOTIONAL PRICING

Project indivisibilities or economies of scale and the lack of temporary, alternative markets for surplus outputs may make it desirable to use promotional pricing schemes resulting in temporary financial losses in order to attract a large number of customers, stimulate greater consumption by existing ones, and thereby expand project utilization at a faster rate. Another reason for promotional pricing of a specific energy resource may stem from the desire to reduce the consumption of alternative energy sources whose use entails substantial negative externalities, as, for example, the overcutting of forest resources.[27]

Promotional pricing is defined as the temporary underpricing of energy sup-

---

[26]This is presently a bitterly disputed issue in the U.S. Pacific Northwest where large blocks of extremely low-priced electric energy (2 to 3 mills per kWh) were originally sold to aluminum smelters under long-term contracts. These contracts are up for renewal now, but other power demands have grown so much by now that high-cost atomic and thermal plants must be added to serve regional demands. Problems have also been encountered in Egypt where low-priced, low-value Aswan Dam energy is used for aluminum production in an export-oriented smelter, while incremental electric energy needs must be satisfied from high-cost fuel-oil based powerplants. Similarly in Ghana, low-priced power from the large Akosombo dam was provided in the late 1960s to a foreign aluminum producer, on a long-term contract. By the late 1970s, Ghana was obliged to continue selling electricity to the aluminum smelter at US¢ 0.9 per kWh while charging much higher prices to local consumers to cover the cost of thermal generation added more recently to meet rapid demand growth. The Ghanaians are continuing to negotiate with the aluminum producer to alter the terms of their long-term electricity contract.
[27]For a discussion of this issue see Chapter 6.

plies to selected customer groups at levels below LRMC (plus user costs, if applicable). Under certain circumstances, such promotional rates may also be set at levels below SRMC.[28]

Promotional pricing schemes are appropriate only for those energy supply systems whose fixed production costs are high, and whose existing market demands are insufficient to absorb output. This could result in high unit costs which, if passed on to consumers, could discourage them from consumption altogether, or at least from increasing consumption at sufficiently rapid rates to justify the initial investment.

A typical example is provided by rural electrification networks. For these distribution costs are usually high because of long distances and low load densities. Technical considerations, however, require minimum investments in terms of line voltage, number and structural strength of distribution poles, conductor size, etc. Almost all of these initial costs (including those of meter reading) are fixed costs, at least up to network capacity. Hence unit supply costs are inversely related to sales,[29] and increased sales reduce LRMC accordingly.

Energy use depends on the use of energy-consuming appliances—in the case of electricity, light bulbs, refrigerators, flat irons, hot plates, etc. If electricity is priced at low rates, more users may be willing to invest in such appliances. This would increase unit sales per customer, thus reducing total unit costs. For example, in Thailand which maintains a vigorous rural electrification program, usually about 50% of all households will sign up initially for electricity supplies when the distribution lines first reach a village. This rate increases to between 75 to 80% after 3 to 5 years. Average per household consumption in newly electrified villages is about 40 to 50% of the average consumption in all rural areas combined. Given the high, initial fixed costs of distribution, it is clear that more rapid increases in demands will tend to lower unit costs, at least until system capacity is reached.

Hence it may be reasonable and economically more efficient to price energy supplies not at the present low-volume, high-unit LRMC, but at the expected average life-time LRMC instead. This has been illustrated in Figure 5.8. Demand schedules in successive years are given by D_t, D_{t+1}, D_{t+2}, and D_{t+3}. With strict LRMC pricing, prices charged would have to be P_1, P_2, P_3, and P_4, resulting in consumption of Q_1, Q_2, Q_3, and Q_4, respectively.[30] As can be

---

[28]If such rates are considered necessary to entice new customers to sign up, and if sales at such rates are strictly limited (i.e., rate structures are inverted), and if it is expected that average consumption will soon exceed the quantity sold at below SRMC.

[29]According to the formula: Total Unit Costs = (Fixed Costs/Unit Sales) + Marginal Variable Costs per Unit.

[30]As drawn in the diagram, prices in the early years could actually be set equal to LRMC. This may often not be the case, i.e., the demand curve may well lie below the LRMC curve at low unit sales (as shown in Figure 5.7, for example).

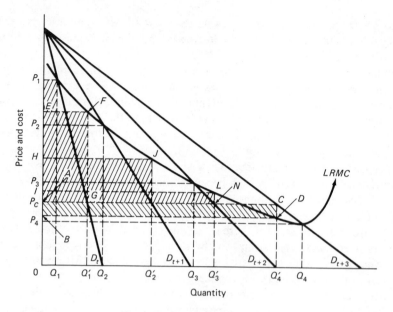

Figure 5.8.　Promotional pricing.

seen, prices charged would be reduced as demand increases and per unit costs fall. The promotional pricing alternative would be to charge a uniform price equal to P_C from the outset. This would increase early annual demands to Q'_1, Q'_2, Q'_3, Q'_4, respectively. Initially, losses would be incurred equal to area AEFG in year 1, AHJK in year 2, and AILN in year 3. To compensate for them, prices charged at or close to full capacity (here assumed to be reached in period 4) would have to be somewhat higher than the LRMC at full capacity. This would have to be just large enough (over the life expectancy of the system) so that its present value equivalent would equal the present value of the initial losses incurred. In Figure 5.8, this difference is represented by area ABDC.

5.8　PRICING ENERGY RESOURCES FACING DECLINING MARKETS

The drastic price increases in hydrocarbon-based fuels have resulted in a general increase in the market values of most other energy resources. However, some of them, in certain regions or countries, nevertheless face steadily shrinking markets. In past decades this was universally true for thermal coal which was replaced in many uses such as transportation or as a household fuel by the more efficient, cleaner, and often cheaper hydrocarbon fuels, or by electricity.

Such trends may well continue because of pricing policies for alternative fuels that try to protect users from the full impact of changing world energy prices, or because of rising incomes that enable users to switch over to a more expensive, but nevertheless preferred, fuel. Furthermore, even at drastically changed petroleum-product prices, the total systems costs of coal, wood, charcoal, or other traditional fuels may be higher than those of petroleum, natural gas, or electricity-based ones. A detailed example of such alternatives has been presented in the case study on Thailand. Another typical example is offered by rail transport—few countries, if any, are presently considering the replacement of steam engines once the existing stock has worn out.

In these situations the question of pricing may become an intricate and complex one because of the interactions of market forces, including consumer preferences, financial costs, economic costs, capital investment decisions, and long-term prospects of future energy costs.

For the producer or supplier of an "inferior" energy resource that faces shrinking markets the necessary pricing decision is well-known from economic theory (even if it is painful): "Disregard long-run marginal costs (LRMC) as well as total average costs and continue production and supply as long as short-run marginal costs (SRMC) are lower or equal to prices; go out of business when price falls below SRMC." This typical microeconomic decision is, however, usually made on the basis of financial costs and prices, which may differ substantially from the real economic opportunity costs (on which the decision should have been based).

The differences in financial and economic costs and prices as well as their distributional impacts are of interest and concern to national energy policymakers. However, private consumers or producers of energy cannot be expected to base their own energy decisions on real economic ones; for them the financial ones necessarily govern. Because of these differences it is necessary at the governmental policy level to establish the actual difference, if any, between the actual market prices and the real economic values of those energy resources that face declining markets, as well as of their substitutes. This may require the shadow pricing of the various factors entering the cost structures of these resources—foreign exchange, labor, externalities such as pollution, etc.—as well as analyzing the costs and benefits of complementarities and/or substitutes of the competing energy systems.[31] If it turns out that the real economic value of the specific energy resource facing declining markets is in fact higher than its market price a case can be made for governmental action in the form of subsidies, regulations, taxes on alternative energy resources, or other

---

[31]See, for example, the differences between the financial and economic costs of LPG for taxis in Thailand as analyzed in the Appendix of the Thai case study (Chapter 10).

direct forms of market interventions. However, care must be taken that the costs of these measures do not, in fact, exceed the net difference in value between the protected energy resource and its next best alternatives.

5.9   THE EFFECTS OF RISING PRODUCTION COSTS

Inevitably, any exhaustible energy resource deposit will eventually reach a stage in its exploitation at which marginal production costs start to increase. This may occur well before physical exhaustion, simply as a consequence of lower yields per unit of effort because of factors such as higher overburden ratios, lower grades of mined resources, narrower seams, higher drilling and/ or pumping costs, the need for the use of secondary and tertiary recovery techniques, etc. This increase in marginal unit costs becomes critical only when two distinct thresholds are reached. The first occurs when LRMC start to exceed the value of output as measured by the costs of supply for alternative or substitute resources, the second when SRMC start to exceed this level. The market response to these two situations is straightforward. When LRMC exceed the unit value of output, further investments will be avoided so that future production decisions are governed only by comparisons between SRMC and market price. At this stage, provided that past planning and investment decisions were based on accurate forecasts, previous investment costs should have been recovered, or should at least be no greater than the present value of the remaining difference between market price and SRMC. When SRMC start to exceed the unit value of output, production should be terminated.[32]

These production decisions are generally based on comparisons between financial costs and revenues. However, these could deviate substantially from economic ones. Government intervention may be called for if either the financial LRMC, SRMC, or the prevailing market prices of the output do not reflect the underlying real economic costs and benefits. Production, including necessary additional investments, should be continued until the social LRMC reach and exceed the real economic value of output; beyond that point new investments should be avoided and the SRMC cost criteria should apply.

It should be noted that these economic benefit and cost calculations should be extended to cover all complementary inputs and outputs as well, to the extent that their activity levels would be affected by changes in the production rates of the energy deposit under consideration. For example, reduction or cessation of output from a given mine may substantially affect the utilization rate of a railroad. These potential costs (evaluated, of course, in shadow-priced economic terms) should be included in the overall evaluation.

---

[32]This assumes that SRMC can no longer be reduced through variations in the rate of output.

Furthermore, output reductions or complete shutdowns will usually affect regional employment levels, utilization rates of existing infrastructure, etc. All of these effects should be taken into consideration in the economic analysis and the evaluation of possible aid measures. Care must be taken, however, not to fall into the trap of continuous, long-term subsidy arrangements. This would certainly be the case if there are other variable energy resources available elsewhere that could be produced at lower economic costs instead. Such considerations also affect the evaluation and shadow-pricing of locally employed labor. In the short run, unemployment may result from a decision to close down an operation. Hence the shadow price of labor would approach zero. However, in the long run, out-migration, even if it requires substantial governmental assistance to be brought about, may be a preferable alternative. The comparison, then, should be made between the net cost of subsidies versus the net costs of resettlement. These trade-offs have to be carefully evaluated on a case by case basis. As a general rule of thumb, however, policies that require continuous, long-term governmental subsidies to maintain submarginal operations in given regions should be avoided, unless it can be shown that there are substantial additional national net benefits to be gained from such a policy. Examples would be the maintenance of a national energy resource production capability, if risks of cut-offs of foreign supplies are high, or expectations that future relative costs and prices will change sufficiently to make the operation economically viable once again, particularly in cases where the net costs of a temporary shutdown are greater than the costs of the subsidy needed to continue operations.

MATHEMATICAL ANNEX (TO CHAPTER 5): MODEL FOR DETERMINING SUBSIDIZED OR LIFELINE ENERGY PRICES IN A DISTORTED ECONOMY WITH A SKEWED INCOME DISTRIBUTION

In this Annex, a general expression for the socially optimal energy price is developed based on a simple model to compensate for distortions and skewed income distribution in the economy. From the general equation, results for optimal energy pricing are derived for cases that reflect:

(a) a perfectly competitive economy (classical result);
(b) economic second best considerations; and
(c) subsidized social prices or lifeline rates for poor consumers.

The supply and demand for a given form of energy is shown in Figure A.5.1, where S is the supply curve representing the marginal cost of energy supply at domestic market prices, and D is the corresponding market demand curve for a specific consumer. Starting with the initial combination of price and consumption (p, Q), consider

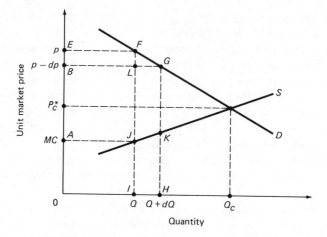

Figure A.5.1.   Supply and demand for a given form of energy.

the impact of a small price reduction (dp) and the resultant increase in demand (dQ), on the net social benefits of energy consumption.

Before evaluating the net social benefit of this price change, several parameters will be defined. First, suppose the marginal cost of energy supply MC is calculated without shadow pricing the inputs, that is, considering them, in market prices. Then a_p is defined as the energy conversion factor (ECF), which transforms the market-priced marginal cost of energy supply MC into the corresponding real economic resource cost. (See annex to Chapter 8). Thus when all the inputs to the energy sector are correctly shadow priced, the shadow priced marginal opportunity cost is MOC = (a_p · MC). Second, a specific social weight W_c is assigned to each marginal unit of consumption (valued in market prices) of a given individual i in the economy. For example, if the energy user is poor, the corresponding social weight may be much larger than for a rich customer, reflecting society's emphasis on the increased consumption of low-income groups. Third, if the given individual's consumption of goods and services other than this form of energy valued in market prices increases by one unit, then the shadow-priced marginal social cost of economic resources used or the shadow cost to the economy is b_i.

As a result of the price reduction, the consumer is using dQ units more of energy, area IFGH, which has a market value of (p · dQ).[1] The consumer's income net of expenditure on energy has increased by the amount $pQ - (p - dp)(Q + dQ)$, however. Assuming none of it is saved, this individual's consumption of other goods and services will increase by the amount (Q · $dp - p$ · dQ)—that is, area BEFG minus area IFGH, also valued in market prices. Therefore, the consumer's total consump-

---

[1]The little triangle LFG may be neglected throughout this analysis because its area is (dp · dQ)/2, where both dp and dQ are small increments.

tion—energy plus other goods—will increase by the net amount $(Q \cdot dp)$ in market prices. This is the traditional increase in consumer surplus benefits. The shadow value of this increased consumption is $W_c \cdot (Q \cdot dp)$, where W_c is the social weight appropriate to this consumer's income/consumption level.

Next, consider the resource costs of these changes in consumption. The shadow cost of increasing energy supply is $(a_p \cdot MC \cdot dQ)$, that is, a_p times area IJKH. The resources used to provide the other additional goods consumed are $b_c \cdot (Q; dp - p \cdot dQ)$, where a_p is the conversion factor for energy production, and b_c is the conversion factor for other goods consumed by the consumer. Finally, the income change of the energy producer must also be considered, but this effect may be ignored especially if, as is likely, the producer is the government.

The total increase in net social benefits caused by the energy price decrease is given by:

$$d\text{NB} = W_c(Q \cdot dp) - a_p \cdot (MC \cdot dQ) - b_c \cdot (Q \cdot dp - p \cdot dQ)$$

Therefore,

$$(d\text{NB}/dp) = Q[(W_c - b_c) + n \cdot b_c - n \cdot a_p \cdot (MC/p)],$$

where $n = (p \cdot dQ/Q \cdot dp)$ is the elasticity of demand (magnitude). The necessary first order condition for maximizing net social benefits, in the limit, is $d\text{NB}/dp = 0$. This yields the optimal price level:

$$p^* = a_p \cdot MC/[b_c + (W_c - b_c)/n] \qquad (5.A.1)$$

This expression may be reduced to a more familiar form, by making some simplifying assumptions.

Case 1: **Classical Results**

Assume a perfectly competitive economy where market prices and shadow prices are the same, and income transfer effects are ignored, that is, there is no social weighting. Therefore, $a_p = W_c = b_c = 1$, and equation (5.A.1) reduces to to:

$$p_c^* = MC \qquad (5.A.2)$$

This is the classical result (discussed in Chapter 4), where net social benefits are maximized when price is set equal to marginal cost at the market clearing point (P_c^*, Q_c) in Figure 5.A.1.

Case 2: **Efficiency Pricing**

Assume that income transfer effects are ignored, because the marginal social benefit of consumption is equal to the marginal social cost to the economy of providing this consumption. There, $W_c = b_c$, and equation (5.A.1) becomes:

$$p_e^* = (a_p \cdot MC)/b_c = MOC/b_c \qquad (5.A.3)$$

This is the optimal marginal opportunity cost based energy price, when efficiency (shadow) prices are used that emphasize the efficient allocation of resources and neglect income distributional considerations.

Note that while the marginal opportunity cost of energy supply (MOC) may be evaluated as discussed in Chapter 4 the conversion factor b_c depends on the type of consumer involved.

For residential energy consumers, b_c represents the resource cost or shadow value of one (market prices) unit of the household's marginal consumption basket. If $b_c <$ 1, then $p_e^* < (MCB_e)$.

Another interesting case illustrates the application of equation (5.A.3) to correct for economic second-best consideration arising from energy substitution possibilities. As an extreme case, suppose we wish to establish the price of grid-supplied electricity. Assume that all expenditures diverted from electricity supplied from the grid will be used to purchase alternative energy that is subsidized by the government. Examples could be kerosene for lighting or diesel for autogeneration. In this case, b_c is the ratio of the shadow priced marginal opportunity cost of this alternative energy to its market price and may be written:

$$b = MOC_{ae}/P_{ae}$$

Thus, from equation (5.A.3):

$$= MOC \cdot (p_{ae}/MOC_{ae}) \tag{5.A.4.}$$

The logic of this expression may be clarified by considering the case when $p_e > p_e^*$. If the consumer spends one unit of currency, in domestic prices, to purchase electricity, the shadow cost of this expenditure is MOC/p_e. If this sum were used to purchase alternative energy, the shadow cost would be MOC_{ae}/p_{ae}. Since $p_e > p_e^*$. from equation (5.A.4) $MOC/p_e > MOC_{ae}/p_{ae}$. Therefore the country is better off if more electricity is used instead of the alternative energy, and p_e should be reduced to p_e^*. Similar reasoning can be used to show that if $p_e < p_e^*$, then p should be increased to the value p_e^*.

If the alternative energy is priced below its shadow marginal cost, that is, $b > 1$, then from equation (5.A.3), $p_e^* < MOC$ also. Therefore, the subsidization of substitute energy prices will result in an optimal electricity price that is below its shadow supply cost.

If it is not possible to determine the consumption patterns of specific consumer groups, then b_c could be defined broadly as the average conversion factor for all electricity users, as discussed in Chapter 4.

Case 3: General Case

Assume equation (5.A.1) is the optimal electricity price when using social (shadow) prices, which incorporate income distributional concerns.[2]

---

[2]In this case MOC should be computed using the so-called social shadow prices. In practice the difference in MOC calculated in efficiency and social prices, however, is likely to be small and may be ignored. That is, the efficiency-priced value of MOC may be used.

Consider the case of a group of poor consumers for whom one may assume: $W_c \gg b_c(n-1)$. Therefore. equation (5.A.1) may be written: $p^* \approx n \cdot \mathrm{MOC}/W_c$. An even greater simplification is possible if it is assumed that $n = 1$; thus,

$$p_s^* = \mathrm{MOC}/W_c$$

For illustration. suppose that the income/consumption level of these poor consumers (c) is one third the critical income/consumption level ($\bar{c}$), which is similar to a poverty line. Then a simple expression for the social weight is:

$$W_c = \bar{c}/c = 3$$

Therefore, $p_s^* = \mathrm{MOC}/3$, which is the "lifeline" rate or subsidized tariff appropriate to this group of low-income consumers.

6
Energy Conservation and Efficiency

6.1 INTRODUCTION

This chapter analyzes the relevance of energy conservation, its technical and economic meaning and its potential as an important component of overall energy management. The range of policies available for the effective implementation of conservation measures is discussed, including pricing, rationing, physical controls, the use of technical devices, legislation, taxes and subsidies, education and propaganda, etc. The discussion begins with a presentation of the technical concepts of efficiency in energy use, followed by an analysis of the economic principles governing conservation, including the potential divergences between optimal private and social energy conservation objectives. A brief assessment is presented of the likely magnitudes of energy savings in various sectors that could result from appropriate conservation measures.

In a period of rising energy prices and potential shortages, the goals of energy conservation seem to be intuitively attractive. Depending on the trade-offs, or "sacrifices," involved, energy conservation can be an important element of demand management and, therefore, a valuable tool for achieving some of the objectives of integrated national energy planning that were discussed in Chapter 3. As a number of analysts have claimed, the potential for energy conservation, at least in the industrialized nations, is rather substantial and often more cost-effective than alternative means for augmenting energy supplies.[1]

In its most general sense energy conservation can be defined as the deliberate reduction in the use of energy below some level that would prevail otherwise. Usually, such reduction requires some trade-offs in terms of comfort, convenience, or the use of additional capital or labor.

Some conservation can be brought about simply by reducing or eliminating certain energy-using activities. Foregoing Sunday pleasure driving, using lower

---

[1]See, for example, Robert Stobaugh and Daniel Yergin, *Energy Future,* Random House, New York, 1979; Marc Ross and Robert Williams, *Our Energy; Regaining Control,* McGraw-Hill Books, New York, 1981.

thermostat settings, and shutting off appliances and lighting fixtures when not directly needed are typical examples. Other conservation measures may require the substitution by either capital or labor. Examples are the recirculation of heat in industrial processing, the energy-saving reductions in the weight of vehicles by better engineering or lighter materials, the use of more sophisticated controls to minimize energy consumption per unit of output or work performed, or the use of improved insulation.

An important subset of effective energy conservation measures consists of the substitution of some form of costly or scarce energy resource by some other that is more readily available. Examples are the use of coal instead of fuel oil in heat processes, the use of natural gas instead of petroleum products for power plants in countries in which gas is plentiful compared to oil, or the use of alcogas (alcohol-gasoline mixtures) instead of gasoline for transportation. In a physical or technical sense (as measured by btu's or kcal's consumed), such substitution may not "save" energy, or may even result in increased energy use per unit of output. In an economic sense, however, such substitution may make perfectly good sense, given the economic scarcity values of the alternative fuels.

The pursuit of energy conservation as a goal raises the issue of up to what point the reduction of energy consumption is socially beneficial or desirable. Common sense indicates that while "wasteful" energy use should be discouraged, there is a limit beyond which conservation measures may become too costly in terms of foregoing other resources or useful outputs, thereby causing more harm than good. In order to define desirable conservation levels more precisely, both the technical and economic feasibilities must be evaluated. As will be argued, technical feasibility is a necessary, but not a sufficient condition for determining the desirability of specific conservation measures. As argued elsewhere in this book, the principle objective of a given policy should be the maximization of the welfare of a society's citizens over time. If aggregate consumption or production (e.g., gross domestic product or GDP) is taken as a proxy for aggregate welfare, then welfare maximization implies the use of scarce resources such as energy, capital, labor, and land in such a way that output is maximized. It also implies that the value of output, at the margin, must be equal to or greater than the costs of the resources utilized in its production.[2] Reductions in the use of energy will contribute to this goal, provided that the added costs of such conservation measures or process changes do not outweigh the value of the energy savings achieved.

Conservation measures usually lead to three types of consequences that are economically significant. First, the reduction or substitution in energy use gives

---

[2]For a detailed exposition of the market conditions and relationships that give rise to the efficient production and consumption of economic goods and services, see, for example, J. Henderson and R. Quandt, *Microeconomic Theory,* McGraw-Hill, New York, 1980.

rise to a net saving, B. Second, the implementation of the conservation measure may involve additional costs, e.g., hardware or operating costs, C_1. Finally, the reduction or substitution in energy consumption may result in added costs C_2 due to foregone benefits to the producer or consumer, as a result of reductions in the quantity or quality of the energy-dependent output. In general, if $B >$ $(C_1 + C_2)$, then the conservation measure is economically justified, i.e., it will improve economic efficiency and, hence, overall economic welfare.

It should be noted, however, that the perceived magnitudes of B, C_1, and C_2 may differ between affected individuals or groups and society as a whole. This may be the result of distorted market prices for some inputs or outputs, differences in the perception of risks between private users and the public (e.g., those resulting from potential supply interruptions), lack of knowledge or foresight by energy users, or strong externalities resulting from the use of specific energy resources. In such cases, the level of conservation considered optimal may differ significantly between individual energy users and society as a whole. This may require the setting of energy prices on the basis of economic shadow prices that take these distortions into account, or may call for the outright regulation of energy consumption and use.

6.2   TECHNICAL EFFICIENCY OF ENERGY USE

The technical efficiency of energy use is usually defined in terms of the first and second laws of thermodynamics.[3] The first law efficiency measures the relationship between total energy inputs and useful energy outputs. The second law efficiency relates to a more subtle concept, one that defines the optimum efficiency as the minimum amount of thermodynamic (heat) differential needed to complete a given task.[4]

The first law in its simplest form states that energy (i.e., chemical, electrical, gravitational, heat, mechanical, nuclear, etc.) can only be transformed from one form to another, but cannot be created or destroyed. The corresponding efficiency of an energy using process may be defined as:

$$e_1 = \text{useful energy output/energy actually input}$$

Application of this efficiency measure requires definition of the boundaries of the system within which the process occurs, and the determination of energy

---

[3]The first and second law efficiencies described here are also loosely called the conservation and thermodynamic efficiencies, respectively. The latter terminology is not used here because of the potential for misinterpretation.
[4]This is strictly applicable only to types of energy users that depend on a flow of heat energy for accomplishing the task at hand.

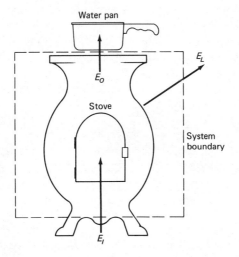

Figure 6.1.   Application of first law efficiency: heating water.

flows across these boundaries. This can be demonstrated by a simple example, the act of heating water in a home by a wood-burning stove as shown in Figure 6.1. Let E_I be the energy input representing the calorific value of the firewood burned; E_O, the useful energy output absorbed by the water; and E_L, the conductive, convective, and radiant heat losses. Then the first law of thermodynamics defines the energy balance as $E_I = E_O + E_L$, and

$$e_1 = E_O/E_I = 1 - E_L/E_I \tag{6.1}$$

Thus the first law efficiency of any process may be determined by correctly identifying all the appropriate energy flows and losses. However, this type of energy "bookkeeping" may be quite complicated for a complex system such as thermal electrical power station involving flows of many forms of energy including heat, electrical, mechanical, and chemical energy.[5]

Energy balance analysis provides a convenient framework for determining the first law efficiencies from primary energy source to final end-use, and therefore may be used as one criterion to analyze the efficiencies of different energy delivery systems. For example, in the case of domestic water heating by either natural gas or electricity, the respective first law efficiencies of the processes are about 0.5 and 0.95. This seems to show that the use of electricity is supe-

---

[5]For details of energy balancing see, for example, John E. Ahern, *The Exergy Method of Energy Systems Analysis,* Wiley, New York, 1980, pp. 24–30.

rior. However, what must also be considered are the efficiencies of the energy production and delivery systems. The first law efficiency for generating electricity from natural gas at the power station is about 35%, and electricity losses in the transmission and distribution networks about 15%, while the losses in the gas delivery system are approximately 30%.

Hence, the first law efficiency of the total chain from gas to hot water via the electric option is given by:

$$e_{1E} = 0.35 \times 0.85 \times 0.95 = 0.283$$

The corresponding efficiency for direct heating of water using piped gas is:

$$e_{1G} = 0.7 \times 0.5 = 0.35$$

Overall, therefore, the direct gas option appears to be superior.

What must be noted, however, is that this simple analysis neglects many other relevant factors such as the convenience of using electricity versus gas, the substitutability of the two energy sources with respect to other noncooking energy uses, the relative costs of the delivered energy, etc. These other considerations may well be dominant, and often economically more attractive to the user.

The second law of thermodynamics seeks to distinguish between energy that is available and unavailable for doing useful work. It states that the entropy (or unavailable energy) of a closed system must remain constant or increase over time.[6] The entropy of a system is closely linked to its state of order. For example, if there are two glasses of water, one at 100°C and the other at 0°C, then by virtue of the temperature difference heat can flow from the hotter to the colder glass. This flow may be used to perform useful work (e.g., using a thermocouple).

If the two glasses of water are mixed together and then separated again each full glass will be at a temperature of 50°C (assuming there are no heat losses). This has eliminated the temperature differential. At the same time it has also eliminated the possibility of any heat flow between the two glasses, and the capability for extracting useful work from the system. The system entropy has increased while its state of order has decreased since there is now only one category of lukewarm water available. However, the total heat content of the system is unchanged since there was no heat loss. The lower entropy of the original system can be regained by the use of external energy to heat one glass and/or cool the other. But if this occurs the system is no longer closed to the

---

[6]This implies that the total entropy of the universe will keep on increasing until a condition of complete thermal disorder known as "heat-death" is reached.

outside world as required by the second law definition. Furthermore, the second law also ensures that the available energy used externally to restore the two-glass system to its original state will always be greater than the available energy which can be reextracted from the system. Thus, even if the original hot-cold system is restored, the total available energy in the system and its external environment would have decreased further, i.e., total entropy has increased.

The second law efficiency relates the minimum input required to perform the task to the maximum useful work that could have been extracted from the fuel used. It can be defined as:

e_2 = Theoretical Minimum Energy Required/Maximum Useful Work Available From Energy Actually Input

In the simple case of a heat transfer from a heat source to a heat reservoir which is to be heated, the theoretical minimum energy required E_M is defined by:[7]

$$E_M = E_O(T_T - T_A)/T_T \tag{6.2}$$

where E_O is the quantity of thermal energy transferred, T_T is the absolute temperature (in degrees Kelvin) at which the heat is transferred, and T_A is the absolute (ambient) temperature of the environment.[8] But $E_o = e_1$. (Maximum Useful Work Avaliable), and therefore using equation (6.1), we may write:

$$e_2 = e_1(T_T - T_A)/T_T = e_1(1 - T_A/T_T) \tag{6.3}$$

Therefore in this simple example, an increase in first law efficiency e_1 implies an increase in second law efficiency, and vice versa, provided the other parameters remain unchanged.[9] If T_T is high, i.e., $T_T \gg T_A$; then e_2 and e_1 are approximately equal, but if the temperature differential is small, e_2 may be much lower than e_1.

As a numerical illustration, consider a process such as fluidized bed coal

---

[7]E_M is defined with respect to the ideal Carnot energy cycle between two temperatures. For details, see John E. Ahern, pp. 9–10.

[8]Degrees Kelvin (°K) = 273° + Degrees Celsius (°C)

[9]Equation (6.3) is not generally applicable and is used here to simply illustrate the relationship between first and second law efficiencies in this special case. e_1 and e_2 are usually linearly independent. Examples have even been quoted where an increase in e_2 is accompanied by a decrease in e_1. See Donald I. Hertzmark, "Joint Energy and Economic Optimization: A Proposition," *The Energy Journal,* Vol. 2, No. 1, pp. 75–88. For a discussion of the various formulae that apply when different forms of energy (such as mechanical, heat, etc.) are involved, see "The Efficient Use of Energy," American Physical Society, Washington, D.C.: U.S. Govt. Printing Office, 1975.

combustion to produce electricity where heat is transferred at a very high temperature. The corresponding first and second law efficiencies are approximately the same, usually lying between 0.4 and 0.45. In contrast, the first law efficiency of a gas furnace used for space heating would lie approximately between 0.75 and 0.8, while the corresponding second law efficiency would be only about 0.05 to 0.1. This large difference is caused by the fact that natural gas which is a high quality fuel with a very high flame temperature (1500–2000°C) is being used to supply low quality heat, i.e., the burning gas heats air or water to an intermediate temperature which in turn transfers thermal energy to the living space. The second law efficiency is low because heat is ultimately transferred at a relatively low temperature difference of a few hundred degrees K.

The exclusive consideration of first law efficiency only may conceal the fact that high quality (or high temperature) energy sources such as fossil fuels are being used for relatively low temperature processes like water and space heating or cooling, industrial process steam production, and so on. Consideration of second law efficiency may permit a better matching of energy sources and uses, so that high-quality energy which may be scarce is not used for performing low-quality work. Some approximate first and second law efficiencies for typical energy uses or processes are given in Table 6.1.

It should be noted that both first and second law efficiencies can be increased by cascading energy use. Thus, a working fluid such as steam may be successively utilized in several processes. As the steam cools along the chain of processes the quality of heat desired is matched with the steam temperature at each stage to provide the best efficiency. This issue is discussed in greater detail in Section 6.6.

6.3 ENERGY ACCOUNTING FRAMEWORKS

Increased concerns about the finiteness of the world's depletable energy resources, together with the shock of the petroleum price rises in the 1970s, have led a number of analysts to conclude that energy should be made the unit of account for all production and consumption processes and activities. Such an energy theory of value would seek to establish the relative values of all other goods and services in relation to a numeraire, or yardstick, based on the energy embodied in them. This would elevate energy to the principal scarce resource, but would neglect the value of all other scarce factors needed in production, mainly land, labor, and capital. In a market economy, price levels are determined by the interaction of supply and demand forces, but in such a single factor theory of value they would be determined by supply only, namely the supply, or use, of one single factor—in this case energy. Such single-factor theories of value are not new; the best known is the labor theory of value devel-

Table 6.1. First and Second Law Efficiencies for Some Typical Energy Uses.

USE	FIRST LAW EFFICIENCY	SECOND LAW EFFICIENCY
1. Electricity Generation or Traction (large scale)	0.9–0.95	0.3
2. Industrial Steam Production	0.85	0.25
3. Fluidized Bed Electricity Generation	0.4–0.45	0.4 – 0.45
4. Transportation (Diesel Powered)	0.4	0.1
5. Transportation (Gasoline Powered)	0.25	0.1
6. Space Heating or Cooling	0.5–0.8	0.05
7. Domestic Water Heating	0.5–0.7	0.05
8. Incandescent light bulb	0.05	0.05

oped by Adam Smith and Karl Marx. Attempts at using energy as the principal unit of account have been made prior to the 1970s as well.[10]

The fatal flaw of any attempt to make energy inputs the major criteria for determining the value of any output or activity is that other resources, as well as time, are needed to produce such output. These resources, as well as the use of more time, are not costless. They have scarcity value and their costs cannot be neglected. If this happens, misallocation of one or all of these other resources is bound to occur. The simple analysis given below illustrates this point, as well as the differences between optimal energy use as defined by second law efficiency and the economic viewpoint.

Consider an industrial activity in which some final output (Q) is produced by employing energy (E,) and other nonenergy inputs (I.) For simplicity, both E and I are considered homogeneous aggregates, although in practice, each is likely to consist of many different subcategories. The activity is described by the production function $Q = Q(E,I)$. Production is said to be economically efficient if the output is produced at minimum cost. This may be represented mathematically by the requirement:

$$\text{Minimize: } p_e e + p_I I, \text{ subject to } Q = \overline{Q}$$

where p_t and p_i are the prices of energy and other inputs respectively, and $\overline{Q}$ is the constant (target) level of output.

The economics of the industrial process is analysed geometrically in Figure 6.2, where $\overline{Q} = Q$ is the isoquant or constant level of final output that can be

---

[10]For a discussion of the history of energy analysis see Frank J. Alessio, "Energy Analysis and the Energy Theory of Value," *The Energy Journal*, Vol. 2, No. 1, January 1981, pp. 61–74.

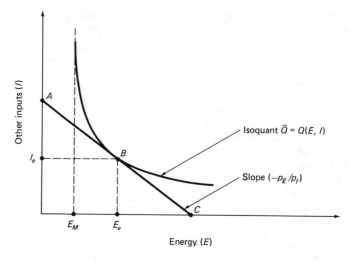

Figure 6.2.   Production isoquant for an energy using activity.

produced by using various combinations of E and I as specified by the production function. The economically efficient or optimal cost-minimizing combination of inputs is (E_e, I_e) at the point of tangency B, between the isoquant and the isocost line AC (which represents different combinations of E and I that have the same total cost to the producer). The isocost line satisfies the equation $p_e e + p_I I =$ constant, and has the slope $-(p_E/p_I)$. We note that AC will become steeper as the price of energy rises relative to other inputs, thus pushing the point of tangency B further up along the isoquant and thereby reducing the economically optimum level of energy input below E_e.

According to the second law viewpoint, the optimal or minimum energy that should be used for production is E_M, represented by the vertical asymptote to the isoquant. This represents the case in which energy is effectively considered to be the only scarce resource. In other words, the costs of all other input may be neglected, so that the line AC tends towards the vertical with an infinite slope. Clearly, in this case, the thermodynamically minimum amount of energy (i.e., second law efficiency = 1) is less than the economically efficient level E_e, but an infinite quantity of (costless) other inputs would also be required. Thus using the economically optimal energy input will generally lead to a second law or technical efficiency considerably less than unity.[11]

---

[11]For details, see: R. Stephen Berry, Peter Salamon and Geoffrey Heal, "On a Relation Between Economic and Thermodynamic Optima," *Resources and Energy, Vol. 1,* October 1978, pp. 125–37.

A subsidiary problem arising out of attempts to elevate relative energy consumption to the sole criteria for ranking is that first and second law efficiencies may not coincide, and, hence, may not always be consistent with each other.[12]

Other problems that arise with the various energy accounting frameworks that have been developed in recent years is that energy itself is not a homogeneous and freely interchangeable commodity.[13] For example, there is a vast difference in the usefulness for specific purposes of say, a million kcal, available in the form of electricity, or of gasoline, or of lignite or peat. Given the often unique matching of specific energy sources and uses it is also not possible to draw up a hierarchical, qualitative ranking order that would define energy resources in terms of consistently declining usefulness. Electricity, for example, is extremely useful for lighting lamps or providing stationary mechanical energy through the use of electric motors, but it is an awkward and costly energy source for independent vehicle propulsion.

Similarly, energy accounting frameworks that claim "inefficiency" for various types of energy production processes, because the net output of useable energy of a given process is less than the total input of energy needed in its production, can be rather misleading. For example, the energy inputs of high fertilizer-pesticide-mechanical energy agricultural production processes may be substantially higher than the outputs of the resulting agricultural production, but the latter, in the form of digestible food, may be far more valuable and useful for the intended purpose than the original energy inputs required for its production.[14] As another example, a given quantity of energy inputs for electric power production provided by coal is far less valuable (has lower economic opportunity costs) than a similar quantity of energy supplied in the form of fuel oil or diesel fuel (at least in North America).

For all of these reasons, energy audits, energy bookkeeping, energy efficiency and energy balance information should be used only as a supplement to the economic assessment of conservation measures, but never as the sole criterion for deciding the appropriate level of energy use for a given activity.[15]

---

[12]For examples see Peter Chapman, "Energy Costs: A Review of Methods," *Energy Policy,* 3, 1975; and Robert A. Herendeen, "The Energy Costs of Goods and Services," *Report No. ORNL-MSF-EP, 40,* Oakridge National Laboratory, Oakridge, Tennessee, 1972.

[13]For a discussion see Donald 1. Hertzmark, "Joint Energy and Economic Optimization: A Proposition," *The Energy Journal,* Vol. 2, No. 1, pp. 75–88.

[14]For an example of such attempts at energy accounting in food production see Bruce M. Hannon, Carol Harrington, Robert W. Howell, Ken Kirkpatrick, "The Dollar, Energy, and Employment Costs of Protein Consumption," *Energy Systems and Policy,* Vol. 3, No. 3, 1979, pp. 227–41.

[15]A general critique of energy accounting is provided in: Michael Webb and David Pearce, "The Economics of Energy Analysis," *Energy Policy,* December 1975, pp. 318–31.

6.4   THE ECONOMICS OF ENERGY CONSERVATION

In a market economy free of distortions the basic rule of energy conservation should be that the total system's cost of any energy-using activity per unit of output, not the energy costs only, should be minimized over time, while at the margin the value of the output produced should be just equal to someone's willingness to pay for it. This, of course, is nothing but the fundamental, long-run marginal cost pricing rule elaborated in Chapters 4 and 5.

The critical conservation condition to be met is that total net benefits associated with any energy using activity must be optimized, not just energy costs. For example, a 40-mile per hour speed limit on freeways may minimize fuel consumption, but it may be economically inefficienct because of the greater time requirements for moving freight and people, which would result in higher labor, transportation-equipment, and freight-inventory costs. Hence, the appropriate amount of energy conservation at the margin is determined by the equality of the value of the energy savings on the one hand and the sum of the costs of the needed additional inputs plus the net reductions in the value of the output produced (if any) on the other, i.e.,

$$\Delta B = \Delta C_1 + \Delta C_2 \tag{6.4}$$

Where ΔB, ΔC_1 and ΔC_2 are the marginal energy savings, marginal additional input costs, and marginal reductions in consumption benefits (or output values), respectively.

It should be emphasized that this optimization should be achieved over the life-expectancy of the activity.[16] This means that resource input mixes (including the share of energy inputs) should be based on expected life-time costs, not just presently prevailing cost relationships. For example, if energy costs are expected to increase relative to other input costs or the value of output over time, greater substitution by nonenergy inputs (i.e., higher levels of energy conservation) is called for. Therefore, we may modify equation (6.4), to derive the following economic criterion which indicates whether a given conservation measure should be adopted:

$$\sum_{t=1}^{n} b_t \left(\frac{1}{1+r} \right)^t > \sum_{t=1}^{n} (c_{1,t} + c_{2,t}) \left(\frac{1}{1+r} \right)^t \tag{6.5}$$

where b_t, $c_{1,t}$, and $c_{2,t}$ are the respective annual energy savings, additional input costs, and losses in consumption benefits or output values, in years $t = 1, 2,$

---

[16]More accurately, over the life expectancy of the fixed input components of the activity.

..., n, and r is the appropriate discount rate. This optimum economic conservation criterion can be illustrated by some simple examples.

Let us consider a particular end-use for energy such as home lighting, and suppose there is a choice of two distinct types of light bulbs, incandescent and fluorescent. For simplicity, we begin by assuming that both have the same economic cost, same lifetime, and provide light output of the same quality. If the fluorescent bulb uses less electrical energy than the incandescent one, then replacing the latter by the former is a conservation measure that results in an unambiguous improvement in the economic as well as technical efficiency. In this case, using fluorescent bulbs instead of incandescent lamps reduces the economic resources expended to provide the desired output, i.e., lighting. Electrical energy has been conserved, with no change in other economic costs and benefits.

Next, let us suppose that the fluorescent bulb is more costly to install. There is a trade-off between the higher capital cost of the fluorescent lamp and the greater consumption of kilowatt-hours by the incandescent bulb. The relevant data to determine whether substitution of incandescent bulbs by fluorescent ones is economically justified, are summarized in Table 6.2. At this stage we distinguish between the economic value (or opportunity cost or shadow price, as discussed in Chapter 5) of a good or service, and its market price. The former is relevant to decision-making from a national perspective and the latter is more appropriate from a private individual's viewpoint.[17]

The national cost (based on economic values) of using the incandescent and fluorescent bulbs over their two year lifetimes are respectively:

$$EC_I = 10.5 + 16 + 16/(1 + r) \text{ and} \qquad (6.7a)$$
$$EC_F = 32 + 4.4 + 4.4/(1 + r) \qquad (6.7b)$$

Assuming an economic discount rate of $r = 0.1$, we find $EC_I = 41.0 > EC_F = 40.4$

What we have done is to compare the energy cost saving of dineros $(16 - 4.4) = 11.6$ per year for two years against the increase in capital costs of dineros $(32 - 10.5) = 21.5$. What we find is that $(16 - 4.4) + (16 - 4.4)/(1 + r) > (32 - 10.5)$, when $r = 0.1$. Therefore the use of fluorescent lightbulbs and the associated reduction in energy consumption will improve economic efficiency as well as technical efficiency.

However, if $r = 0.2$ is used as a variant, then $EC_I = 39.8 < EC_E = 40.1$. In this case the conservation measure is no longer beneficial. This reduction in the relative value of conservation measures will always occur with increases in

---

[17]For an analysis of the divergence between national and private optimization see section 6.5.

Table 6.2. Physical and Economic Data to Assess the Economic Efficiency of Energy Conservation for Lighting.

		INCANDESCENT BULB	FLOURESCENT BULB
Installation	Economic Value (Opportunity Cost)	10.5	32
Cost (Dineros)	Market Price	18	36
Physical Energy Consumption (kWh per year during 2 year lifetime)		40	11
Value of Energy consumption (Dineros per year during 2 year lifetime)	Economic Value (Marginal Opportunity cost)[a] _____ Market Price[e]	$\downarrow \dfrac{16}{12} \downarrow$	$\downarrow \dfrac{4.4}{3.3} \downarrow$

[a]Dineros 0.4 per kWh   [b]Dineros 0.3 per kWh

the discount rate, because increases in initial investment costs are traded off against the future cost-savings realized by conservation. This finding has important policy implications. Energy users who confront high opportunity costs of capital (as for example, in many developing countries) will find capital-cost-intensive energy conservation measures relatively less attractive than users who have access to low-cost sources of capital. This means that economically "optimal" conservation measures may differ significantly among different countries or user classes.

The analysis of optimal economic rates of conservation becomes more complex under conditions of projected relative price changes for energy inputs, and technically determined, fixed energy-capital-labor inputs over the life expectancy of given energy-using capital assets (such as a truck, or boiler, or smelting oven). While it is usually possible to vary the physical mix of inputs (at least within certain ranges) in the design stage of new equipment or production processes, this usually not possible (or at least quite difficult and costly) once the equipment has been installed. Hence, equipment decisions relating to conservation should be made beforehand.

It can be shown that under these conditions the "optimum" economic rate of conservation (or energy utilization) will lie between the most efficient (lower) rate prevailing in the early life stage of the asset and the most efficient (higher) rate that would be most appropriate towards the end of its life. Furthermore, the optimum "average" rate will be the lower, the higher the applicable discount rate is.

To accommodate such relative price changes in the evaluation of the benefits from energy conservation, equation (6.5) has to be changed to:

$$\sum_{t=1}^{n} b_t \left(\frac{1}{1+r}\right)^t (1 + g)^t > \sum_{t=1}^{n} (c_{1,t} + c_{2,t}) \left(\frac{1}{1+r}\right)^t \quad (6.6)$$

where g is the expected average annual rate of relative price increase of the energy input.

The effects on maximum allowable energy conservation expenditures of such relative price changes, and the interaction with the level of the appropriate rate of discount have been illustrated in Figure 6.3. The underlying data of the figure assume that at constant relative real prices between energy and other inputs and outputs, the maximum allowable annual average or levelized costs of the conservation program $C_1 + C_2$ would be equal to 100 dineros per year, given a discount rate of 5% and a life expectancy of the equipment of twenty years. This would be equivalent to total lifetime savings of energy of 1,308.50 expressed in present values. At this point, the benefits from the lifetime savings of energy would just be equal to the lifetime costs of the conservation effort, in present value terms. If the applicable discount rate rises to 10%, the maximum allowable annual average costs of conservation are reduced to 71.5 dineros or a present-value-equivalent of 822.20.

If real energy costs are expected to increase at an average annual rate of 3%, the equivalent maximum allowable annual levelized costs of conservation would rise to 128.1 dineros in the 5% discount rate case and to 87.8 dineros in the 10% case. As can be seen from the graph, in year 0 the optimum annual energy savings is 100 dineros in the 5% case which is lower than the annual levelized conservation cost of 128.10 dineros. In contrast, in year 20 the optimal annual energy savings rate is 180.60 dineros per year, well above the annual average conservation cost. The values of energy savings and levelized annual conservation costs would be equal approximately in year 7 in the 10% discount rate case, and in year 8.4 in the 5% case. In other words, although the btu value of energy savings is constant over the 20 years, the 3% per year increase in real energy costs steadily raises the value of conserved energy.

The policy conclusion which can be drawn from this analysis confirms that in the case of fixed input-output relationships and expected relative price rises of energy inputs, the input-output relationships should be chosen in such a way that the expected lifetime savings in the value of energy, expressed in present values, should be greater than, or at most, equal to the additional costs incurred in bringing them about. This also means that in the early years the average annual costs of conservation will exceed the value of energy savings, while in later years the opposite will be the case.

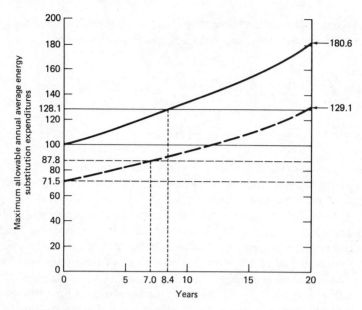

Figure 6.3. The effects of variations in discount rates and rising relative energy costs on maximum allowable energy substitution expenditures.

6.5 THE EFFECTS OF MARKET IMPERFECTIONS

So far the analysis has been based on the national viewpoint, using values for all inputs and outputs (including those for energy) that reflect the shadow or economic opportunity costs. However, market prices differ from shadow prices because market imperfections, particularly in the pricing and availability of energy, abound throughout most countries of the world. Legal barriers, capital rationing or lack of credit facilities, inappropriate foreign exchange rates, price controls, externalities, and many other factors interfere in the normal functioning of the energy market. Other complications arise from a lack of knowledge of available alternatives, uncertainties with respect to future costs and prices and the lock-in effects of long-lived facilities and equipment. All of these factors tend to distort rational choice patterns, with the result that private energy use and conservation patterns diverge substantially from those found to be optimal from a national viewpoint (i.e., based on economic efficiency criteria).

To illustrate the effects of these divergences, let us return to the simple light bulb example of Table 6.2 in the preceding section. As we have seen, at an economic discount rate of $r = 0.1$ the use of fluorescent light bulbs proved to be superior in economic terms.

The private costs (based on market prices) of using incandescent or fluorescent lighting respectively are as follows:

$$PC_I = 18 + 12 + 12/(1 + r) \qquad (6.7a)$$
$$PC_F = 36 + 3.3 + 3.3/(1 + r) \qquad (6.7b)$$

At a discount rate of $r = 0.1$ (e.g., market interest rate based on commercial bank rates), $PC_I = 40.9 < PC_F = 42.3$. This means that a rational consumer would prefer to use incandescent light bulbs, because this is the cheaper option. At any higher discount rate the advantage of the incandescent system over the fluorescent one increases further. Thus, as a consequence of the divergence between market prices and actual underlying economic costs, economically inefficient energy use and conservation decisions are made by users.

Such cases of inefficient energy utilization patterns resulting from inappropriate energy pricing policies are common throughout the world. In the U.S., for example, part of the domestically produced crude oil (the so-called "old" oil) was price controlled from the early 1970s to early 1981, and natural gas prices are still subject to a complex system of gradually declining price controls that are targeted to be phased out by 1985.[18] These policies resulted in domestic price levels for petroleum products and natural gas that were well below replacement cost levels.[19] As a consequence, more crude oil based energy was consumed (e.g., through the purchase and use of heavier, fuel-inefficient cars), and fewer investments in energy-saving substitutes were made (e.g., installation of insulation, or accelerated substitution of coal for natural gas).

In Canada, two of the major goals of the Trudeau Government's National Energy Policy are:

(a) a blended or "made-in-Canada" price of oil consumed in Canada, an average of the costs of imported and domestic oil, which will rise gradually and predictably but will remain well below world prices and will never be more than 85% of the lower of the price of imported oil or of oil in the U.S.

(b) Natural gas prices which will increase less quickly than oil prices.[20]

---

[18]For an analysis of U.S. natural gas price regulations see Keith C. Brown, *Regulation of the Natural Gas Producing Industry* (Resources for the Future Series), Johns Hopkins Press, Baltimore, 1972.

[19]For example, the average costs of natural gas to US electric utility companies in December 1980 amounted to $2.33 per MCF, while imported gas from Mexico or Canada at the border was about $4.47/MCF. *From:* US Department of Energy, *Electric Power Monthly,* January 1981, Table 69; and Brian L. Scarfe, J. L. Helliwell, and R. N. McRae "Energy Policy and the Budget," *Canadian Public Policy,* Vol. VII, No. 1, Winter 1981, p. 7.

[20]Brian L. Scarfe, et al., p. 6.

The adverse consequences for energy conservation resulting from such a policy, which is but a continuation of Canadian petroleum and natural gas pricing policies in force since the 1973 oil crisis, have been discussed in more detail in the case studies.[21]

In developing nations, such price distortions, often justified on equity grounds, are even more common. For example, a recent study of energy issues in South and Southeast Asia has shown that retail prices for kerosene—an important fuel for lighting and cooking—amounted to only 49% of normative prices[22] in Burma, 32% in Indonesia, 30% in Sri Lanka, 69% in Pakistan, but as much as 124% in the Philippines. Therefore, it is perhaps not surprising that in Indonesia rural per capita consumption of kerosene in 1978 was about 2.8 times higher than that in the Philippines. In Pakistan, it is estimated that as much as 25% of total kerosene consumption is accounted for by illegal diversions for blending with high-priced gasoline (see also Sri Lanka case study).[23]

In Indonesia, which has long followed the policy of subsidizing diesel fuel and fuel oil prices, diesel-powered power-plants accounted for 23%, and fuel-inefficient gas turbines for another 33% of total electricity generating capacity in 1979.[24]

Inappropriate pricing of energy resources is not the only reason for inefficient energy conservation decisions. In many developing countries, the lack of foreign exchange resources forces governments to maintain strict import controls. As a consequence, it is often impossible for large energy users to import new, more energy-efficient equipment in order to replace existing ones, even though they are usually able to secure their share of high-cost imported fuel supplies in order to keep their existing fuel-inefficient equipment operating. In countries in which fuel prices are subsidized at the same time, there is little incentive for such equipment owners to press for appropriate changes in import policies.

A special problem exists with regard to energy consumption patterns of households that depend on the collection and utilization of traditional fuel resources such as firewood, charcoal, and dung. As many studies have shown, the vast majority of these households depend on rather primitive cooking appli-

---

[21]Average Canadian crude oil prices, a blend between lower domestic well-head prices and oil imported at world market prices was equivalent to US $16.28/per bbl. in November 1980, compared to a world market price of $32.90/per bbl. The delivered natural gas price at Toronto City gate was equivalent to $2.45/MCF in November 1980.

[22]"Normative" prices were defined as market prices required to cover the actual costs of imported crude plus current refining and distribution charges plus pre-1973 excise tax rates. *From:* T. L. Sankar and G. Schramm, *Asian Energy Problems,* Praeger, New York, 1982, Table 6.5.

[23]Actual 1979 gasoline prices in Pakistan amounted to 105% of the normative price, while the corresponding ratio for kerosene was only 69%. *From:* T. L. Sankar and G. Schramm, *Asian Energy Problems* ibid. Table 6.5.

[24]T. L. Sankar and G. Schramm, *Asian Energy Problems,* Table 4.10.

ances which are usually little more than open fires. These are highly inefficient in terms of energy utilization and usually use only about 5% of the inherent heat energy of the fuel. Heavy population pressures, dwindling firewood resources resulting in sharply increased costs of fuel-wood gathering as well as increased soil erosion, and reduced availability of crop residues from new short-stem, high yielding crop varieties, all combine to make this one of the foremost and serious energy problems in the majority of developing countries.

The use of simple cooking stoves constructed of locally available materials at out-of-pocket costs barely exceeding $5 to $10 has improved energy efficiency by a factor of 4 to 5 in laboratory tests, and would perhaps do so by a factor of 2 or better in actual day-to-day household use.[25] However, while improved stoves have been developed in many countries, it has proved difficult to ensure their widespread use. The reasons for failures are many and usually have little to do with relative fuel efficiencies, but more with local customs, lack of effective extension services, and other socioeconomic factors. Nevertheless, this is an area in which major savings could be achieved at relatively low materials costs.

6.5 POLICY OPTIONS

In addition to appropriate pricing there are a wide variety of direct and indirect policy measures that can be taken to bring about desirable levels of energy conservation. Among them are direct regulations of energy uses, regulations of the use of energy consuming equipment and appliances, mandatory standards, mandatory information requirements about energy consumption rates, taxes and subsidies, appropriate infrastructure investments for energy saving facilities (e.g., better roads, railroads, marine shipping facilities), education and propaganda, and others.

To analyze some of the effects of such conservation-oriented policies let us first return to our light bulb example of the preceding sections. As we have found, existing market prices have made it more attractive for users to opt for the incandescent light bulb system. To resolve this difference between optimal economic and private market choices, the first option policy makers might consider could be to raise the market price of electricity from 0.3 dineros per kWh to its economic value of 0.4 dineros per kWh. We now have: $PC_I = 48.5 > PC_F = 44.4$ and rational electricity consumers will make the correct decision in favor of fluorescent lighting. In addition, setting the electricity price equal

[25]R. S. Dosik, D. Hughart, and P. Krahe, *Renewable Energy*, International Bank for Reconstruction and Development, Energy Department Note No. 53, Washington, D.C., February 19, 1980, p. 10.

to its marginal opportunity cost will also establish electricity consumption for nonlighting purposes at optimal levels.

Suppose that public resistance or other social pressures make it impossible to raise electricity prices. Let the economic value of an incandescent bulb be its cost of production or producer price, while the imposition of a government tax of 7.5 dineros determines the market price. Similarly, assume that an import duty of 4.0 dineros represents the difference in the c.i.f. import cost (32 dineros) and the market price of fluorescent bulbs. Instead of raising electricity prices, an alternative policy option might be to raise the tax on incandescent light bulbs to 9.5 dineros, making the market price 20 dineros. In this case, $PC_I = 42.9 > PC_F = 42.3$, which encourages the desirable consumer decision.

Reducing the duty on fluorescent bulbs to 2 dineros and lowering the retail price to 34 dineros would also yield a favorable result, since now: $PC_I = 40.9 > PC_F = 40.3$. Some combination of the tax increase and lowering of duty could also be used. From a general economic viewpoint, and ignoring effects outside the lightbulb market, reducing the import duty would be preferable to raising the producer tax because the former action reduces the divergence between market price and economic opportunity cost of fluorescent bulbs whereas the latter has the opposite effect and increases the market distortion in the price of incandescent lightbulbs.

Next, returning to our light-bulb example, assume that the tax on incandescent light bulbs cannot be increased because the legislation affects a much larger class of related products. Similarly, suppose that the import duty on fluorescent bulbs cannot be reduced because it would undercut the price of a high-cost local producer and drive him out of business. In this instance, some final options left to the energy policymaker might be to legislate that all incandescent lightbulbs be replaced by fluorescent ones, or to give a direct cash subsidy to consumers who adopt the measure, or to mount a major public education and propaganda campaign to bring about the required change.[26]

To conclude, we summarize the steps that an energy policymaker should take before adopting and implementing any given conservation measure. First, using economic opportunity costs consistent with the national viewpoint, he should establish whether the benefits of such an action exceed the costs. If this is the case, then the same test should be repeated, using market prices relevant to the appropriate consumer group, in order to establish whether a rational consumer would adopt the conservation measure. If this is not the case, changes

---

[26]The evidence concerning the effectiveness of education and propaganda as an energy conservation policy tool is mixed. See for example: James M. Walker, "Voluntary Responses to Energy Conservation Appeals," *Journal of Consumer Research,* Vol. 7, June 1980, pp. 88–92; and A. E. Peck and O. C. Doering, "Voluntarism and Price Response: Consumer Response to the Energy Shortage," *The Bell Journal of Economics,* Vol. 7, Spring 1976, pp. 287–292.

in energy prices, taxes, or import duties on equipment, subsidies to consumers, legislation, and other policy options may have to be used to implement the conservation technique. In general, price changes that reduce the divergence between market prices and opportunity costs should be preferred. However, care should be exercised to ensure that these policy actions do not have adverse repercussions in other energy as well as nonenergy markets. We discuss next some complications that could arise in the assessment and application of conservation measures.

If the useful lifetimes of technological alternatives are different, then economic comparisons become somewhat more complicated. This would be the case in our earlier example if the lifetime of incandescent bulbs were to be only one year while that of fluorescent lamps might be 3 years. Two alternative approaches could be used to overcome this difficulty. In the first, the investment costs of each alternative would have to be annuitized over its lifetime at the appropriate discount rate and the associated energy consumption and other recurrent costs for one year would be added on. Then the total costs for each option would be compared. The second method would compare the full costs of each alternative over a much longer period, say 20 years, including the costs of periodic replacement of worn-out equipment.[27] The two methods should give consistent results, assuming the same values are used for parameters such as the discount rate.

Another type of difficulty associated with changes in the benefits of consumption arises if either the quality or quantity of the end-product of energy use is different for the two alternatives. As an example of the first effect, consider a comparison of electric versus kerosene lamps for lighting. In addition to the differences in equipment and fuel costs, the cost-benefit assessment of the two options should also include a term to recognize that electricity is likely to provide lighting of a superior quality.[28] While the quantification, in monetary terms, of this qualitative superiority will be difficult, one measure might be the willingness-to-pay of the customers for the different forms of lighting, usually represented by the area under the relevant demand curve (see Figure 4.1 and explanation in Chapter 4).

Specific conservation measures such as rationing have a quality effect that must be taken into account. Let us consider the physical rationing of gasoline. In this case, the cost, or welfare loss, to the consumer due to the reduction in the miles he can travel in his car must be added to the cost of implementing

---

[27]Strictly speaking the discounted scrap value of equipment left at the end of the 20 year period should also be deducted from the costs stream associated with the corresponding alternative.

[28]The ease and convenience of using a fuel, the danger from its use, its social acceptability, and so on, are all factors that may affect the consumer's choice. See, for example, the case study on the Thai tobacco industry in Chapter 10.

the rationing scheme and then compared with the benefits of reduced gasoline supply. Once again, the willingness-to-pay of gasoline users would be the appropriate measure of the foregone consumption benefit. However, in the long-run, gasoline consumption could also be reduced by the introduction of a more fuel efficient automobile engine without (perhaps) requiring a reduction in the miles traveled. This shows that a reduction in energy consumption does not always imply a reduction in consumption benefits. A major focus of the appropriateness of conservation policies should be the service derived from the energy use.

Finally, the costs and benefits associated with externalities should be included in the economic cost-benefit comparison of alternatives. For example, improvements in technical efficiency or fuel substitution measures may give rise to pollution, as in the case of conversions from oil burning to coal-fired electric power plants. These additional "external" costs should be explicitly evaluated in the analysis.

The effects of sunk costs must also be recognized. Thus, if an oil burning generating-plant already exists, the initial comparison must allow for the fact that the system's costs of the oil option have no associated capital costs until the plant is physically fully depreciated.

6.6   CONSERVATION OPPORTUNITIES IN COMMERCIAL ENERGY PRODUCING AND USING SECTORS[29]

As discussed in Chapter 2, Energy in the Economy, aggregate economy-wide measures such as the ratio of energy use to value added or GDP are not very helpful in determining specific policies to improve the energy situation. The practical application of energy conservation policies requires the disaggregate analysis of the technical, economic, and behavioral relationship underlying the various types of energy conversion and end-use. In this section, we attempt to summarize the principal practical possibilities for energy conservation in the near term in several selected commercial energy producing and consuming sectors of the economy, based on the recent experience of both industrialized and developing countries. Since the emphasis here is on conservation, mention of improvements in efficiency generally refers to increases in the first law efficiency of energy use that will yield actual reductions in energy inputs required to perform a particular task. It is also understood that the desirability of adopt-

---

[29]General references which explore in greater detail some of the aspects discussed in this section include: Richard C. Dorf, *Energy Resources and Policy*, Addison-Wesley Publ. Co., (Reading, MA, 1978); *Energy Programs and Policies of IEA Countries*, 1977 Review, OECD, Paris; ibid. 1978 Review; ibid., 1979 Review; ibid., 1980 Review; John C. Sawhill (ed.), *Energy Conservation and Public Policy*, Prentice-Hall, Inc., Englewood Cliffs, NJ, 1979); and Sam Schurr, et. al., *Energy in America's Future*, Johns Hopkins Univ. Press, (Baltimore, 1979).

ing any conservation measure must depend on the economic criteria discussed in the previous section, and on the social, political, overall physical, and other constraints peculiar to the country concerned.

Transportation

The principal purpose of a transportation system is to physically convey people or goods from one location to another. Therefore, any measure that increases the payload in terms of energy used per passenger-mile or ton-mile would help the conservation effort. Important methods of achieving this result include: (a) changing from more to less energy intensive transport modes; (b) increasing the technical efficiency of energy use of given modes of transportation; and (c) changes in behavior and overall systems effects. We will examine each of these three aspects below.

Table 6.3 summarizes some typical characteristics of the chiefly used transport needs. The energy intensity figures for a given mode will vary widely depending on the geographic characteristics, goods transported, behavioral characteristics, and so on. However, even on a rough basis, it may be seen that switching modes could provide substantial savings. For example, a change from passenger-car travel to mass transit modes such as bus or rail, or a transition from freight-trucks to freight-trains would reduce fuel consumption. In a period of rapidly rising petroleum prices, conservation in the principally liquid fuel using road transport mode should have a high priority. In particular, savings in traffic congestion costs in urban areas and also reductions in air pollution may be significant. However, behavioral, physical, and other impediments such as the unwillingness of motor car owners to use public transport, or the inaccessibility of railway stations for freight hauling will make it difficult to effect these changes.

The technical efficiency of energy use may be increased by introducing more fuel efficient engines, improving the quality of roads, electrifying railways, and so on. A combination of price, legislation, and government investments or other initiatives can be used effectively.

Behavioral changes include car-pooling, getting people to live closer to their place of work, using alternative methods of communication like telephones where possible, and so on. In many cases people may be very resistant to adopt these changes in lifestyles, many of which affect the whole socioeconomic system rather than the transportation system alone (e.g., urban living patterns, travel and migration patterns, etc.). In the developing countries some of the developmental changes themselves may encourage shifts to more energy intensive and modern modes, e.g., use of motor vehicles instead of walking, bicycles, or animal drawn carts. Therefore, the desirability of any conservation policy must be assessed against the overall economic criterion discussed earlier. Coor-

Table 6.3. Typical Energy Use Characteristics of Principal Transport Modes (1975-80).

MODE	ENERGY INTENSITY		SHARE OF ENERGY CONSUMPTION[1]	
	Btu/PASSENGER-MILE	Btu/TON-MILE	INDUSTRIALIZED COUNTRY[3]	DEVELOPING COUNTRY[4]
Walking/Bicycle/Animal[2]	300–500	—	<.01	.05
Pipeline	—	300–600	.05–.1	<.05
Water	—	400–700	<.05	.05–.1
Railway	2,000–3,000	500–900	<.05	<.05
Bus	1,000–1,500	—	.01–.1	.05–.1
Truck	—	1,500–2,000	.1–.2	.35–.45
Car	2,000–4,000	—	.4–.5	.25–.35
Air	5,000–9,000	15,000–25,000	.05	.05

[1]As a fraction of total commercial energy consumption.
[2]Noncommercial energy use given as a fraction of commercial energy total.
[3]*Source:* Author's estimates and OECD, IEA, and USDOE data.
[4]*Source:* Author's estimates and "Energy in the Developing Countries," Washington, D.C., August 1980.

dinated use of price and nonprice tools is again important. For example, public exhortations alone are unlikely to be effective.

Building, Lighting, Space Heating, and Cooling

Three factors affect the consumption of energy in buildings: (a) behavioral characteristics and attitudes of occupants; (b) energy using equipment installed; and (c) architectural design practices and material used. Keeping living and working space lighted and cool are the chief concerns in the tropics where most of the developing countries lie. In these areas the use of air-conditioning is growing rapidly for commercial buildings such as businesses, tourist hotels, etc., and to a lesser extent for residences of upper income urban and expatriate groups whose numbers are fortunately small in this respect. In the industrialized countries, which are located more in the temperate and colder zones, lighting and both space heating and cooling are required.

The principles that govern the application of energy conservation practices remain the same. Thus the full range of policy tools including increases in energy prices, legislation, tax incentives, and so on will help to implement conservation measures. However these policies and their consequences must be simple and easily comprehensible. Educating occupants of buildings (especially domestic residents) concerning simple conservation practices, making them aware of new but readily available energy saving devices, and explaining the consequences of new pricing structures and taxes should have a high priority.

The so-called "information gap" is particularly critical in this area of conservation.

Changes in behavior and attitudes often take time to occur. For example, admonitions to switch off unused lights or set back thermostats may take years to sink into the public consciousness especially if these requests run up against false but commonly held beliefs. Thus in Indonesia, where a fixed charge life-line electricity tariff was in effect, low income households were found to be using their lightbulbs as many as 16 hours a day. This occurred despite an energy conservation campaign, because it was widely believed that switching lights on and off would reduce their lifetime more than if the bulbs were kept continuously lighted, while the electricity bills were fixed and unrelated to kilowatt-hour consumption.

In other cases, people are unwilling to give up comforts they have grown accustomed to, unless some costs are involved. Thus requests to set back thermostats in many industrialized countries have not been effective except when accompanied by increased energy prices, and only after the effect of increased energy bills was felt. Public resistance may also occur in developing countries where people who are just beginning to enjoy the benefits of economic growth resent being asked to cut back on consumption of electricity for lighting, air-conditioning, and so on. Pricing of energy at economic opportunity cost is very useful in all these cases, because the consumers can get the correct price signal and then choose how much energy to use on the basis of willingness-to-pay. Even then responses to price changes could be slow because these adjustments in energy usage patterns may imply major expenditures for purchasing new energy using equipment as discussed below. In brief, behavior changes that facilitate conservation may be realized only slowly, and policymakers should fully investigate local attitudes and idiosyncracies with respect to energy use patterns.

Improvements in energy using equipment and appliances in buildings are an obvious target for conservation programs. Thus occupants must be made aware of the many opportunities for replacing inefficient equipment such as furnaces and air-conditioners, for retrofits or improvements to existing equipment, or for simply improved operation and maintenance procedures. Appropriate financial incentives should be provided to reinforce the message. Technical advice, energy audits, and guidance could be provided by the government, especially in the case of large buildings, e.g., some of the modern heating and cooling systems are very effective, using an array of techniques ranging from heat pumps to computerized control of equipment in different parts of the building. Legislation on minimum energy efficiency standards for equipment and appliances is also helpful.

As mentioned earlier, the high cost of replacing old equipment may delay consumer response to higher energy price signals. For example, the lump sum cost of a new refrigerator may be a significant fraction of income, thus causing the consumer to wait until his old inefficient refrigerator is worn out before replacing it. Replacing incandescent light bulbs with fluorescent fittings is another energy saving measure whose costs may deter a poor consumer. This phenomenon is particularly significant in developing countries where incomes are low and the fuel may be relatively cheap or subsidized (relative to its shadow price). For example, several programs are under way to replace open hearth fires whose first law efficiency is only about 5% with simple stoves that are over four times as efficient. These have had limited success, and only when government officials provided the improved stoves free or at a subsidized price, coupled with a strong promotion campaign. In all these cases, financial incentives or subsidies on new equipment could be most effective. Thus energy saving improvements in equipment will be realized quicker in general, with the combined use of higher prices, legislation, public education, and equipment subsidies.

Architectural design, building practices, and construction materials used is the third area in which energy conservation gains may be made.[30] Policymakers may make a significant impact by altering building codes and implementing legislation relating to minimum efficiency standards. The orientation of buildings, location of windows, type of glass used, and other architectural features can improve heat losses. Use of improved insulating materials and high standards of construction to avoid flaws or gaps in insulation are also helpful.[31] If buildings are not completely enclosed as in some developing countries, proper design will promote natural air-circulation. Use of simple local materials such as brick tiles (instead of asbestos), higher ceilings, and installation of fans are often substitutes for air conditioning in the tropics. Much can be achieved in the way of conservation by a well informed and imaginative architect.

Industry

Improvements in the efficiency of industrial energy use cover such a broad range of techniques that only the general principles can be touched on here.

---

[30]See for example: G. Dallaire, "Designing Energy Conserving Buildings," *Civil Engineering*, April 1974, pp. 54–8.

[31]There are practical limits to conservation gains that may be achieved in this respect. Thus, improving construction practices or using better building materials will become increasingly costly and beyond a certain point they would not justify the reduction in heat losses, according to the economic criterion described earlier. Again, a perfectly air-tight and insulated building may be ideal from the viewpoint of energy efficiency, but would be stifling and uncomfortable to occupants, impose health hazards due to constantly recirculated stale or polluted air, and so on. In brief, savings due to energy conservation must be weighed against both the quantifiable and sometimes nonquantifiable costs incurred.

Contrary to widespread belief, many industries are unaware of increases in efficiency that can be realized quite simply and are extremely cost effective. Because of the concentrated nature of industrial energy users, both governments as well as energy suppliers and utility companies can be particularly effective in legislating improvements, counseling, providing energy audits, and helping consumers carry out technical improvements. Four broad areas for conservation are: (a) waste heat recovery and cogeneration; (b) other retrofits and improvements in operation; (c) major changes in manufacturing processes and production methods; and (d) recycling and recovery of waste materials. We will briefly examine each topic in turn.

Most large industries use energy for heating. A significant fraction of this thermal energy is expelled into the external environment at temperatures well above ambient conditions, usually in the form of hot gases, steam, or water. This waste heat can be harnessed in a number of ways, thus improving the overall efficiency of energy use in a plant by as much as 30%. In completely integrated or total energy systems (also called cogeneration systems), fuel would be used to generate electricity, yield process heat for industrial use, heat buildings in the area (i.e., district heating), provide hot water, process solid and liquid wastes, and so on. A central concept in this type of system is that the overall efficiency of energy use for the total plant be maximized rather than the efficiency of any single component or subsystem, such as electricity generation.[32] In fact the energy efficiency of certain components may have to be reduced below what it would have been if it was operating on its own. The gains in other parts of the system more than compensate for this loss, and thus overall efficiency improves. For example, the exhaust heat from an electric power generator in a cogeneration scheme would be extracted at a somewhat higher temperature than a stand-alone unit, with consequent loss of power output. However, the waste heat could be used much more efficiently in another task because it is available at a higher temperature.

The best potential for development of cogeneration or total energy systems occurs when new industrial plants are being set up. Many examples of cogeneration already exist in Europe, and the U.S. is not far behind.[33] There is also considerable scope for such schemes in large industrial estates being set up in many developing countries. In many cases, legal and institutional barriers to agreements that facilitate the exchange of energy between different entities such as utility companies, industries, and municipalities, appear to pose greater difficulties than technical constraints. Energy policymakers should take action

---

[32]See, for example, P. Bos, et al., *The Potential for Cogeneration Development in Six Major Industries by 1985,* FEA Report, Resource Planning Associates, Cambridge, MA, 1977.
[33]See, for example, L. Schipper and A. J. Lichtenberg, "Efficient Energy Use and Well-Being: the Swedish Example," *Science,* Dec. 3, 1976, pp. 1001–12.

to smooth out these problems. In the case of existing industries, where major changes in plant layout are not possible, there is still scope for more limited use of waste-heat. For example, an industrial plant using steam for heating could also run an auxiliary generator to produce electricity, or hot exhaust gases could be used for drying industrial materials, preheating incoming air, and so on.

The effectiveness with which waste heat can be put to use depends on several factors including the temperature of the exhaust and the degree of compatibility with the quality of heat required in the particular application (as discussed in Section 6.1, under second law efficiency), the availability of sufficient quantities of waste heat to achieve economies of scale, and the distance between the source of the waste heat and its final use. In general, the higher the temperature of the exhaust fluids, the more efficiently it can be used based on the thermodynamic concept of cascading (see section 6.1). Some industrial activities that produce waste heat at different temperatures are presented in Table 6.4.

The second broad area for encouraging conservation is in the improved operation of existing plant and by appropriate retrofits. Adapting old equipment for cogeneration is of course also a form of retrofitting, but this has been discussed earlier. There are many other ways in which the energy efficiency of industrial processes may be improved. For example, the thermal insulation or lagging of boilers and pipes carrying heating or cooling fluids may be increased, the mixing of fuel and air could be improved to provide better combustion, heat transfer can be enhanced, and so on. More sophisticated techniques such as computerized control of industrial processes can also increase conservation. In many cases, detailed energy audits by external experts can pinpoint these improvements, identify new energy efficient devices that are readily available, and demonstrate the cost effectiveness of such conservation practices to the industrialists concerned.

The third aspect of conservation involves major changes in industrial activity. Shifts in technology and production processes most often occur in response to changes in the relative prices of energy and other inputs such as capital and labor. Thus, an increase in the real cost of energy would favor a shift towards a less energy intensive technology. We note that autonomous technological improvements can also occur quite independently and sometimes contrary to price changes, due to inventions that improve the efficiency of industrial processes. For example, in the case of aluminum smelting, which is one of the most energy intensive industrial processes, the electricity requirements for smelting a kilogram of aluminum from bauxite ore decreased from about 26 kilowatt-hours in 1940 to about 17 kilowatt-hours by 1975 despite an overall decrease in the real price of electricity. With the added impetus of higher energy costs, a new chlorine process now being introduced will decrease energy consumption

ENERGY CONSERVATION AND EFFICIENCY 203

Table 6.4. Sources of Waste Heat by Temperature Range.

HIGH TEMPERATURE		MEDIUM TEMPERATURE		LOW TEMPERATURE	
SOURCE	TEMPERATURE (°F)	SOURCE	TEMPERATURE (°F)	Source	Temperature (°F)
Nickel refining furnace	2500–3000	Steam boiler exhausts	450–900	Process steam condensate	130–190
Aluminum refining furnace	1200–1400	Gas turbine exhausts	700–1000	Cooling water from:	
Zinc refining furnace	1400–2000	Reciprocating engine exhausts	600–1100	• Furnace doors	90–130
Copper refining furnace	1400–1500			• Bearings	90–190
Steel heating furnaces	1700–1900	Reciprocating engine exhausts (turbo charged)	450–700	• Welding machines	90–190
Copper reverberatory furnaces	1650–2000			• Injection molding machines	90–190
		Heat treating furnaces	800–1200		
Open hearth furnace	1200–1300	Drying and baking ovens	450–1100	• Annealing furnaces	150–450
Cement kiln (Dry process)	1150–1350	Catalytic crackers	800–1200	• Forming dies	80–190
Glass melting furnace	1800–2800	Annealing furnace cooling systems	800–1200	• Air compressors	80–120
Hydrogen plants	1200–1800			• Pumps	80–190
Solid waste incinerators	1200–1800			• Internal combustion engines	150–250
Fume incinerators	1200–2600			Air conditioning and refrigeration condensers	90–110
				Liquid still condensers	90–190
				Drying, baking, and curing ovens	200–450
				Hot processed liquids	90–450
				Hot processed solids	200–450

Source: W. M. Rohrer and K. Kreider, "Sources and Uses of Waste Heat," *Waste Heat Management Guidebook*, NBS Handbook 121, Government Printing Office, Washington, D.C., January 1977, p. 5.

to about 10 kilowatt-hours per kilogram of aluminum.[34] More generally, the energy inputs per dollar of industrial value added in the U.S. decreased from about 110,000 btu to 80,000 between 1945 and 1975. Appropriate price signals regarding higher energy costs and incentives to encourage research and development to improve the energy efficiency of major industrial production processes are important tools in the conservation specialist's arsenal.

While the bulk of manufactured products is derived from the processing of raw input materials, the recycling of previously manufactured but discarded materials may be less energy intensive. However, the difficulties of collecting and sorting waste materials prior to recycling could significantly increase processing costs especially where labor inputs are required to do this. Some typical figures for energy saving through recycling are given in Table 6.5.

Electric Power

Electric power is a relatively mature energy sub-sector where conservation techniques are well developed, and this permits us to clearly illustrate the links between energy conservation, demand management and pricing. The three principal opportunities for conservation arise in the (a) generation; (b) transmission and distribution; and (c) efficiency and patterns of end use of electricity.

Conservation gains in the production of electricity may be achieved through efficiency improvements in individual generating plants or with respect to the whole power system.[35] In the former category, efficiency improvements in generation technology have occurred steadily over the last fifty years or more. Many of these advances have stemmed from economies of scale as unit sizes of generators have increased. For example, the largest steam units of about 200 MW available in 1930 had first law efficiencies of less than 20%, while the largest unit size for today's thermal plant is about 1,500 MW and they operate at much higher steam pressures and temperatures to achieve first law efficiencies of about 35–40%. New technologies such as fluidized bed combustion and magnetohydrodynamic generation could improve this figure substantially in the future, but thermodynamic laws will limit ultimate efficiencies to around 55%.

---

[34]Richard A. Charpie and Paul W. McAvoy, "Conserving Energy in the Production of Aluminum," *Resources and Energy,* Vol. 1, September 1978, pp. 21–42. A minimum value for energy use of about 8 kilowatt-hours per kilogram of aluminum based on thermodynamic limits has been estimated in: Elias P. Gyftopoulos, J. Lazaros, J. Lazaridis and Thomas F. Widmer, *Potential Fuel Effectiveness in Industry,* Ballinger, Cambridge, MA, 1974.

[35]Switching some generation where possible from fossil fuels to renewable and new sources such as solar wind and wave power can conserve depletable energy resources. The possibilities are extremely country specific and depend on resource endowments.

Table 6.5. Energy Savings From Recycling (1978).

| MATERIAL | SHARE OF ENERGY COSTS IN VALUE OF OUTPUT | ENERGY USE FOR PROCESSING (kWh/kg) | | ENERGY SAVING (%) | CHIEF CONSTRAINTS |
		From RAW MATERIAL	FROM WASTE MATERIAL		
Glass	0.35	2.3	2.3	None	Collection
Paper	0.35	1.9	0.95	50	Separation
Steel	0.35	13.4	6.6	50	Impurities, separation, and collection
Plastics	0.05	13.0	0.58	95	Technology commercially unavailable

In the case of large modern hydroelectric generating units, the hydrostatic or potential energy of stored water is already converted into electrical energy at efficiencies exceeding 90% and the scope for improvement is somewhat more limited. Retrofitting of existing thermal and hydro plant, and improving operating and maintenance procedures to at least bring them up to original design standards is another focus for conservation. The cogeneration type arrangements discussed earlier under industrial energy conservation will also improve efficiency. While introducing new plant is likely to be highly capital intensive, the alternative of upgrading of existing units is so case-specific that high inputs of skilled manpower will be required. The cost effectiveness and desirability of these various options will have to be established case by case according to the overall economic criterion outlined in section 6.2.

Systemwide improvements in the efficiency of producing electricity can also be achieved by correctly matching the available technology to the pattern of demand. As explained in greater detail in Chapter 8, Introduction to Investment Planning, meeting a certain shape of load duration curve at the least possible cost requires optimal long-range planning and operation of the power system. For example, the least cost generation planning of an all thermal system implies that steam or nuclear plants should be built for base load duty (i.e., operating at least 6,000 out of 8,760 hours per year) because their fuel costs are low, although their investment costs are high. The same logic dictates that gas turbine units which have low capital costs but high fuel costs should be used for peak period operation, usually about 2,500 hours per year. Similarly, in optimal system operation and load dispatch, the available generating plant is used sequentially, starting with the newer base load units that are cheapest to run and ending with old or peaking units that have the highest fuel costs. Generally, electric utilities offer scope for improvements in system effi-

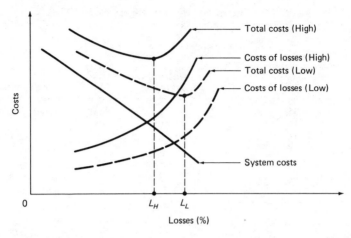

Figure 6.4. Optimal trade-off between losses and system costs for electricity transmission and distribution systems. Total costs = costs of losses + system costs.

ciency, especially in many developing countries where engineering-economic optimization of system planning and operation is neglected or poorly done.

Conservation in the delivery of electric power is achieved by reducing technical losses in the transmission and distribution (T&D) networks. These losses may be as high as 30% of gross generation in some developing countries, although norms in industrial countries are about 10%. The determination of optimal or desirable loss levels is based essentially on the trade-off between the increased capital costs of augmenting T&D capacity, and the corresponding savings in both kilowatt and kilowatt-hour losses. The rapid increases in the costs of electricity supply in the 1970s indicate that the levels of losses previously considered desirable are likely to be unacceptably high today. As shown in Figure 6.4, the optimal trade-off occurs when total costs defined as the costs (all in present discounted value terms over a long period of about 10–20 years) are minimized.[36] The desirable loss level shifts from L_L to L_H as energy costs rise. Losses due to theft can also be significant, reaching levels of 10–15% in some developing countries. U.S. norms are about 2–3%. Such losses may be reduced by appropriate improvements in legislation and management of the power utility.

---

[36]We note that as system losses decrease, the quality of electricity supplied may also improve (e.g., better voltage and fewer supply interruptions or outages). Therefore, some adjustment to total costs defined above may be required to account for the accompanying change in consumption benefits due to improved quality. The magnitude of consumption does not change due to loss reduction because these losses were not consumed originally, in any case.

Energy conservation at the end-use stage may be achieved by two principal methods: improving the technical efficiency of energy using devices and appliances, or changing the shape or characteristics of the load through demand management techniques. The first aspect, increasing the energy efficiency of devices and appliances, has already been discussed in the sections on energy conservation in buildings and industry. The second aspect uses demand management as a tool to conserve energy and increase the economic efficiency of energy use. In the electricity sector, the term load management is used synonymously with demand management. Because of the relative sophistication and maturity of this sector, load management techniques are well developed and fall into two basic categories: soft load management which relies on prices, financial incentives, and public education to achieve voluntary changes in consumer electricity user patterns; and hard load management which seeks to realize the same result by actual physical control of consumer loads.

As explained in greater detail in Chapter 4 (in the section on peak load pricing), and also in the Sri Lanka and Thailand case studies, it is more costly to supply electricity during certain seasons or hours of the day known as peak periods than during other off-peak periods. Therefore changing the shape of the power utility's load curve by shifting electricity consumption from peak to off-peak periods will reduce the costs of supply and conserve energy. Charging higher prices (equal to long run marginal costs) during peak time-of-day or seasonal periods signals to the consumer that he should try to switch at least some of his load to off-peak periods.[37] Separate capacity and energy changes may also be used. Providing financial incentives to retrofit old equipment, improve the power factor (by adding a capacitor bank), and so on, are also soft demand management techniques.

The clearest example of hard load management is the case of interruptible loads, where certain industrial consumers receive cheap power but may be cut-off or shed at short notice when the total system load approaches the available capacity (during peak periods). Domestic loads such as water heaters can also be controlled through ripple control or radio by the utility. Within the next decade, advances in solid state hardware and metering will allow greater control and switching of individual loads through use of microprocessors and so on.

---

[37]For details of peak load pricing, see Mohan Munasinghe, *Principles of Modern Electricity Pricing, Proc. IEEE,* Vol. 69, March 1981, pp. 332–48. Shifts from peak to off-peak consumption result in an increase in the load factor. However, from an economic efficiency point of view, the objective of load management is not to achieve a load factor of 1.0, but to see as far as possible that the price or consumers' willingness-to-pay (i.e., benefit) equals the marginal cost of supply, at all times.

6.7 CONSERVATION OF TRADITIONAL FUELS AND ENVIRONMENTAL EFFECTS

While shortages of traditional fuels are widespread, use efficiencies of these fuels are also generally appallingly low. A substantial percentage of users worldwide simply place a cooking pot on stones over an open fire (which costs them little). However, this process will utilize only about 5–8% of the heat energy of the woodfuel. Even quite primitive stoves will increase fuel-use efficiencies to about 15%, or by a factor of 2 to 3, while modern efficient wood stoves may utilize as much as 50–60% of the available heat energy of the fuel.

Laboratory and field tests have shown that improved cookstoves, constructed out of local materials for between $5–$20, may save one half or more of the original fuel needed. Stoves may be made of metal by local artisans and sold in local markets. Alternatively, they may be made of clay and sand by local masons, who could be taught initially by government extension services or private voluntary organizations. Researchers in at least thirty less developed countries are currently developing cookstoves adapted to local cooking habits and are attempting to spread their use.

In terms of material costs and resulting savings in energy use, the widespread introduction of improved cookstoves is one of the most cost-effective ways to reduce the mounting pressures on traditional energy resources. It can yield quicker results, if properly implemented, than efforts at reafforestation which require a minimum of 6–8 years from planting to harvesting. The major problem involved in their widespread introduction and adaptation is that of social acceptability and appropriate organization. Therefore, at least initially, costs to the promoting organization are likely to be high.

In regions with high densities of animal populations (e.g., cattle or pigs), biodigesters may provide a useful and cost-efficient source of energy. The low-btu gas produced by such biogas units can be used for cooking, lighting, or the operation of spark-ignition engines. Major problems affecting the widespread introduction and utilization of these digesters are the high initial costs— between $400 to $600 for a family-size installation under South and Southeast Asian conditions; the need for continuous operation and continuous maintenance and supervision of the system to keep it operating; and the costs of collecting the dung to feed the digester. Therefore, digesters have had their most notable success in areas in which animal density per family is high (6–10 animals per family), and where animals are stabled, which drastically reduces collection costs. Pig manure is more suitable than cow manure. This may help explain the notable success of digesters in China, where pigs are common, versus Pakistan and India, where only cattle manure is available and animals usually are not stabled. An added advantage of the use of digesters compared to the direct burning of dried manure as fuel is the fact that the fertilizing properties of residual material from biogas digesters are retained and even enhanced.

Much has been written regarding the relationship between energy and the environment. It should be noted that the economic criterion for conservation described in section 6.3 seeks to include environmental gains and losses wherever these can be identified, in the cost-benefit analysis, so that the desirability of a conservation measure is judged in relation to its environmental impact. This is important because different energy conservation practices can either improve or worsen the environment.

For example, in the case of automobile exhaust emission, pollution limiting devices will generally decrease fuel efficiency. Therefore, there is a conflict between the conservation and environmental objectives, requiring the trading-off of various gains and losses. On the other hand, reducing the use of woodfuel in certain developing countries may prevent severe deforestation problems that may lead to soil erosion, loss of vegetation, reduced watershed potential, downstream siltation and flooding. Thus conservation of fuelwood is consistent with the ecological goal, but may involve other costs such as substitution of a more expensive commercial fuel for firewood. Once again, an economic analysis that takes these various external factors into account provides the basis for rational decision making. Where quantification of environmental costs and benefits is difficult, judgment must be used.

While many energy supply activities as well as energy uses produce important negative environmental effects some supply activities also create significant external benefits. Well-known examples are those related to multiple outputs from hydropower projects, such as irrigation and/or potable water supplies, flood protection, and navigational or recreational benefits.[38] The value of these joint outputs has long been recognized because of their clearly identifiable and measurable outputs.

However, there exists another group of joint benefits from a different type of energy supply project—woodfuel plantations—that take on increasing importance in those parts of the world where most of the population depends on indigenous biomass resources to satisfy basic cooking and heating fuel requirements. Few systematic attempts have been made to evaluate these external benefits, even though the available evidence suggests that they may be large—larger even than the on-site benefits of the plantations themselves.

[38]There is a large literature on the subject available. See, for example, Otto Eckstein, *Water Resources Development,* Harvard University Press, Cambridge, MA, 1961; or Arthur Maass et al., *Design of Water Resources Systems,* Harvard University Press, Cambridge, MA, 1962.

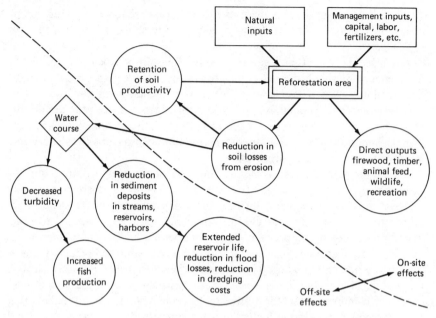

Figure 6.5. Joint outputs, on-site & off-site, of a reforestation project.

While the latter may often be marginal, measured in terms of conventional benefit-cost analysis, inclusion of these extended benefits may well prove that such projects have a far higher rate of return than initially estimated.

Figure 6.5 presents a flow diagram of the likely on-site and off-site effects of a reforestation project. The direct outputs are likely to be firewood and/or charcoal, timber, animal feed, and, perhaps wildlife and recreation. A second, and most important, consequence of the reforestation of denuded land is the reduction in soil losses. As many studies have shown, tree growth on previously unprotected sloping land will dramatically reduce water run-off rates.[39] As a consequence, soil losses from erosion are reduced. This has one important on-site effect: Soil productivity is maintained over much longer periods of time and the output of the land (in the form of usable biomass) is enhanced. Such

---

[39]For example, a study on reforestation showed that peak flow rates of run-off from storm water decreased by 75% over a 6 to 8 year period. *From:* Lloyd L. Harrold, "Effect of Vegetation on Storm Hydrographs," in *Biological Effects on the Hydrological Cycle,* Proceedings of the 3rd International Seminar for Hydrology Professors, Lafayette, Ind., 1971.

benefits from "avoided losses of land productivity" may, in some cases, be large enough to justify reforestation as a soil erosion control measure alone.[40]

In addition, however, there are usually important off-site benefits. Eroded material from an unprotected site would usually enter some water course. This could lead to increased turbidity and decreased fish production. More importantly, it would lead to sediment buildup in streambeds, reservoirs, shipping channels, and harbors. Such sediment deposits lead to increased flooding,[41] decreased reservoir life,[42] and high costs of dredging.[43] Taken together, the reduction in these losses as a result of a reforestation project designed primarily for firewood production may well outweigh the direct, on-site benefits.

Unfortunately, few studies are available that provide estimates of the actual magnitude of these on- and off-site interrelationships. The reason for this dearth of evidence can be found in the difficulties of linking the consequences of reduced run-off and sediment production from a given site with downstream effects that are often hundreds or even thousands of miles away.[44] While we are able to detect the aggregate relationships—the drastic reduction of tree cover in the Himalayan Mountain chains is undoubtedly the major cause of the huge flood losses in the Ganges plains—it is difficult to tie individual sites to specific downstream effects. Much more effort will be needed to establish these linkages so that the resulting benefits can be taken appropriately into account.

Other problems are endemic to reforestation projects. First, the issue of common property resource ownership—most of the original forest stands were denuded because the fuel source was owned by everyone in common. Second, the benefits from avoided off-site losses must be counted in the benefit-cost evaluation of the reforestation efforts; this would allow for higher investments per unit of land, but no revenues could be collected to actually pay for these investments. The public sector would have to assume the burden, thus produc-

---

[40]This was the case for high-cost terracing measures designed to protect the output of basic, rainfed crops in Mexico. See: Gunter Schramm, "A Benefit-Cost Model for the Evaluation of On-Site Benefits of Soil Conservation Projects in Mexico," *The Annals of Regional Science,* Vol. XIII, No. 2, July 1979, pp. 19–28.

[41]See also Chapter 2.

[42]Increased sediment inflow at the Tarbela Dam, Pakistan's largest hydroproject, is estimated to have reduced total reservoir life by more than 20 years compared to original estimates.

[43]Dredging costs due to increased sedimentation in the harbor of Buenos Aires exceeded 5 million dollars a year in the mid-1970s. *From:* R. Cano, "Argentina, Brazil and the La Plata River," in: A. Utton and L. Teclaff (eds.), *Water in a Developing World,* Westview Publ. Co., Boulder, Colorado, 1978.

[44]For a detailed discussion of the complex issues involved see Daniel J. Basta and Blair T. Bower, eds., *Analyzing Natural Systems,* Resources For The Future, Washington, D.C., 1982, Chps. II & IV.

ing future benefits to downstream users. However, such investments would pay little political dividends downstream because of the very nature of the benefits, i.e., "avoided" damages that would remain invisible to the recipients. Third, the gestation period for such projects is long, usually at least 6–8 years, even for fast growing species. If local fuel availability is already strained, newly planted trees may not survive to maturity because of illicit cutting that usually cannot be prevented by policing or fines.[45]

---

[45]In the Upper Mangla Dam watershed protection area in Pakistan, for example, over 60% of the indigenous population had fines for illicit cutting outstanding. Needless to say, such fines are rarely paid. *From:* West Pakistan Water and Power Authority, *Mangla Watershed Management Study,* Lahore, 1961.

7
Energy Demand Analysis and Forecasting

7.1 INTRODUCTION

Energy planning is impossible without a reasonable knowledge of past and present energy consumption and likely future demands. These consumption patterns are significantly affected by energy prices. Any demand analysis and demand forecast, therefore, must take explicit account of relative and absolute energy prices, because prices not only affect choices among alternative energy sources, but also choices between the use of energy versus other alternative inputs such as capital and labor, or choices between energy and nonenergy consuming activities.

However, prices of specific energy resources are only one set of parameters that affect use. Others, such as availability, reliability of supply, uniformity of quality, convenience in use, technical and economic characteristics of energy-using equipment and appliances, population growth, income, rate of urbanization, as well as social habits, acceptability, and knowledge by potential users are as important, or even more important, than price in determining demand. Hence any analyses of past and present consumption patterns and forecasts of future demands have to take these other factors explicitly into consideration.

Nevertheless, prices may substantially affect energy consumption patterns, at least in the long run, since substitutes, in the form of alternative energy resources, or capital, or labor, can be found, or consumption habits may be changed. For sector planning purposes, therefore, estimates are needed of the short- and long-run price elasticities of demand for every specific energy resource. Such estimates are usually based on some form of econometric analysis. However, for a number of reasons which are discussed in greater detail below, such aggregate statistical measures are usually unsuitable for demand forecasts in developing nations. Even in industrialized nations they are subject to wide margins of error. Hence direct sector and activity-specific estimates are likely to provide more reliable results.

Demand forecasts could be made either on the basis of statistical evaluations and projections of past consumption trends, or on the basis of specific micro-studies. The former approach is appropriate in industrialized nations in which data coverage is excellent and energy-consuming activities are ubiquitous, complex, as well as mature, so that changes from observed trends are slow. In most developing nations, although trend-line extrapolation is common, a micro-survey-research type approach will usually be more useful because it will yield more reliable results. This is so because statistical data are often lacking, or of poor quality. Furthermore, sectoral demand changes resulting from specific policies, such as rural electrification programs or the establishment of new industrial plants, for example, can be very substantial relative to existing demand. Such program- or project-specific effects on future demand usually cannot be forecast on the basis of observed past or present consumption data. However, such case by case investigations must necessarily be limited to surveys of the larger energy consumers or energy-related development programs. Both for reasons of costs and time, forecasts for sectors such as urban or rural households, commercial activities, or transportation, for example, must normally be based on statistical data analysis, although specific factors such as changes in relative prices, disposable income, rate of urbanization, or of sectoral production or output must be specifically considered as determinants of future sectoral energy demands.

7.2   THE IMPORTANCE OF ENERGY DEMAND FORECASTS

There are three interrelated reasons for the importance of accurate energy demand forecasts. The first is that the timely and reasonably reliable availability of energy supplies is vital for the functioning of a modern economy. The second is that the expansion of energy supply systems usually requires many years. And the third is that investments in such systems generally are highly capital intensive, accounting, on average, for some 30% of gross investments in most countries. If forecasts are too low, energy shortages may develop whose costs are usually a large multiple of the volume of energy not supplied; but if forecasts are too high large amounts of capital with high opportunity costs might be uselessly tied up for long periods of time. Either of these consequences is costly to an economy; far more costly than the resources that may have to be marshaled to undertake detailed and reliable demand studies that could help to avoid such errors.

A number of recent studies have shown how costly such errors can be. If supply shortages develop as a consequence, more expensive foreign energy supplies may have to be imported, emergency equipment may have to be installed, rationing may have to be introduced, and forced outages may occur. In Thailand, for example, the net value of refined petroleum product imports rose from 15 percent of unrefined crude oil imports in 1972 to 21 percent in 1977 because local refinery capacity was not expanded in time to meet rapidly growing demands.[1] In the late 1970s in both Sri Lanka and Colombia, delays in expand-

---

[1] See also the case study on Thailand, chapter 10.

ing low-cost hydro/powerplants forced the installation of small size thermal powerplants with high fuel costs to bridge the gap.[2] Also in Colombia, restrictions had to be imposed on gasoline sales in 1979 because of rapidly rising high-cost imports, even though the country had been a net exporter of crude oil until 1975/76 and its prospects for additional domestic oil discoveries are rated high.

Rationing or forced interruptions of energy supplies usually lead to substantial losses in output and consumer welfare. This is so because the cost of energy, relative to the value of output or services received from energy use, is low. This is apparent from Table 12.10 which shows that in the USA, for example, electricity costs to industry amounted to 1.2 percent on average of the value of output in 1979. These percentage ratios are reasonably typical for developed as well as developing countries, although the recent petroleum product price increases will have increased them for some of the latter. Hence, while the cost of energy is quite low relative to the value of output, without energy no output could be produced.

In a detailed case study of the costs of outages of electricity supply in Brazil, Munasinghe found that in 1976 the net loss in the value of industrial output from forced outages amounted to about $1–6 per kWh lost, while for residential consumers the estimated outage cost per hour during the evening period was approximately equal to the income earning wage rate of the household affected.[3]

Overestimates of future demand may be equally costly, if expansion plans are based upon them. For example, the electric generating plant expansion program for a government-owned South American utility system was originally based on a demand growth forecast of some 6.5% per year throughout the 1980s. This called for the addition of some 275 MW of hydroelectric capacity by 1986 and 1988, which in turn was used to justify a large, integrated water development scheme that included important irrigation components as well. However, the overall development, costing in excess of 1.5 billion dollars, could not be justified on the basis of the irrigation component alone, because its expected rate of return excluding the power component was at best about five percent per year. However, more recent, and more thorough studies of future electric power demand showed that the original estimates were far too high. As a consequence, the first stage of the hydroplant component of the transfer scheme will probably not be needed for at least another four years. Evaluated at an opportunity cost of capital of at least twelve percent, this results in an approximate net loss to the national economy of some 350 million dollars,[4]

[2]See Sri Lanka Case Study, chapter 9.
[3]Mohan Munasinghe and Mark Gellerson, "Economic Criteria for Optimizing Power System Reliability Levels," *The Bell Journal of Economics*, Vol. 10, Spring 1979 (pp. 353–365); and Mohan Munasinghe, "The Costs Incurred by Residential Electricity Consumers Due to Power Failures," *Journal of Consumer Research*, March 1980.
[4]Net of agricultural net benefits.

which, expressed in present value terms, is a sum equivalent to some 90% of the country's annual public investment budget. This is a staggering loss indeed.

These examples clearly show the importance of accurate demand forecasts. The need for forecast accuracy is the greater, the larger the time horizon is for new energy supply installations. Thermal power plants may need 4 to 6 years to completion,[5] atomic power plants 8 to 12 years, and hydro power plants some 5 to 8 years. Oil and gas fields from initial seismic surveys through exploratory drilling to first production may need 5 to 15 years, refineries 3 to 6 years, and forest plantations for firewood some 5–30 years.

We may conclude that the need for forecast accuracy is directly related to the size, cost, complexity, and irreversibility of projected supply components. In relatively small and simple economies, in which marginal additions to demand are either satisfied from readily available local resources or through incremental imports of widely available finished or semifinished energy products (gasoline, diesel, fuel-oil, coal, or diesel power plants) elaborate, long-term demand forecasts are not so important.[6] However, in countries where decisions have to be made with respect to large additions to existing energy systems that will require many years to completion, or where new energy sources are to be introduced (e.g., natural gas) accurate demand forecasts are of crucial importance, and substantial planning resources should be allocated for their preparation and continued updating.

7.3   FORECAST METHODOLOGIES

Various methodologies can be used to make energy demand forecasts. The most important ones are:

1. Trend analysis
2. Sector-specific econometric multiple-correlation forecasts
3. Macroeconomic or input-output based forecasting models
4. Surveys.

1.   Trend Analysis

Trend analysis is the most commonly used approach. It consists of the extrapolation of past growth trends assuming that there will be little change in the

---

[5]Although high-cost gas turbines or diesel power plants can usually be commissioned on an emergency basis within 1 to 2 years, with the result that many utility systems use them excessively because of initial forecasting errors.

[6]This does not negate the need for them for predictions of import requirements, balance of payments effects, etc. However, for these purposes general trend analysis based on economic aggregate data may suffice.

growth pattern determinants of demand such as incomes, prices, consumer tastes, etc. These trends are usually estimated by a least square fit of past consumption data or by some similar statistical methodology. Depending on the availability of data, they may be estimated either on a national basis for a given energy source (e.g., gasoline) or they may be broken down by region, by consuming sector (e.g., households, commercial enterprises, industry, transportation, etc.), or by both. Frequently, ad hoc adjustments are made to account for substantial changes in expected future demands due to specific reasons. For example, this may take the form of projecting on a case by case basis the expected demands of new industrial plants or other economic activities. This combination of overall trend projection together with survey-research type specific adjustments is used, for example, for power planning in Colombia, Peru, and Thailand,[7] to name but a few countries.

The main advantage of this approach is its simplicity. Forecasts can be based on whatever data are available. The major disadvantage is that no attempt is made to explain why certain consumption trends were established in the past. The underlying, and usually unstated, assumption that whatever factors brought about consumption changes in the past will continue unchanged in the future as well is, of course, a rather limiting one in a world in which relative energy prices are changing at rapid rates, and in a direction opposite to that observed until just a few years ago.

2. Econometric Multiple Correlation Forecasting

Econometric forecasting techniques are usually somewhat more sophisticated, and, in theory, hold out the promise of greater forecast accuracy. Past energy demand is first correlated with other variables such as prices and incomes, and then future energy demands are related to the predicted growth of these other variables. However, these methods are frequently nothing more than a special form of trend analysis if the projections of the selected determinants themselves are based on historical trends in turn. The other problem that they encounter is that of data availability. Usually it is difficult, if not impossible, to obtain the required time series that are needed to produce statistically acceptable (i.e., statistically significant) results. Not only are data series often incomplete, they are also subject to changes in definitions over time and, even more frequently, subject to substantial errors. Furthermore, the need for "proving" statistically "significant" results is such that long periods of time have to be covered, periods long enough to have experienced significant changes in the underlying

---

[7] In Thailand, one of the specifically enumerated sectors is rural power demand which is based on estimates of the progress of their electrification program.

structure of the economy. Hence, the confidence in the results must be low, even if they pass the test of statistical significance. However, the advantages of econometric studies are that they can take a number of important, demand-determining variables, such as price, income, number of vehicles, etc., explicitly into account.

The formulation of a typical residential energy demand model would be based on consumer theory in economics.[8] The direct utility function of a consumer which indicates the intrinsic value he derives from the consumption of various goods may be written:

$$U = U(Q_1, Q_2, \ldots, Q_n; \underline{Z})$$

where Q_i represents the level of consumption of good i in a given time period (e.g., one year) and $\underline{Z}$ is a set of parameters representing consumer tastes and other factors.

The set of prices $P_1, P_2, \ldots, P_n$, for these n consumer goods, and the consumer's income I, define the budget constraint:

$$I \geq \sum_{i=1}^{n} P_i Q_i$$

Maximization of the consumer's utility U subject to the budget constraint yields the set of Marshallian demand functions for each of the goods consumed by the household:

$$Q_i = Q_i(P_1, P_2, \ldots, P_n; I; \underline{Z}) \text{ for } i = 1 \text{ to } n. \tag{7.1}$$

Consider the demand function for a particular fuel (e.g., gas). Then equation (7.1) may be written in the simplified form:

$$Q_g = Q_g(P_g, P_e, P_o, P; I; \underline{Z})$$

where the subscript g denotes gas, while subscripts e and o indicate the substitute forms of energy, electricity and oil (e.g., for cooking), and P is an average price index representing all other goods.

Next, assuming that demand is homogenous of degree one in the money variables (i.e., prices and income), we may write:

$$Q_g = Q_g(P_e/P, P_o/P, P_o/P; I/P; \underline{Z}) \tag{7.2}$$

---

[8]See for example: W. Nicholson, *Microeconomic Theory*, Second Edition, Dryden Press, Hinsdale, Ill., 1978, Chap. 4.

Thus, starting from consumer preference theory, we may arrive at a demand function for a given fuel which depends on its own price, the prices of substitutes and income, all in real terms. The effects of other factors, $\underline{Z}$, such as quality of supply, shifts in tastes, and so on can also be explicitly considered. More specific examples of the effects of supply quality on energy demand are discussed in Chapters 8 and 13.[9]

The final specification of an equation such as (7.2) could vary widely.[10] Q_g could be household consumption or per capita consumption; the demand function could be linear or linear in the logarithms of the variables or in the transcendental logarithmic form and could include lagged variables; and $\underline{Z}$ could include supply side constraints such as access to supply, and so on.

Analogously, the industrial demand for energy may be derived from production function theory in economics. For example, consider the output of a particular firm of industry over a given time period:[11]

$$X = F(K, L, M, Q_1, \ldots, Q_n; \underline{S})$$

where K, L, and M represent the inputs of capital, labor and other nonenergy materials respectively; Q_i is the input of the ith form of energy, and $\underline{S}$ is a set of parameters that represents other factors such as shifts in technology, industrial policy, and so on.

The problem posed in production theory is the minimization of the costs of producing a given quantity of output X, given exogenous prices of inputs.[12] In principle, the solution yields, as in the household case, a set of energy demand functions:

$$Q_i = Q_i(P_K, P_L, P_M; P_1, \ldots, P_n; X; \underline{S})$$

---

[9]For an example from the telecommunications sector, in which quality of supply is a key explanatory variable, see Mohan Munasinghe and Vittorio Corbo, "The Demand for CATV Services in Canada," *Canadian J. of Economics,* Vol. 11, August 1978, pp. 506–520.

[10]See, for example, Lester D. Taylor, "The Demand for Energy: A Survey of Price and Income Elasticities," in: *International Studies of the Demand for Energy,* W. D. Nordhaus (ed.), North Holland, Amsterdam, 1977; and Robert S. Pindyck, *The Structure of World Energy Demand,* MIT Press, Cambridge, MA, 1979.

[11]To avoid the problem of different types of capital, labor, etc., it is necessary to assume weak separability of the inputs K, L, M and $E_i = (Q_1, Q_2 \ldots Q_n)$, so that each may be represented as an aggregate, with a distinct, composite price index, i.e., P_K, P_L, and P_M. Furthermore, in order to be able to determine the minimum cost mix of capital, labor, materials, and energy, it is assumed that the individual energy inputs, i.e., oil, gas, electricity, coal, etc., are subject to homotheticity. This makes it possible to use a two-step process. First the mix of fuels that make up the energy input are optimized, in the second step, the optimum growths of capital, labor, other material inputs and energy are chosen. This, for example, was the approach used by Pindyck, *The Structure of World Energy Demand,* Chap. 2.

[12]See, for example, R. W. Shepard, *Cost and Production Functions,* Princeton Univ. Press, Princeton, N.J., 1953.

As before, we may use one of the nonenergy input prices as numeraire, (e.g., P_K) and rewrite Q_i in normalized form:

$$Q_i = Q_i(P_L/P_K, P_M/P_K, P_1/P_K \ldots, P_n/P_K; X/P_K; \underline{S})$$

The demand for energy at time t is therefore a function of its own price, the prices of energy substitutes, the prices of nonenergy inputs, and other factors $\underline{S}$. Many different choices of variables and specifications of demand function may be used.[13]

In this way demand functions for various fuels could be developed for other end-use sectors such as transport, agriculture, and so on. The demand function may be estimated by standard econometric techniques. The estimated equations then could form the basis for future demand forecasts.

The main difficulties with this approach are:

1. The mechanistic nature of the econometric equations and their extrapolation into the future, which often fails to capture structural shifts in demand growth. Such structural shifts are particularly important in developing countries as a result of the introduction of new technology (e.g., tractors instead of draft animals, auto manufacturing instead of assembly operations, or rural electricity instead of kerosene lamps).
2. The difficulties of separating out short-run and long-run effects in the analysis of changes in the structure and level of prices.[14]
3. The lack of an adequate data base to make accurate regression estimates.[15] In developing countries this problem is particularly severe

---

[13]See, for example, Robert S. Pindyck, *The Structure of World Energy Demand.*
[14]It was this difficulty that led Pindyck to his massive study of comparative energy demand patterns, utilizing combined time-series-cross-sectional data for some nine industrialized, as well as for a number of developing countries. However, Pindyck was unable to determine what "long-run" really means; as he states: " ... our results tell us nothing about how long the long run is, and in particular what the adjustment speeds of the various elasticities are. ... The determination of these adjustment speeds requires the estimation of demand models that are explicitly dynamic. We have outlined some dynamic versions ... but we have not estimated them ... the estimation ... did not prove fruitful." Robert S. Pindyck, *The Structure of World Energy Demand,* p. 120 ff.
[15]A typical comment about data availability is the following: "With respect to energy use, there are reports on electricity consumption and generation for the country as a whole, by regions and, to some extent, by sector of the economy. There appears to be little attempt to break sectoral consumption down into subsectors corresponding to discrete industrial activities, i.e., steel, cement, fertilizers, etc. With respect to fossil fuel use (coal, coke, and petroleum-backed energy resources) statistics are almost nonexistent in publicly published form, ... in terms of its usefulness for energy policy analysis, the data available in published (and centralized) reports is clearly inadequate." *From: Joint Peru/United States Report On Peru/United States Cooperative Energy Assessment Report on the Industrial Demand Sector of Peru,* prepublication copy, Lima, Peru, June 1979, p. 6.

because energy price and consumption data by consuming sector are usually lacking, income data are unreliable or nonexistent and do not account for changes in income distribution, and statistical data of output and sales of specific industries are often unobtainable because of their confidentiality if the number of producers is small. Even in industrialized countries with far better data coverage, demand studies should be based on specific processes, rather than broad classifications such as "iron and steel production," or "household use."[16]

4. The inherent limitation of estimating procedures that concentrate on energy prices as demand determinants, but do not account for the prices, availability, life expectancies and replaceability of the energy-using appliances and equipment that must be utilized with alternative energy sources.[17] This problem is particularly acute under the assumption of homotheticity for all alternative energy sources.[18]

5. The problem that for almost the whole post-World War II period until the early 1970s real costs of commercial energy sources were falling, while they have been rising steeply, albeit irregularly, since. Econometric data series, unless they rely on cross-sectional data only (not a practical procedure in developing countries), necessarily utilize the data of these past periods of falling prices. These are unlikely to yield reasonable estimates of future energy demands because of the sharp reversal in relative price trends.

6. The fact that specific energy resources are often allocated by governmental fiat, or determined by such factors as availability or reliability of supply, rather than observable market price.[19] Another important factor, particularly in developed countries, may be quality.[20]

---

[16]This need for more detailed breakdowns by use classification has been realized in a number of studies. See, for example, Joel Darmstadter, Joy Dunkerley, and Jack Alterman, *How Industrial Societies Use Energy,* Johns Hopkins Press, Baltimore, 1977; Gunter Schramm, "The Effects of Low-Cost Hydropower On Industrial Location," *The Canadian Journal of Economics,* Vol. II, No. 2, May 1969, pp. 210–229; See also: The Conference Board, *Energy Consumption in Manufacturing,* Ballinger Publ., Cambridge, MA, 1974.

[17]For a useful example of the importance of these factors see the case study of the Thai tobacco industry in chapter 10.

[18]This is so because energy users will usually base their use decisions on total systems costs rather than energy costs only (e.g. total costs of electric versus total costs of diesel-powered irrigation pumps, or tractors versus draft animals).

[19]This is the case, for example, in Thailand where the choice of fuels for electric power production is determined by governmental policy rather than free market prices (see the case study on Thailand). However, it is also true in countries such as the USA, where, for example, electric utility companies were forced by governmental orders to convert from oil and gas to coal, and, more recently, from oil to natural gas.

[20]An example is the choice of high- versus low-sulphur coal in utility and industrial cases.

7. The problem that demand elasticities, even if they were estimated accurately, are likely to change significantly themselves, rather than remain constant, if price changes of energy are large. This is so because for most activities the ability to substitute other resources for energy (including alternative energy resources) is limited.[21]

Because of these problems, it is not surprising that the results of various empirical studies of energy demand functions vary by substantial margins even in the industrialized nations. For example, some of them found that in industry, capital and energy are substitutes for each other, while other studies concluded that they are complementary to each other.[22] Estimates of the own price elasticity of aggregate energy use range from -0.3 to -0.8. For specific energy sources the range is much larger. For example, for the U.S., long-run price elasticity estimates for electricity range from -0.5 to -1.2, and for oil they range from -0.22 to as much as -2.82.[23]

One cross-country study of energy demand in the industrialized market economies indicated that during the period 1950–70, the elasticities of total and commercial energy demand with respect to gross domestic product (GDP) were 1.78 and 1.55 respectively, while the corresponding price elasticities were -0.055 and -0.026.[24] Another recent paper gave total energy demand elasticities with respect to GDP in the range 1.3 to 1.4 for a group of developing countries during 1970–76.[25] Other estimates made for developing countries are likely to be less reliable.[26]

What we must conclude, then, is that most of these econometric studies of energy demand aggregates are probably of limited value in industrialized countries and practically useless for developing nations. More specific studies, of clearly defined subsectors such as urban residential energy demand by city size and income, for example, may be more promising and useful, even if the required disaggregation reduces the more mechanical aspects of statistical significance.

---

[21]In the production of aluminum metal, for example, the potential range between high and low power consuming potlines is about 30%. Once all substitution has taken place, the price elasticity becomes necessarily zero.

[22]All data from Robert S. Pindyck, *The Structure of World Energy Demand*, Chap. 5.

[23]For example, if we assume an increase in the real cost of oil by 4% p.a., a price elasticity of -0.22 would reduce demand to 92% after 10 years, but to as low as 30% given an elasticity of -2.82. The latter is obviously unrealistic.

[24]M. Beenstock and P. Willcocks, "Energy Consumption and Economic Activity in Industrialized Countries," *Energy Economics, vol. 3,* October 1981, pp. 225–32.

[25]Ben-Zion Zilbfarb and F. Gerald Adams, "The Energy-GDP Relationship in Developing Countries, *ibid,* pp. 244–48.

[26]See Robert S. Pindyck, *The Structure of World Energy Demand,* Chap. 7 for pooled estimates for Greece, Spain, and Turkey, and for Mexico and Brazil.

3. Macroeconomic and Input-Output Based Models

In the United States, Canada, and Europe a number of large, multisector energy planning models have been developed in recent years.[27] These models are either additions to already existing general purpose macroeconomic models, or they were specifically constructed to analyze energy supply and demand relationships.

Basically, an energy model connects a number of energy variables with various driving variables. A simple model, for example, would relate how total energy use and prices result from a driving force on the demand side (say, consumer income) and a constraint on the supply side (say, the cost of production of energy). More detailed models would distinguish between the various end uses of energy (heating, processing, transportation, etc.) and several energy sources such as petroleum fuels, natural gas, hydro, coal, wood, etc., each with its own production constraint. The relations must be estimated on the basis of empirical information. The major sources for these empirical estimates are technological information relating to past and future production and utilization possibilities, and econometric estimates of past demand, supply and other economic behavior relations (production by activity sector, income, number of vehicles, rate of electrification, etc.). Much of the technical information consists of physical quantities of inputs and outputs with input quantities sometimes aggregated into cost estimates.

If the models are to yield determinate answers, the number of independent relations and of binding constraints must at any time equal the number of energy variables projected.

Figure 7.1 provides a compressed view of the characteristics of a number of energy models that were developed for the United States. Three groups of "driving" variables are identified in these models: (a) policy variables, (b) realization variables, and (c) blends of the two. Variations of the policy variables chosen allow estimates to be made of policy impacts, and variations of realization variables permit sensitivity analyses that trace the differences in policy variables when alternative facts of nature pertain. Blend variables represent a mixture between policy and realization variables. Typically, these models consist of a large number of variables that are related to each other through sets of behavioral equations and/or constraints. Figure 7.2 presents the flow diagram of one of them that was developed by Data Resources, Inc.

It contains about 200 econometric equations estimated on annual data for 15 years. It projects final demands for and consumption of energy in four fuel sectors: coal, electricity, petroleum, and natural gas. A submodel of the utility

---

[27]For a detailed, comparative analysis of six of the models designed for the U.S. see National Academy of Science, *Energy Modeling for An Uncertain Future,* Washington, D.C., 1978.

sector, in turn, allocates the generation of electricity between hydroelectric, nuclear, and fossil-fuel generated electricity. Demands are forecast for 13 geographic regions of the United States and in each of four final consuming sectors: commercial, industrial, residential, and transportation. Projections of energy prices were taken from other sources, i.e., the Federal Energy Administrative model, judgments about Federal Power Commission regulatory rulings on natural gas prices, and projections of OPEC oil prices.

The major advantages of such comprehensive modeling efforts are that they yield internally consistent estimates of future demand and supply relationships, and they make it possible to test the effects of various assumptions about policy and realization variables.

However, the construction of such models requires not only a very substantial amount of intellectual resources, but also quite detailed and reliable sets of statistical data. The latter are usually unavailable in a consistent and com-

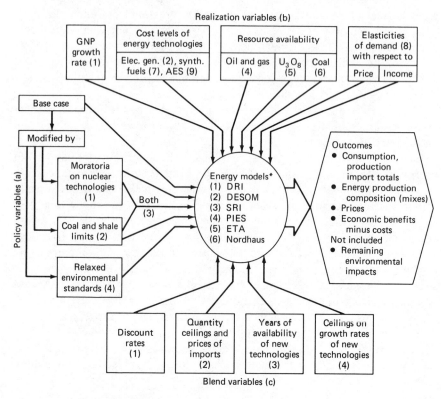

Figure 7.1.   Compressed view of the interrelationships of the driving variables, and outcomes of six U.S. energy models.

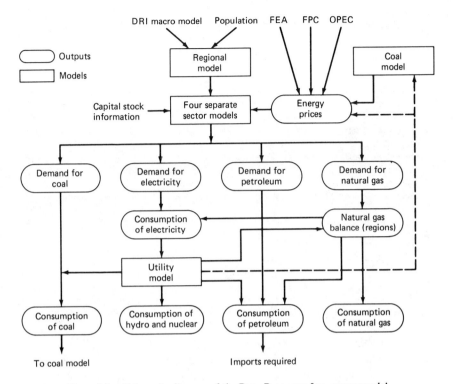

Figure 7.2.   Schematic diagram of the Data Resources Inc. energy model.

prehensive form in developing countries. Hence what has been said about the data problems of econometric demand forecasts in the previous section applies even more forcefully to the construction of comprehensive energy models.

Models, for all their intricacies and sophistication, have another drawback. By trying to model and optimize a whole economy with all its complex inter-relationships, specific details may still be insufficient for reasonably accurate forecasting purposes. This becomes apparent from a comparison of the results of the six U.S. energy models analyzed in the National Academy of Science study. With all of the models normalized in terms of basic assumptions, projected production and consumption rates for the year 1990 vary significantly among the various models. This can be seen from the data in Table 7.1 which show, for example, differences in projected total energy consumption of between 96.2 and 114.5 quads of energy[28] and for electricity generation of

_____

[28]Quad is equivalent to the energy content of 500,000 bbls of oil, or 10^{15} btu.

Table 7.1. Comparison of Selected Outputs of Primary Energy And of Electric Energy for 1990 From Six U.S. Energy Models.

	INTERMEDIATE-RUN MODELS		LONG-RUN MODLES			
	FEA	DRI	DESOM	ETA	NORDHAUS	SRI
Domestic energy production	93.4	—	93.6	81.5	81.1	94.7
Coal production	25.3	25.2	20.5		28.1	29.8
Natural gas	22.2	—	31.6	46.6	21.1	19.3
Oil and natural gas liquids	29.4	11.6	24.9		5.2	24.0
Nuclear energy	12.3	—	11.3	10.1	26.7	16.7
Solar heating and cooling	0.0	—	0.1	0.0	0.0	0.8
Other	4.2	4.7	4.9	4.6	0.0	4.2
Oil and gas imports	21.1	—	18.9	23.3	15.1	14.2
Total energy consumption	114.5	107.9	112.5	104.8	96.2	108.9
Electricity generation (10^{12} kWh)	4.3	4.1	3.1	3.3	2.8	4.1

Source: National Academy of Science, *Energy Planning for an Uncertain Future,* Washington, D.C., 1978, p. 47.

between 2.8 and 4.3 quads. These ranges are quite large and of little help for sector-specific planning purposes.

Over time, as data coverage improves and a better understanding is gained about the interrelationship of the various policy and realization variables and energy use, energy models may become useful as tools for energy planning purposes. For the time being, however, even if they can be constructed at all, their role as forecasting and planning tools will be quite limited.[29]

4. Surveys

Given the limitations of the other forecasting methodologies, surveys potentially provide a direct and reliable tool of demand analysis and forecasting. In essence, surveys consist of a list of more or less sophisticated questions that are put to energy users in order to measure and record their present consumption and future consumption plans. The basic types of questions that might be asked are the following:

1. How much energy (of each type) do you use per month/year?
2. What do you use it for?

---

[29]See, for example, the description of Mexico's "Energeticos" model, which is a 213-row linear optimization model. Energy demands for this model were simply based on historical trend projections. See G. Fernandez de la Garza and A. S. Manne, "Energeticos," in: Louis M. Goreux and Alan S. Manne, *Multi-Level Planning: Case Studies in Mexico,* North Holland/American Elsevier, New York, 1973 pp. 233–275.

3. How much do you pay per unit of energy used (by energy type)?
4. What do you produce or sell and what is the value and value added of each major product line?[30]
5. Can you identify specific energy uses with specific outputs?
6. What is your net income (for households)?
7. What are your future expansion plans, and their timing, and what additional energy requirements do they imply?
8. What additional energy-using appliances or equipment are you planning to acquire in the foreseeable future (identify)?

The major problems afflicting surveys are the following:

(a) They require substantial amounts of time.
(b) They can be costly.
(c) They require skilled interviewers.
(d) Energy users may be unable to provide the information asked because they themselves do not know.
(e) Energy users may be unwilling to provide the information for competitive reasons or because of fear of the consequences of revealing the information, etc.
(f) Energy users may wittingly or unknowingly give inaccurate answers.
(g) Future energy use plans may be vague, or too optimistic/pessimistic.

The major drawbacks are costs on the one hand, and ignorance, or unwillingness to provide the information, on the other. Because of costs, surveys must generally be limited to major energy consumers such as medium- to large-size industrial plants, mines and smelters, large transportation companies, utility companies, important governmental energy users (e.g., armed forces), etc. Fortunately, these enterprises and activities usually account for a large percentage of total energy consumption in developing countries. A recent 3 week survey in Peru, for example, covered a total of nine subsectors, which, in turn, were estimated to account for some 50% of the industrial sector demand in 1976 (projected to rise to about 70% by 2000) and close to 100% of the mining sector.[31]

In the industrial sector, energy consumption per unit of output is substantial for only a few activities. In the USA, for example, in 1967, six industrial groups used some 128 btu per dollar of output, while the rest of all other industries

---

[30]This, for multiproduct enterprises, is usually a difficult if not impossible question to answer because many specific uses result in multiple outputs. Typical examples are multimetal ore, processing and smelting operations, or multiproduct steel fabricating plants, to name but a few.
[31]The nine subsectors were mining, nonferrous metals production, iron and steel, cement, oil refining, petrochemicals, chemicals, fertilizers, and agriculture (major crops). The mining sector in Peru is particularly important, accounting for some 56% of Peru's total exports in 1976. *From: Joint Peru/United States Report on Peru/United States Cooperative Energy Assessment*, p. 3.

combined used only 21 btu. Overall, these six industrial groups accounted for some 80% of total U.S. industrial energy consumption,[32] but for only 36% of total value added. Similar, although usually somewhat less concentrated industrial consumption patterns exist in developing countries.[33] Industrial surveys of energy utilities, therefore, should first of all concentrate on those activities that are known to be high users of energy per unit of output, and second on those industrial plants and activities that are large relative to the country's industrial sector as a whole. Second, surveys should attempt to evaluate the energy-use implications of new economic development programs, such as irrigation, industrial settlements, mining and hydrocarbon developments, and rural or urban electrification programs, etc. However, care must be taken (and seasoned judgment employed) to assess the realism of these specific projected development programs. Not infrequently, these represent more the dreams of somewhat overoptimistic development planners or politicians than the cold reality of realizable objectives.[34]

The use of surveys will be less practicable for analyzing the consumption of ubiquitous sectors such as households, farms, and small-scale commercial and industrial enterprises. For these sectors survey costs are generally high, as well as time consuming. They also require the hiring and training of a large number of enumerators—not an easy task in most developing countries.

However, energy consumption surveys could be usefully combined with others, such as general population censuses or similar inquiries (such as income studies, etc.). On occasion, useful information may be available from already existing surveys, as for example, the FAO Timber Trends Study of Thailand.[35] A number of similar studies have been undertaken in other countries as well.[36] However, because of their costs such surveys will usually be undertaken only as one-time efforts. Because results usually do not become available until quite

---

[32]These six groups are food and kindred products (6%); paper and allied products (7%); chemicals and allied products (18%); petroleum and coal products (15%); stone, clay and glass products (7%); and primary metals (27%). *From:* The Conference Board, *Energy Consumption in Manufacturing*, Tables 1.1 and 1.3.

[33]See, for example, the industrial consumption data for Peru quoted in footnote 31 above.

[34]For example, a few years ago one of the authors reviewed the electric power expansion plans of a government-owned electric utility. Projected growth rates for the following 5 years exceeded 20% p.a. each year. These rather unusual projections were based on ambitious plans of the planning ministry involving projected new power users such as smelters, chemical plants, etc.— almost all of which turned out to be based more on wishful thinking than reality. However, when the power planners were confronted with the evidence, their reply was that they had to use the official forecast of the politically dominant planning ministry since they would lose their job if they did not. In actual fact, a review three years later revealed that growth rates had been less than 5% p.a.

[35]De Backer and K. Openshaw, *Timber Trends Study, Thailand: Detailed Description of Surveys and Results,* FAO, Report No. TA 3156, Rome, Italy, 1972.

[36]Arnold de Backer and Pringle, *Present Wood Consumption and Requirements in Kenya,* Report No. TA 1503, FAO, Rome, 1962.

some time after the actual data collection phase, the information may already be outdated. However, this is problematic only if substantial changes in some consumption-related factors have taken place in the meantime, such as drastic changes in prices,[37] or availability, or availabilities of competing energy resources.[38]

Another limitation is that such surveys usually cannot elicit any useful information from respondents about future energy consumption intentions. Demand projections, therefore, must be developed on the basis of estimated trends using more or less sophisticated methods that take account of the various identifiable energy demand determinants such as disposable income, household formation, etc.

In general, the need for surveys of these ubiquitous demand sectors and the required degree of accuracy of the results[39] will depend on the importance of the information thus acquired. If significant and potentially costly decisions must be based on it, requirements for accuracy will be high; but if the information is needed only for statistical record keeping, relatively crude estimates may suffice.

7.4   VERIFICATION

Whatever methods are used to estimate sectoral energy consumption, verification of the results should always form an integral part of the analysis. Such verification, at least in aggregate terms, is usually possible if reliable data exist either about total supply or total consumption or both. The latter will usually be the case for electricity and natural gas, because supply and consumption form closed systems with metering at both ends. The only troublesome aspects are autogeneration by nonpublic enterprises, leakage, transmission-distribution losses, and outright theft. The latter is a serious problem for electric utilities in

---

De Backer and K. Openshaw, *Present and Future Forest Policy Goals. A Timber Trends Study, 1970–2000,* A report prepared for the Government of Thailand Report No. TA 3156, FAO, Rome, 1972.

K. Openshaw, *Present Consumption and Future Requirements of Wood in Tanzania,* Technical Report 3, FAO, Rome, 1971. *A Timber Trends Study: Detailed Description of Survey and Results,* FO-SF/TAN 15, Project working document, FAO, Rome, 1971.

K. Openshaw, *The Gambia: A Wood Consumption Survey and Timber Trend Study, 1973–2000,* Report to the Overseas Development Administration, London, 1973.

R. Revelle, "Energy Use in Rural India," *Science,* Vol. 192, 4 June 1976.

K. F. Wiersum, *The Fuelwood Situation in the Upper Bengawan Solo River Basin,* INS 72/006, Project working paper, FAO, Rome, 1976.

[37]This was the case in the Thai tobacco curing industry (chapter 10). In the fall of 1978, the industry was rapidly converting to diesel fuel as primary energy source. By late 1979, the drastic price increases following the Iranian crisis forced the industry to return to fuel wood and lignite instead.

[38]Such as the introduction of electricity, replacing kerosene.

[39]Which will affect sample sizes, regional coverage, and efforts to verify the results.

many countries.[40] For petroleum products and coal, reasonably accurate supply data can usually be obtained from production, refinery, and import and export statistics. Data will usually be poor or nonexistent for both supply and consumption of indigenous energy resources such as wood, charcoal, dung, crop residues, etc.

Verification starts from the necessary equality between supply and consumption per time period. Total production, plus imports minus exports, minus net additions to storage, minus autoconsumption, processing, and transportation losses have to be equal to total consumption. This flow from the supply to the demand side has been shown in Figure 7.3 which depicts the basic supply-demand relationships of the petroleum products sector. In most countries reliable supply data are available for the supply side, except perhaps for net changes in storage, autoconsumption, and conversion and transportation losses. However, since the number of entities involved in these activities usually is small, these data could probably be obtained and assembled with relatively little effort. On the demand side, reliable consumption data will usually be available only for a few enumerated sectors or activities, although these may account for a substantial percentage of total consumption. After these have been accounted for, estimates must be made for the remaining sectors. These should be checked for reasonableness against sample surveys, statistics such as numbers of cars, buses, and trucks times average miles times miles per gallon, number of households times average consumption, etc. Consumption, plus losses, plus net additions to inventories, added over all consuming sectors has to be equal to total available supply.

Verification problems and estimation errors may arise from illegal diversions of specific energy sources to alternative uses. This, for example, is common in countries in which kerosene is heavily subsidized because of its importance as cooking and lighting fuel for the poor. As a consequence, kerosene is often illegally diverted to other uses such as additives to motor fuels or for use in privately-owned diesel generators. Furthermore, little, if any, information (except, perhaps, installed capacity) is usually available about privately owned generating plants or other stationary power sources such as diesel-driven irrigation pumps. Another problem is that in some regions consumption data are biased because of smuggling to neighboring countries, or because of substantial sales to temporary visitors.[41]

---

[40]In the late 1960s the rate of unexplained losses (i.e., theft) in the Port au Prince, Haiti utility system exceeded 50% of total generation, after allowance had been made for normal distribution losses. From G. Schramm, "Le Secteur de L'Energie Electrique," in: *Haiti, Mission D'Assistance Technique Integree,* App. I, OAS, Washington, D.C., 1972.
[41]This is the case all along the Mexican-U.S. border where U.S. visitors take advantage of the lower Mexican gasoline and diesel fuel prices.

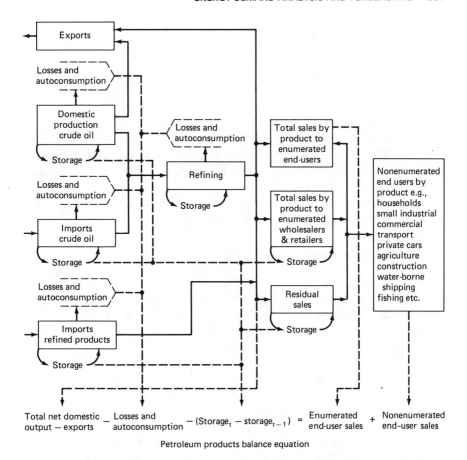

$$\underset{\text{output} - \text{exports}}{\text{Total net domestic}} - \underset{\text{autoconsumption}}{\text{Losses and}} - (\text{Storage}_t - \text{storage}_{t-1}) = \underset{\text{end-user sales}}{\text{Enumerated}} + \underset{\text{end-user sales}}{\text{Nonenumerated}}$$

Petroleum products balance equation

Figure 7.3. Supply/demand accounting framework for petroleum products.

Another means to check estimated consumption data for their reasonableness is to compare them with published consumption data for similar activities in other countries. Detailed data have become available in recent years in most industrialized nations, and institutions such as the OECD, the U.S. Department of Energy, and others regularly publish detailed consumption data for specific activities and production processes.[42]

---

[42]See, for example, The Conference Board, *Energy Consumption in Manufacturing;* Gordian Associates Inc., *The Steel Industry,* NATO/CCMS Study No. 47, 1974; Gordian Associates Inc., *Potential Energy Conservation of Energy,* U.S. Fed. Energy Adm., Washington, D.C., June 1974.

Verification is not limited to the analysis of past consumption data. Projections for the future can also be checked, at least for internal consistency with other projections. For example, projected consumption of gasoline (which may have been estimated on the basis of trend analysis of past consumption only) could be checked against the projected number of vehicles in circulation, their likely annual mileage and their expected consumption per mile or kilometer. If the technical characteristics of the vehicles are expected to change, this may imply significant changes in future gasoline consumption—a development that would not be apparent from either simple trend analysis or econometric projections that use as determinants only the estimated number of vehicles in circulation.[43] Similar internal consistency checks are possible taking into account new developments such as rural electrification, or irrigation developments based on power-driven pumps, or new industrial development projects.

7.5   ELECTRIC POWER

The electric power sector is one of the most important elements within the broader energy framework. Fortunately, demand analyses and load forecasts for this sector have a better foundation than those of other energy sources because of the availability of accurate consumption data from individually metered customers.[44] These provide a more detailed and reliable base for future projections than the more aggregate data of domestic disappearance rates that usually provide the main source of information for other energy sources. However, forecast errors, on the other hand, are usually more serious because of the long lead times for new equipment installations, their high initial capital costs, and the difficulty of replacing electricity by other energy sources, except for cooking, heating, and domestic lighting purposes; but even for these uses different appliances are needed. Forecasts that turn out to be too low may lead to costly, short-term solutions for alleviating capacity shortages, such as the installation of high-cost generating plants (diesels or gas turbines), or to rationing or forced supply interruptions. In some areas lack of sufficient capacity has effectively shifted industrial expansion and economic growth. Overestimates, on the other hand, will lead to the tying-up of scarce investment capital and other resources. Figure 7.4 indicates schematically the various types of losses

---

[43]For the U.S., for example, a recent study has estimated that the total fuel consumption of private passenger vehicles will have fallen to between 52 to 69% of 1976 consumption by the end of the century inspite of an estimated increase in total mileage driven by some 38%. *From:* Sam Schurr, et al., *Energy in America's Future,* Johns Hopkins Press, Baltimore, 1979, Table 5.5.

[44]However, these data may not always be arranged in a useful form for analytical purposes. This is apparent from the observations of the U.S.-Peruvian Energy Study quoted above. Often utilities are unable to identify the specific customers that are connected to specific feeder lines. Hence, factors such as daily demand patterns for specific user groups cannot be determined.

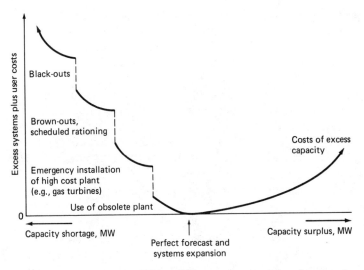

Figure 7.4.   Effects of forecast errors on system & user costs.

that can result from forecast errors. Load forecasting, therefore, plays a critical part in the electric systems planning process.

The main elements of load forecasting which are relevant to the system planning process are reviewed in this section, and further details regarding these aspects may be found in the cited references.[45] Unlike in the traditional engineering context where demand is synonymous with capacity or power, we use the terms demand and load forecast interchangeably to indicate the magnitude as well as the structure of the requirements of both electrical power and energy (i.e., kW and kWh respectively), except in instances where it is necessary to make the appropriate distinction. The structure of demand includes disaggregation by geographic area, by consumer category, and over time, as well as characteristics such as load factor, diversity factor and system losses.

A knowledge of loads at the disaggregate level is required because the demand characteristics vary by type of consumer, by geographic area, and over time, and therefore the properties of the aggregate demand (e.g., at the system level) may be quite different from the characteristics of the individual loads. Since demands by individual consumer groups and/or by region may grow at different rates, aggregate demands will also change. Disaggregate loads are very important in the system planning process. Whereas for generation plan-

---

[45]See, for example, *The Methodology of Load Forecasting,* 1970 National Power Survey, Federal Power Commission, Washington, D.C., 1970; and R. L. Sullivan, *Power System Planning,* McGraw-Hill International Book Co., New York, 1977, Chap. 2.

ning purposes the system may be modeled as a point source feeding a single lumped load, in the design of transmission networks, the characteristics of demand by region and by major load center (e.g., a city) become important. Ultimately, for distribution grid planning, a detailed knowledge of demand at each load center is required. Furthermore, the sum of many disaggregate demand forecasts, estimated by using a variety of different techniques, can often serve as a useful check on an independently made global forecast for the same region.

Before we discuss the other characteristics of demand, it would be helpful to examine the basic relationship between power and energy. The commonly used unit of energy in heavy current electricity applications is the kilowatt-hour (kWh). The rate of energy per unit time is called power (or sometimes, capacity), which is measured in kilowatts (kW). Therefore, it follows that energy in kWh is the product of power in kW and time in hours. The distinction between power and energy is important because the same amount of energy may be delivered within a small interval of time at a relatively high rate of power flow, or over a longer period at a lower power level. This point is best exemplified in nuclear fission, by the contrast between the practically instantaneous explosion of an atomic bomb, and the controlled release of power from a nuclear reactor.

The load factor (LF), which is the ratio of average to maximum (or peak) kW over a given interval of time, may be calculated at various levels of aggregation, e.g., for a single customer or the whole system, on a daily or annual basis, and so on. It is important, because the size or capacity (and therefore cost) of power system components are determined to a great extent by their capability to handle peak power flows. Since practically all customers require their maximum kW only during a short peak period during the day, the LF is also a measure of the intensity of capacity use. A load forecast may be made either in terms of the peak power, or the total energy consumed, during a given period (e.g., one year), with the conversion from one unit to the other being made by the use of the LF. In general, the kW peaks of different customers will not occur simultaneously. The diversity factor for a group of consumers is a measure of the divergence of spreading over time of the individual peak loads, and permits the computation of the combined peak load for the group in terms of the disaggregate peak values.

The total amount of both kW and kWh generated at the source will be greater than the corresponding values consumed because of losses in the system. Generation losses are mainly of the station use type, i.e., for driving various auxiliary equipment at the power plant itself. Transmission and distribution network losses are basically resistive losses, including transformer losses. Another type of loss which is important in some countries is theft, and allowances must be made for it in the demand forecast. Since it is also a type of electricity consumption, ideally, the corresponding user benefits and outage costs should also be considered in the optimizing process; in practice this would be extremely difficult to do, because little is likely to be known about the consumers and ethical considerations may make it difficult for the planning authority to give special considerations to such illegally appropriated benefits.

Load forecasts are made for time periods of varying duration.[46] Very short-term demand projections are made on a daily or weekly basis for purposes such as optimizing system operation, and scheduling of hydro units, while short-run forecasts ranging between one and three years are used in hydro reservoir management, distribution system planning and so on. The time horizon for medium-term demand projections is about four to eight years, which corresponds to the lead times required for major transmission and generation projects. Such forecasts, therefore, are especially useful to determine the next steps in building these facilities. Long-range demand projections, usually of a duration of 10 to 35 years, are extremely important in long-run system expansion planning. The type most relevant to the topic addressed in this study is the long-range forecasts. All the methods discussed earlier, trend analysis, econometric-multiple correlation techniques, macroforecasts, and surveys are used for this purpose.

All the methodologies discussed so far have dealt with deterministic load projections, where uncertainty is usually accounted for by parametric shifts in the forecasts to test for sensitivity. Recently, more sophisticated techniques involving the use of matrix methods (e.g., input-output matrices, to capture the interrelationships between productive sectors), as well as stochastic processes (e.g., representing sectoral growth of probabilities in terms of a Markov model), have permitted explicit consideration of probabilistic effects.[47] However, the resulting long-range demand forecasts are still deterministic. Once again, for sensitivity testing purposes, the basic assumptions may be varied to provide low-, medium- and high-demand forecasts, covering a range of relevant values of load growth.

Special care must be taken in cases in which substantial tariff changes are expected. While, for reasons discussed elsewhere in this report, existing esti-

[46]See F. D. Galiana, "Short Term Load Forecasting," *Systems Engineering for Power: Status and Prospects,* NTIS no. CONF-750867, USERDA Conference, Henniker, New Hampshire, August 1975, pp. 105–114; and K. N. Stanton, "Medium Range, Weekly and Seasonal Peak Demand Forecasting by Probability Methods," *IEEE Transactions on Power Apparatus and Systems,* Vol. PAS 71, May 1971, pp. 1183–1189.

[47]See, for example, *Nuclear Power Planning Study for Indonesia,* IAEA, (Vienna, 1976); N. R. Prasad, J. M. Perkins, and G. Nesgos, "A Markov Process Applied to Forecasting. Part I—Economic Development," *IEEE Power Engineering Society Summer Meeting,* Vancouver, July 1973; and N. R. Prasad, J. M. Perkins, and G. Nesgos, "A Markov Process Applied to Forecasting. Part II—The Demand for Electricity," *IEEE Power Engineering Society Winter Meeting,* New York, January 1974.

mates of price-demand elasticities for both the short- and long-run are suspect, price changes, particularly if they are substantial in real terms, are likely to have a significant effect on consumption. How significant will depend on many local factors such as the ability of users to pass on cost increases to customers, the price and availability of substitute energy sources (and their total costs including needed changes in equipment), the effect of these tariff changes on real income, etc. No generalization is possible. Table 7.2 lists some of the major factors that are likely to determine the demand for electricity. Estimates of future changes in these determinants are needed to project demand.

We conclude this section with the following summary. There is no single, universally superior approach to load forecasting. The more sophisticated techniques tend to obscure the basic methodological assumptions as well as weaknesses in the data, and therefore the simpler techniques may be more appropriate in situations where the available information is known to be unreliable. Whenever possible, survey approaches should be used to forecast the demand of large users. This may account for a substantial percentage of the total load. Frequent updating of this information, to account for unexpected changes is an additional important requirement. The determinants of demand for the remaining, ubiquitous sectors have to be carefully selected, because they vary from case to case, and the level of disaggregation chosen will depend on the data, as well as on the ultimate purpose of the projection. Demand forecasting is a dynamic process in the sense that the procedure must be repeated often, in light of improved information and techniques of analysis. The building up of a reliable data bank is an invaluable tool in this process of constant revision. Finally, it should be noted that in some cases institutionally imposed constraints which suppress demand, such as peak lopping or nonconnection of customers, and the presence of captive generation, may lead to a substantial gap between the potential (or underlying) and the achievable (or realistic) demand.

7.6 NATURAL GAS

The analysis and forecasting of natural gas demand raise some special issues. While the measurement of existing or past consumption is relatively simple because supply networks are closed systems with metering at both ends, forecasting future demand is not because of the relative ease of substitution by other energy sources. Most larger industrial utility and institutional gas users are equipped to switch to alternative fuels, such as fuel oil or coal, if this proves to be more advantageous.[48] This applies to the major uses of gas as a source of

---

[48]Canadian gas utilities found that out to their chagrin in recent years. When gas prices were revised by government order to 85% of the also regulated price of oil on a btu-content basis demand growth fell drastically. Worse still, a developing imbalance in fuel oil supplies in Ontario

process heat or boiler fuel, although not to its use as chemical feedstock. However, in most countries use of gas for these purposes is minor. It also does not apply to residential or small-scale commercial users because these usually do not own multiple-fuel appliances.

A second problem for the demand analysis of gas is that its supply requires heavy front-end investments, and that these investments are subject to substantial economies of scale. Once installed, such supply networks have a rather long life expectancy. Hence, secure, long-term markets are needed to write off these investments over periods that approximate their useful life in order to reduce unit capital costs, and large throughputs for customers are required for the same reason. Competing fuels, such as fuel oil or coal do not face the same difficulties because they can be supplied in discrete quantity if needed, providing more flexibility and reducing fixed costs.

Residential markets, which represent a major component of gas consumption in northern latitudes, are almost nonexistent in the developing world. This is so because most of the latter are located in subtropical or tropical zones, eliminating the need for residential heating, which is the major domestic source of consumption in the former. Low income levels in developing countries prevent the use of air conditioning units in all but a few households, and gas demands for other appliances, except cooking, are low as well. As a consequence, the fixed costs of house connections are usually prohibitive because potential consumption levels per connection are too low.[49]

Another problem that increases fixed costs results from the fluctuating nature of demand. In countries with substantial space conditioning needs for heating and/or cooling,[50] peaks are seasonal, although gas field production preferably should be kept constant the year around. As a consequence storage and larger distribution pipeline diameters are needed. All of these factors militate against the supply of natural gas to small-scale users.

Presently prevailing thinking about the real value of gas reinforces this problem. Since the beginning of the so-called energy crisis in 1973/74 general

----

led to depressed prices for the latter with the result that most larger industrial gas users switched to oil. As a consequence, gas prices in Ontario had to be lowered to 65% of the oil equivalent price. Presently ongoing negotiations about the extension of gas supplies in Quebec assume that subsidies will lower gas prices to the same level in Quebec as in Ontario. *From: Financial Post,* Toronto, Dec. 1, 1979, p. 30.

[49]There are a few exceptions here and there. In Baranquilla, Colombia, for example, residences in a new government-financed housing project were uniformly equipped with gas connections from the nearby offshore gas fields. Equipping existing residential areas is much more costly because of higher retrofit construction costs and fewer than 100% hook-up rates. See also the discussion on residential gas use in Bangladesh in Chapter 11.

[50]Use of natural gas for air conditioning only is uncommon because of relatively high equipment costs. It is used only when whole structures (houses, stores, office buildings, etc.) are to be fully air-conditioned.

Table 7.2. Major Demand Determinants of Electricity.

A. Residential Sector
1. Number of households (or population)
2. Real family income weighted by income distribution
3. Price of electricity
4. Connecting charges
5. Availability of service (urban, rural)
6. Reliability of service
7. Cost and availability of electricity-using fixtures and appliances
8. Availability and cost of consumer credit
9. Costs of alternative energy sources (kerosene, LPG, natural gas, charcoal, firewood, coal)
10. Degree days below or above reasonable comfort range[1]

B. Commercial Sector
1. Sales or value-added of nonsubsistence commercial sector
2. Price of electricity
3. Connecting charges
4. Costs of electricity-using appliances relative to those using alternative fuels for space conditioning
5. Costs of alternative fuels for space conditioning including costs of appliances
6. Reliability of service
7. Working hours of various types of commercial establishments

C. Industrial Sector
1. Price of electricity (demand and energy charges by type of service)
2. Type of industry and its electric intensity[2]
3. Degree of market power of individual industry
4. Relative price of alternative energy sources[3]
5. Reliability of supply
6. Comparative total costs of autogeneration
7. Availability

D. Agricultural Sector
1. Price of electricity
2. Availability
3. Reliability of service
4. Supply costs of alternative energy systems, e.g., diesel-driven irrigation pumps and milling and processing equipment, kerosene for lighting, etc.
5. Comparative total costs of autogeneration

E. Public Sector
1. Price of electricity
2. Per capita revenue of municipal governments
3. Number and size of public schools
4. Expenditures on armed forces and police (i.e., size of forces)
5. Types of public water supply (e.g., deep-well pumps)

---

[1]These types of load (heating and cooking) are limited to middle and high income earners. In regions of more severe winters (e.g., Korea, Turkey) electric heating loads may be lower than in others with more benign, but occasionally cold weather (e.g., altiplanos of North-Central-South America) because of cost and more permanent heating facilities based on alternative fuels.

[2]Electric energy intensity can be checked against consumption data of similar production processes in other countries.

[3]Only for industrial activities for which alternative fuels or energy sources represent technologically feasible sources of supply; the implied substitution would likely require complete equipment substitution for already existing plants.

238

agreement has been reached among energy pricing analysts as well as governments that cost-of-supply-based gas pricing schedules are no longer justified. Instead it is argued that gas prices should be raised to the going prices for petroleum fuels on a btu-equivalent basis. While this appears to be eminently reasonable from the point of view of opportunity-cost accounting, it also eliminates any incentive to smaller-scale users to switch to gas from alternatives such as fuel oils, LPG, or others. This is so because the initial connecting costs of natural gas are much higher than those for other fuels so that total costs of supply are necessarily higher, if btu-equivalent prices are charged for the gas. In countries in which natural gas supplies are relatively ample and petroleum products must be imported (such as Pakistan, Thailand, and Colombia, for example) it could be argued that gas prices should be set at prices such that total supply costs (including fixed costs) are somewhat lower, or at least not higher, than the total supply costs of other alternative energy sources.[51]

Particular problems in demand forecasting are encountered if natural gas is to be introduced into new regions and markets. On the one hand, a large and costly supply network has to be constructed, but on the other, no preexisting market exists. In earlier times, when natural gas was not considered as a valuable, exhaustible fuel with high alternative use values now or in the future, gas was priced at low promotional rates that often only reflected its marginal supply costs, excluding the costs of exploration which usually were written off against the costs of higher-valued crude oil exploration.[52] As a result, gas prices, on a btu-equivalent basis, were frequently so much lower than prices of competing fuels that development of markets through displacement of alternative fuels was highly successful. Now, with prices set close to or equal to fuel oil prices market development can be a major problem unless captive markets can be tapped. Such captive markets may be developed by government fiat, if, for example, electric utilities are ordered to switch from alternative fuels to natural gas. This, for example, is the case in Thailand, where the government-owned natural gas pipeline will initially sell almost all of its available throughput to the government-owned utility company.

Where such government-enforced consumption of natural gas is impractical,

---

[51]Such a rule would have to be subject to some lower limit to avoid the installation of high-cost, uneconomic distribution systems. One such rule could be to set prices equal to the present value equivalent of the future value of gas in alternative uses plus long-run marginal costs of supply. (See Chapter 11).

[52]This was the prevailing policy in the United States where the Federal Power Commission, as the protector of the public weal, for many decades followed a utility-type cost-plus pricing rule in natural gas rate making cases. This policy has been a major contributing factor to the lack of gas exploration and decline in U.S. reserves in recent years. For a detailed discussion of U.S. regulatory policy, see Paul W. MacAvoy (ed.), *Federal Energy Administration Regulation,* American Enterprise Institute for Public Policy Research, Washington, D.C., 1977.

long-term contracts with large-scale users are usually required to assure the financial viability of a new gas supply network. In order to arrange such contracts sufficient demand must exist or must be created. This is a serious problem in gas-rich regions and countries such as Algeria, most of the Persian Gulf oil producing states, Trinidad and Tobago, Bolivia, Indonesia, the northern Australian shelf, and the Northslope of Alaska, as well as Mexico. While locally natural gas is plentiful, local markets are too small or nonexistent to absorb potential output.

If deposits consist of gas only, this creates no overwhelming problem since the gas can be left in the ground until markets are developed. However, if the gas is associated gas that must be produced together with oil and cannot be reinjected, then often flaring becomes the only alternative. To find markets elsewhere, either expensive, long distance pipelines have to be built,[53] liquefaction plants installed to supply overseas markets,[54] or chemical feedstock plants developed which in turn have to sell much of their output in world markets, because domestic markets usually are too limited to absorb the output capacity of large-scale, economically efficient plants.[55] However, given the very substantial supplies of natural gas existing in many inconvenient, remote places throughout the world the temptation to convert them into chemical feedstock is large, and a potential, large oversupply of such chemicals a constant threat. Future market demands for their output at stable prices therefore are quite uncertain, at least in the intermediate and long run.

Demand forecasts for natural gas can be conveniently divided into four groups: The first is domestic captive demand that is secured through long-term contracts at predetermined prices. If such contracts are backed by the coercive power of government little risks or uncertainties remain, except for potential variations in consumption (below some agreed-upon maximum ceiling) resulting from changes in demand for the gas customers' outputs.

The second consists of projected future, domestic demand by industrial, commercial and residential sectors that are free to make choices between natural gas and alternative fuels within a market in which prices of alternative fuels are subject to free market forces. Under these circumstances the demand

---

[53]See the case study on Alaska's gas pipeline in Chapter 12.

[54]Because of the high cost of liquid gas processing and transportation systems, delivered gas prices are inherently high, with the result that the wellhead price of the gas must be kept low to remain competitive with alternative supplies or alternative fuels in offshore consumer markets.

[55]The new "world-scale" natural-gas-based plants of Alberta Gas Ethylen Co. Ltd. in Alberta, Canada, for example, have a capacity of 1.2 billion pounds per year, far in excess of domestic Canadian needs. As its owners state: "Output here will remain in excess of domestic needs; clearly we need export markets." From "Feedstock for Plastics to Pour Forth in Alberta," *Christian Science Monitor*, Nov. 29, 1979, p. 11. See also "Canadian Producers Could Grab Half U.S. Petrochemical Market," *The Financial Post*, Oct. 27, 1979, p. 14.

for gas may vary significantly as a result of price changes of alternative fuels, and these changes can be quite abrupt for industrial users that are equipped to switch readily to alternative fuels. The effects of such relative price changes on domestic, small-scale industrial and commercial users will be less because of the need for the installation of alternative appliances. However, longer-term trends in new equipment-customer markets may be strongly influenced by such relative price changes.[56]

The third consists of offshore or long-distance markets that must be reached through long pipelines or liquefaction installations. Because of the high capital costs of such systems, demand is usually guaranteed through long-term sales contracts that assure to the financing institutions repayment of the costs of development and a reasonable rate of return. In these cases, demand uncertainties are minimized once the supply contracts have been signed. However, the question whether such markets can be developed is usually highly uncertain, as the unsuccessful attempts to bring more Algerian gas into the United States, the unsettled future of the proposed Alaska gas pipeline, and the long drawn-out negotiations about Mexican gas sales to the U.S. clearly show.

The fourth, finally, consists of projected markets for gas-derived chemicals and fertilizer products. These are probably the most uncertain markets of all, given the large, potential competition from similar producers around the world and the relatively low freight costs for the chemical feedstocks and finished outputs produced. Before such ventures are started, detailed market studies are needed, and long-term sales contracts with reliable customers may be a necessity. However, such long-term sales contracts almost always are subject to some adjustment clauses that relate to prevailing world market prices—with the result that quantities sold may be assured, but not the prices.

Table 7.3 lists some of the major demand determinants for natural gas.

7.7   COAL

Coal, until the early twentieth century, was the major industrial and domestic fuel in the industrialized world. However, since then it has been rapidly displaced by the far cleaner, easier-to-handle petroleum fuels and natural gas. In recent years, the use of coal was largely limited to the production of steam for thermal powerplants or industrial boilers, or, in the form of coking coal, for the

---

[56]In Eastern Canada, for example, the rapid increase in the prices of petroleum-based fuels and natural gas on the one hand, and the relatively stable price for hydroelectricity have led to a rapid expansion of electricity use for domestic heating purposes in recent years. In 1978, for example, some 27% of all housing units in Montreal, Quebec were electrically heated, and more than half of all new houses in the five eastern Canadian provinces are equipped with electric heating systems. From: "More Homes Switch to Electric Heat as Oil Rises," *Financial Post*, Toronto, Nov. 10, 1979, p. 1.

Table 7.3. Major Demand Determinants of Natural Gas.

A. Residential and Commercial Sector
1. Price of delivered gas
2. Price of alternative fuels
3. Connecting charges for natural gas (together with (1))
4. Availability
5. Real family income weighted by income distribution[1]
6. Degree-days below reasonable comfort levels
7. Degree-days (and humidity) above reasonable comfort level
8. System costs of alternative (electric) air conditioning units
9. Cost of gas-using appliances relative to those utilizing other energy sources
B. Industrial and Electric Utility Sector
(I) Process and Space Heating Requirements
1. Price of delivered gas
2. Price of alternative fuels
3. Availability
4. Long-term price stability and supply reliability
5. Laws and regulations affecting gas utilization
6. Demands for industrial and electric utility outputs
(II) As Chemical Feedstocks and Fertilizers
1. Domestic demands for outputs of conversion plants
2. Relative costs of competing domestic or imported products
3. World market prices for excess outputs that must be exported
4. Price of gas to processing plant
5. Costs of industrial conversion
C. Export Sector
1. Potential delivered costs of gas in importer's country relative to alternative sources of supply (gas or competing fuels)
2. (Political) security of supply to importing country

[1]Equally important will be the local distribution of high and medium income residential areas as well as their rate of development. New natural gas connections may be quite feasible economically if they are systematically built into new middle to upper class housing developments, so that high density of connections can be achieved.

smelting of iron ores. A few developing countries maintain some railroad systems with coal-fired steam-engines as well (India and the Peoples Republic of China, for example).

The price explosion of hydrocarbon fuels in recent years has made coal once more an attractive alternative source of energy. At $25 per barrel of fuel oil, a ton of lowly lignite could cost as much as $60 on a btu-equivalent basis, and a ton of high-quality steam coal as much as $115. Needless to say that such a comparison is misleading because of the much higher handling and equipment costs involved in the utilization of coal. Nevertheless, with mining costs of coal running anywhere between $5 to $40 per ton in major coal producing areas the cost differential to oil is such that a strong resurgence of coal utilization throughout the world is a virtual certainty.

Table 7.4. Approximate Calorific Values of Various
Grades of Coal.

	GJ/ton[1]	btu/lb[1]
Anthracite	33	14,300
Bituminous coal (average)	29	12,500
Subbituminous coal	25	10,600
Brown coal and lignite	15	6,300
Peat	8	3,400
Coke	28	12,000

Source: David Crabbe and Richard McBride (eds.), *The World Energy Book,* MIT Press, Cambridge, MA, 1979.
[1] 1 GJ (Gigajoule) is approximately equal to 0.95 million btu; 1 ton = 1,000 kg = 2,208 lbs.

To analyze and forecast the demand for coal it is useful to divide it into several subgroups. As can be seen from Table 7.4, carboniferous, or coal, resources occur in a wide variety of physical characteristics and quantities. The table presents a rather simplified division in terms of calorific values only. From a consumption point of view, a more useful division might be one that divides existing deposits into coking coal, steam coal and lignite. The line of division between these three major groupings is quite fuzzy, with many deposits, for example, more or less suitable for use as coking coal.[57]

Coking coal commands premium prices substantially in excess of prices for steam coal. F.O.B. export prices from Western Canada and Australia, for example, range between $35 to $60 per ton for shipment to Japan. As another example, Poland finds it economically worthwhile to export its coking coal from the Silesian mines to Cuba, Argentina, and Japan, while Australian coal is sold both in Japan and Great Britain.

Exports of steam coal are all but nonexistent at the present time. However, Australia, Western Canada, the U.S., Poland, and Colombia, to name only a few major producers or potential producers, are all poised to start steam coal exports in a major way.[58] Given present prices of hydrocarbon fuels, their large-scale replacement by coal as the principal boiler fuel in the future appears to be only a question of time.[59]

---

[57]Characteristics useful for coking are low volatile matter, low mineral (i.e., ash) content, sulphur content of less than 1%, and high structural strength (to resist caking under the high pressure conditions of blast furnaces). Ongoing research has widened the types of coal that can be utilized, as well as reduced the amount of coke needed per ton of iron produced.

[58]For example, Korea expects to import 4 million tons of steam coal this year, 9 million in 1981, and 27.1 million in 1986. Japan's imports will rise from 1.5 million tons in 1979 to 15 million in 1985. *From: Wallstreet Journal,* Dec. 19, 1979, p. 14.

[59]Subject, however, to environmental constraints, to transportation costs, as well as to the problem of scale-economics which militate against small-sized coal-fired boiler plants.

Table 7.5. Representative U.S. Transport Costs for Coal.

TYPE	DISTANCE ASSUMED (MILES)	COST PER TON-MILE[1] (¢/TON-MILE)
Unit Train	300	1.46
Conventional Train	300	2.70
River Barge	300	0.62
Slurry Pipeline	273	1.25
Ship	n.a.	1.04

Source: Great Lakes Basin Commission, *Energy Facility Siting in the Great Lakes Coastal Zone: Analysis and Policy Options,* Ann Arbor, MI, Jan. 14, 1977, Table 33.

A major problem confronting the wider utilization of coal is transportation costs. Table 7.5 shows rates typical for the United States for various shipping modes. What they indicate, for example, is that transportation in ordinary railroad hopper cars will double the price of coal sold at $20 per ton at the mine pits at a shipping distance of some 750 miles. Because of these high costs it is usually prohibitive to ship low-btu coal, such as lignite, over any great distance. Mine-mouth utilization, usually in electric powerplants, is the only economically feasible form of use.

Demand forecasts for coking coal are usually relatively simple. There are few, but rather large customers, and the negotiating of long-term sales contracts with them usually precedes mine development.[60] Markets, therefore, are generally assured before development.

The major sources of demand for steam coal are electric powerplants, industrial plants that need heat, and, in some countries or regions, small-scale commercial, industrial, and household sectors.[61] While some of these uses may already be well established so that projections of future use can be made from an existing base, potential future markets may be substantially larger and broader given the rapid increase in costs of previously preferred alternative hydrocarbon fuels.

The potential, future demand will depend on the projected industrial and utility needs for processing heat, and high pressure steam, the cost of delivered

---

[60]Coking coal can also be used as boiler fuel or for domestic purposes. However, if there exists a market for the coal as coking coal, prices are usually too high to use it for thermal purposes only. In the central highlands of Colombia, for example, many small-scale, family-type subsistence mines produce coking coal in very small quantities that can only be used for domestic and nearby industrial purposes as steam coal because of a lack of markets and the high cost of transportation.
[61]Coal can also be used as a fuelstock for chemical plants, however, this use is relatively rare today because of the lower costs of natural gas or fuel-oil-based plants.

coal relative to fuel oil, natural gas, and LPG, the rate at which new plants capable of using coal will be built or existing ones can be converted, and the additional costs in equipment, operation, and maintenance of using coal instead of alternative fuels.

The major users in most countries are likely to be electric utilities. The demand of these utilities will depend on the underlying demand for electricity and the total delivered cost of electricity from coal-fired powerplants relative to the cost of electricity from other potential sources such as hydro. However, it may also depend on governmental policy which may enforce the use of coal in new power generating plants.[62] On occasion, time pressure to bring new capacity on line may lead to the construction of coal-fired plants, even though other alternatives that have longer construction time horizons promise to have lower generating costs.[63] In some countries, the potential demand for steam coal may be met from imports, if domestic energy resources at reasonable costs are not available. Such potential imports may come from coal-rich, low-cost producing areas such as South Africa, Australia,[64] Western Canada, Alaska, the Guajira Peninsula of Colombia, and other regions with large, low-cost coal deposits close to tidewater locations.

Domestic and small-scale commercial/industrial demands for coal may already exist or develop in regions in which alternative fuel sources are scarce and expensive. Coal may become a substitute for more traditional fuels such as charcoal and firewood, if the latter's supplies become scarce.

Export demands for coal will generally be developed under long-term sales contracts. Scale economies, the need for special handling and dock loading equipment, unit trains or slurry pipelines, etc. usually require that quantities sold in export markets must be large, usually amounting to several million tons per year. Hence, future demand generally depends on the successful conclusion of such (long-term) sales contracts.

The market for lignite is more restricted than that for steam coal because of the much lower heat value of lignite coals. Lignite is mostly used for the production of electricity in thermal powerplants specially equipped to handle the particular type of lignite available. Because of the low btu/ton ratio, lignite-fired powerplants usually are built close to the mine site. The economics of such plants is affected by the higher equipment, operating and material handling costs of lignite versus other fuels, and the distance of such mine-mouth powerplants to load centers (i.e., transmission costs). As in the case of steam coal

---

[62]This is the case in the U.S.A. for example.
[63]This is the case in Colombia at the present time where two small coal-fired plants are being installed in Cartagena and Baranquilla to fill short-term supply shortages caused by delays in the completion of a large hydropower scheme.
[64]Both of which already hold contracts to supply steam coal to Japan.

plants the demand for lignite is determined by the underlying demand for electricity and the costs of obtaining the latter from alternative sources.

A secondary market for lignite may develop for industrial, commercial, and domestic purposes if other fuels are scarce and expensive. Because of the generally low quality of raw lignite, such markets are often served through the intermediates of lignite briquetting plants.[65] Such plants usually produce some by-products as well, such as tar oils. Table 7.6 summarizes major demand determinants for coal.

7.8 PETROLEUM PRODUCTS

Petroleum products and their uses are at the heart of the so-called energy crisis today. Their convenience in use, relative cleanliness, and easy transportability have made them the preferred and, at least until recently, lowest-cost energy source for a wide variety of uses. In this century, most of our transportation systems and equipment, and much of our industrial as well as commercial and domestic energy uses have become strongly dependent on the use of liquid hydrocarbons. For the majority of them, suitable substitutes are simply not available at present or are only available at system utilization costs that exceed those of petroleum-based-ones by substantial margins. Even the most strenuous efforts and policy measures designed at developing substitute energy sources and energy-using systems will require decades to make a decisive impact. In the meantime, many of our most vital activities will remain utterly dependent on the availability and use of petroleum products regardless of their price.

However, the demand for them is not entirely inelastic. Within limits, substitutes are available. But the ease of utilizing them instead of petroleum varies widely among different uses. Some of these substitutes may be far more expensive in terms of total energy systems costs;[66] others may require substantial time periods until conversion to their use could be accomplished.[67] In any case, wholesale conversions from crude-oil-based petroleum product uses to such substitutes will usually require many years of relentless and costly efforts, as can be seen from their so far limited success in most industrialized countries since 1973/74.[68]

There are five major types of petroleum substitutes available. The first is simply the possibility to do without or to do with less. Curtailing pleasure driv-

---

[65]For an example see the case study in the Thai tobacco industry.

[66]e.g., the use of hydrocarbons from coal gasification.

[67]e.g., the substitution of coal for fuel-oil-fired powerplants.

[68]Between 1974 and 1979, total OECD petroleum imports increased from an average of 23 MMB/day to some 27 MMB/day, while total free world production increased from 49.3 to 52.3 MMB/day. *From:* U.S. Dept. of Energy, *An Analysis of the World Oil Market, 1974–1979*, Washington, D.C., July 1979, Table 1, Fig. 3.

Table 7.6. Major Demand Determinants of Coal.

A. Coking Coal
1. Mining costs
2. Shipping costs to potential customers
3. Export taxes or other levies, if any
4. Physical characteristics of coal
5. Demand for raw iron from blast furnaces
6. Costs of delivered coking coal from alternative suppliers
7. Considerations of security of supply to customer
8. Costs of alternative reducing fuels (e.g., natural gas in Mexico)

B. Steam Coal
  I. Electric Utilities and Industrial Steam Plants
1. Mining Costs
2. Quality (physical characteristics and heat rate)
3. Transportation costs to user's plant
4. Costs of alternative fuels
5. Additional equipment operating and handling costs to user compared to other fuels
6. Governmental regulations affecting fuel choices by users
7. Total costs of electricity from steam-power plant delivered to load centers
8. Costs of electricity from potential alternative sources delivered to load centers
9. Growth in demand for (a) electricity
                      (b) industrial output of plants using coal

  II. Residential and Small-Scale Industrial and Commercial Sectors:
1. Delivered costs at retail
2. Costs and availability of alternative fuels (charcoal, wood, LPG, fuel oil, kerosene, natural gas, etc.)
3. Number of households
4. Household income
5. Degree days below reasonable comfort level
6. Consumer habits

---

ing in private automobiles or changing thermostat settings are typical examples. The second consists of the substitution by other energy sources. Examples are the use of coal, natural gas, or wood in the production of process heat or steam, the use of ethylene instead of gasoline, or the substitution of hydro for fuel oil based thermal power plants. The third consists of the substitution of capital for energy. Better insulation, recovery of waste heat, or replacement by more energy-efficient equipment fall into this category. Closely related to the substitution by capital is the fourth source, namely technological change. Systematic reductions in vehicle weights, use of cogeneration equipment instead of straight gas or steam turbines in power and steam production, or the use of electronic ignition instead of pilot lights come to mind. Fifth, labor can be used as a substitute for petroleum-based energy. Animal-pulled plows instead of tractors, hand loading instead of the use of fork-lift trucks, hand shovels and pickaxes to replace bulldozers or tractors in construction are some of them.

Because of the potential availability of the substitutes, the demand for petro-leum-based energy resources can never be completely price inelastic, provided price changes are large (as they certainly have been in recent years). Because time is often needed to switch-over, the price elasticity is usually greater in the long-run than in the short-run. This is confirmed by most of the demand studies that have been undertaken in recent years.[69] Unfortunately, as has already been pointed out in the preceding chapter, few, if any of these analyses (at least those that relied mainly on econometric estimates) have been able to determine with any degree of confidence how long the so-called long-run really might be[70] and what the likely short- or long-run price estimations might be over varying potential ranges of price changes.[71] Given the drastically changed environment with respect to petroleum product prices, availability, and security of supply, observations of past consumption trends can provide only a limited and gen-erally unreliable guide for estimating future consumption trends.

A more appropriate approach appears to be to take explicit account of the specific variables that are likely to affect future demand. Among these vari-ables are specific government policies that affect energy use and prices, petro-leum product and alternative energy resource prices, potential investments in alternative energy supply systems, potential technological change, the cost and feasibility of energy conservation measures, etc.[72] In other words, demand pro-jections should be based on estimates of the technological, socioeconomic, and managerial feasibility of utilizing either petroleum-based or alternative energy sources, the potential for reducing petroleum energy inputs per unit of output or unit of consumption and the likely effects of governmental policies related to energy using activities.[73] Such estimates are likely to yield more reliable projections than those that are anchored in past relationships of economic activities, relative prices, and petroleum product consumption patterns.

The increase in the cost of petroleum products by about an order of mag-nitude within less than seven years[74] has been a major factor in the economic

---

[69]See the comparison of energy demand projections and elasticity estimates contained in R. S. Pindyck, "The Characteristics of the Demand for Energy," in: John C. Sawhill (ed.), *Energy Conservation and Public Policy,* Prentice-Hall, N.J., 1979, Table 1.

[70]See Robert Pindyck, *The Structure of World Energy Demand,* p. 120.

[71]While numerical elasticity estimates have been produced in profusion, their most notable char-acteristic is their substantial variation from one study to the next.

[72]This approach, for example, has been used for the projection of future demand for gasoline in the U.S.A. by S. H. Schurr et al., *Energy in America's Figure.*

[73]Major government-affected variables could be energy prices, taxes or subsidies, direct alloca-tion decisions (e.g., the mandated use of specific energy resources), import, export, or use restric-tions for energy resources in energy using equipment (e.g., automobiles or air conditioning equip-ment). For a striking example of the effects of automobile import restrictions on the demand for gasoline see the case study on Sri Lanka.

[74]In 1971 average crude oil prices per barrel in Thailand were US$2.46; by the end of 1979 they had risen to over US$26.00. *Source:* National Energy Admin., Thailand.

dislocation and inflationary pressures in many countries, as well as in the rapid and dangerous increase in the external indebtedness of most petroleum importing developing countries.[75] For these countries, accurate demand forecasting for petroleum product uses, given a range of potential prices and availabilities, is obviously of major importance for their economic well-being and growth. However, accurate demand forecasts are equally needed for countries that are more or less self-sufficient in petroleum products as well as those that are modest exporters with limited crude oil reserves at the present time.[76] For them, some of the important policy questions are how much oil should be exported now to earn needed foreign exchanges given perceived domestic needs tomorrow, what domestic prices should be charged given world market prices, how domestic demands may change if domestic prices are changed, and what the negative effects of such domestic price changes (i.e., increases) may be in economic or political terms.[77]

While there is basically only one source of petroleum products at present—i.e., crude oil—the products derived from crude oil serve a wide variety of different markets and different needs. Hence demand analysis must be subdivided into the various, use-determined subsectors such as transportation, power production, industrial process heat, chemical feedstocks, construction, agricultural uses, armed forces, etc. Within these broad groupings, further subdividing may be necessary to reflect distinct and independent uses (e.g., fuels for tractors, irrigation pumps, and processing machinery in agriculture).

The demands for petroleum products are demands that are derived from other primary activities or human wants and needs. The driving demand determinants for petroleum products, therefore, are the demands for these primary outputs such as, for example, the transporting of foods or people from A to B, the provision of heat for processing or cooking, etc. Hence demand forecasts have to be based on forecasts of the levels of these petroleum using activities, subject to the possible use of substitutes now and in the future. Given the rapid changes in the technology of energy using systems and appliances that have been stimulated by the drastic changes in petroleum prices, special attention

---

[75]Between 1970 and 1977 the external public debt as a percentage of GNP rose from 20.3 to 25.0% for the 37 least developed countries of the world. *From:* World Bank, *World Development Indicators,* Washington, D.C., June 1979, Table 13.

[76]As, for example, Canada, Peru, Ecuador, Argentina, Malaysia, or Egypt. In some of these countries (e.g., Canada and Venezuela) a subsidiary policy question is whether available lower-cost, conventional petroleum resources should be exported now given that much higher-cost ones are likely to be available in the future (i.e., heavy oils, tar sands, oil shale, or costly off-shore or frontier deposits).

[77]Political costs may be more serious than economic ones as the riots in Bogota, Colombia and Quito, Ecuador in response to fare increases in public transport attest to. In Thailand, a 1979 announced 56% increase in electric utility rates necessitated by increased fuel oil prices had to be hastily withdrawn when serious student riots broke out as a result.

must be placed on the potential feasibility and time-path of introducing such new or improved technologies. As existing equipment wears out or becomes inefficient because of high energy costs (e.g., gas turbines or airplanes) it will be replaced by others, subject to constraints such as investment capital availability, technological know-how, etc. New plants and new or expanding activities, unfettered by the deadweight of sunk investments, will tend to utilize substantially more energy efficient installations and equipment. Because much of this new equipment as well as the lay-out and design of new plants is imported from industrialized nations, which have felt the brunt of high petroleum prices for some time, the transfer of such energy-saving technologies should be rapid. Unfortunately, such rapid transfers are unlikely to affect existing plants and equipment which, because of the chronic investment capital shortage in most developing nations, have a tendency to be kept in operation far longer than in industrialized nations. Such existing tendencies and constraints have to be taken explicitly into account in the projection of future demands.

As pointed out above, demand forecasts have to be prepared separately for each separate use. For some of them, mainly industrial users, a survey-research approach will yield the best results. Cement plants, pulp and paper mills, steel mills, fuel-oil-based powerplants, chemical producers that use petroleum as feedstocks, refineries, etc.—all of them will be in a better position to forecast their likely future demand than anyone else. However, cross-checks should be made to compare their projected consumption patterns per unit of output with those elsewhere, particularly in comparable plants in industrial nations. If substantial differences are discovered, further investigations may be warranted to find out why and if changes and improvements could be brought about.

Unfortunately, the majority of petroleum product uses originate in ubiquitous, widely-spread activities such as transportation, agriculture, construction, etc. For these sectors, forecasts have to be based on use categories, such as private cars, taxis, commercial transport, railroads, water-borne shipping, aviation, armed forces, heating and cooking, lighting (i.e., kerosene), etc. Within these sectors, specific and basically independent subsectors must be distinguished, such as the use of diesel fuels in agriculture for deepwell pumps for irrigation and/or water supply, agricultural tractors, produce trucks, drying sheds, or individual generator sets. Diesel will also be used in commercial transport, for industrial or isolated power plants, for construction equipment, and for boats and ships. Total projected diesel fuel demand then is the sum of all of these individual demands, but the demand determinants for each individual subsector will, in most cases, be quite independent from each other.

Table 7.7 indicates some of the likely demand determinants for various types of petroleum product uses. It also indicates in a separate column the most likely substitutes that may affect demand in the specific sector. Obviously, these substitutes, their availabilities and costs, are themselves determinants of demand for the given fuel in question. The importance of the various determinants will vary from country to country and often from region to region or place to place.[78] Country, region, and time-specific studies are needed to isolate the more important determinants and to estimate their likely effects on future demand.

7.9 TRADITIONAL FUELS

Traditional fuels comprise a wide variety of materials—all of them of organic origins—that serve as the primary energy source for over one half of the world's population. They include firewood, charcoal, dung, and crop residues. The demand for them is entirely regional or local because of high transportation costs relative to their value, with the possible exception of charcoal. While most of the traditional fuels are used in households or small-scale commercial enterprises (e.g., restaurants), some of these biomass fuels are also used for industrial purposes (e.g., bagasse in sugar factories, wood residues in sawmills and pulp and paper mills, and charcoal in steel mills).[79]

Perhaps the most important characteristic of these traditional fuels is that in most regions only a small percentage enters commercial channels, i.e., are bought and sold. Most users gather them themselves. For them, their costs are represented by the time required to collect and transport them to the family's residence.

As various studies have pointed out in recent years, shortages of these traditional fuels are developing in many countries of the world.[80] These shortages are particularly severe in regions in which alternative, commercial fuels are not available, or only at costs that are beyond the reach of most of the people living there. Continuing population growth and ever-diminishing availability of fuel resources such as wood or crop residues then forces increased use of animal dung, which in turn reduces its availability as natural fertilizer. This, in turn, decreases crop yields and increases erosion. Such increased shortages, ulti-

---

[78]For example, natural gas is an obvious substitute for fuel oil, but only in locations in which it is available, just as electricity may easily replace kerosene for lighting, but only if a powerline is nearby.

[79]One large steel mill in Monlevade, Brazil uses a 1957-built 50 km long aerial ropeway to transport 40 tons/h of charcoal from its forest kilns to the smelter.

[80]See, for example, David Hughart, *Prospects for Traditional and Nonconventional Energy Sources in Developing Countries,* World Bank Staff Working Paper No. 346, Washington, D.C. July 1979; Erik Eckholm, "The Other Energy Crisis: Firewood," *World Watch Paper No. 1,* Worldwatch Institute, Washington, D.C., 1975; Dale Avery, Gunter Schramm, and Kenneth Shapiro, "Production Systems," in: Kenneth Shapiro, et al., *Science and Technology for Managing Fragile Environments in Developing Nations,* Univ. of Michigan, School of Natural Resources, Ann Arbor, Sept. 1978, pp. 148–225.

Table 7.7. Major Demand Determinants and Potential Substitutes for Petroleum Products.

DETERMINANTS	MAJOR POTENTIAL SUBSTITUTES

(I)   *GASOLINE*

   (A) *Passenger Vehicles (automobiles):*

(1) Number of vehicles in circulation	(1) Cost and convenience of public transport
(2) Number of families with incomes above specified minimum (for car ownership)	(2) Cost of taxi fares
(3) Growth of modern sector GDP	(3) Cost of diesel-fueled vehicles and cost of diesel fuel
(4) Costs of new cars	(4) Cost of LPG conversion and cost and availability of LPG
(5) Automobile license fees or taxes	(5) Cost of kerosene (for gasoline dilution)
(6) Price of gasoline	
(7) Average annual mileage per vehicle	
(8) Import restrictions on new automobiles or assembly plants	
(9) Cost and availability of spare parts	
(10) Road conditions	
(11) Degree of traffic congestion	
(12) Technical characteristics of existing and future automobile stock (e.g., gas mileage)	
(13) Tax rules re deductibility of automobile expenses	
(14) Number of taxis in circulation	
(15) Technical characteristics of taxis (i.e., gas mileage)	
(16) Average annual mileage of taxis	
(17) Public policies toward government ownership and use of passenger automobiles	
(18) Armed forces demands (No. of vehicles, gas mileage, annual use)	
(19) Number of motorcycles in circulation	
(20) Number of persons or families with income above specified minimum for motorcycle ownership	

   (B) *Buses*

(1) Number of gasoline-driven buses in circulation	(1) Diesel buses
(2) Technical characteristics of	(2) Trucks used for passenger transport

DETERMINANTS	MAJOR POTENTIAL SUBSTITUTES
buses (e.g., gas mileage) including age	(3) Bicycles
(3) Average annual mileage per vehicle	(4) Number and costs of taxis, private cars, or motorcycles
(4) Public subsidies, if any	(5) Increase in local employment opportunities
(5) Costs and availability of new buses	(6) Existence of alternative passenger transportation modes (railroads, boats)
(6) Costs and availability of spare parts	
(7) Fares per passenger or passenger mile	
(8) Costs of gasoline	
(9) Degree of traffic congestion	
(10) Urban layouts (i.e., distance between residential, industrial, and commercial areas)	
(11) Miles of nonurban road network	
(12) Road conditions	
(13) Population growth	
(14) Per capita income	
(15) Privately owned buses including school buses, armed forces, company-owned, etc.)	
(16) Public ownership of bus systems and available budgets for their maintenance and expansion	

(C) *Gasoline-Driven Trucks*

(1) Number of trucks in circulation	(1) Diesel-operated trucks
(2) Technical characteristics of vehicles (e.g., gas mileage) including age	(2) LPG-operated trucks
(3) Average carrying capacity per vehicle	(3) Rail transport, convenience, costs
(4) Costs and availability of new trucks	(4) Other transport modes (boats, animal-driven vehicles, etc.)
(5) Cost and availability of spare parts	
(6) License fees and vehicle taxes	
(7) Level and characteristics of regulated tariffs, if any	
(8) Real gross domestic product (GDP) by sector (i.e., mining, agriculture, industrial output, etc.)	
(9) Public sector-owned vehicles (post office, municipal services, armed services)	

Table 7.7 (continued)

DETERMINANTS	MAJOR POTENTIAL SUBSTITUTES
(D) *Other Gasoline Using Equipment* 　(1) Construction machinery 　(2) Stationary machinery (pumps, processing equipment, construction equipment, generator sets, etc.) 　(3) Boats 　(4) Gasoline-using aircraft 　(5) Gasoline-driven agricultural machinery 　(6) Real GDP by sectors which tend to use gasoline-operated equipment	(1) Cost and availability of electricity and electric-driven equipment (2) Cost of alternative diesel- or LPG-operated equipment
(II) *DIESEL FUELS* (A) *Automobiles and Taxis* 　(1) Overall cost and availability of diesel-driven vehicles relative to gasoline-driven ones 　(2) All other determinants as for gasoline-driven vehicles. (See I, A)	(1) Total costs and availability of gasoline- and LPG-driven vehicles (2) Public transport (3) Cost of kerosene (as fuel admixture)
(B) *Buses and Trucks* 　(1) Availability and cost of diesel-driven buses and trucks 　(2) All other determinants as for gasoline-driven buses and trucks (See I, B and C)	(1) Total costs and availability of gasoline-driven vehicles
(C) *Construction Machinery* 　(1) Level of heavy construction activity in private and public sectors 　(2) Cost of equipment 　(3) Equipment availability (possible import restrictions	(1) Gasoline-driven equipment (2) Cost of unskilled construction labor
(D) *Agriculture* 　(1) Number of farm tractors 　(2) Technical characteristics of tractors (size, hp, fuel consumption per hour) 　(3) Cost of tractors 　(4) Government or bank supported farm mechanization programs 　(5) Tractor availability 　(6) Farm size 　(7) Average farm income	(1) Animal-driven farm implements (2) Cost of unskilled farm labor (3) Total cost of gasoline-operated tractors (4) Cost of kerosene (as fuel substitute) (5) Light fuel oil (as fuel substitute)

DETERMINANTS	MAJOR POTENTIAL SUBSTITUTES
(8) Cost of diesel fuel	(6) Gasoline- or electricity driven
(9) Diesel-driven water pumps	water pumps
(10) Other stationary farm	(7) Biogas driven pumps
machinery	(8) Wood, charcoal, straw, etc., for
(11) Diesel-driven generator sets	drying operations
(12) Diesel-oil fired drying kilns and	(9) Gasoline-driven harvesting
sheds	machinery
(13) Harvesting machinery	

(E) *Misc. Diesel-driven Machinery*

(1) Cost of diesel generator sets	(1) Cost of electricity supplies
(2) Cost of diesel fuel	from alternative sources
(3) Public support programs for	(2) Costs of alternative gasoline-
noncentral power station	driven machinery and
supplied rural electrification	equipment
(4) Industrial and mining	(3) Steam- or electricity-driven
developments in isolated areas	locomotives
without electric power supply	
(5) Mining machinery and	
equipment	
(6) Diesel use by armed forces	
(7) Diesel-operated boats and ships	
(8) Diesel-locomotives	
(9) Public sector demands for	
diesel-driven equipment (e.g.,	
road maintenance)	

(III) *LPG*

(A) *Residential Sector and Commercial Sectors*

(1) Price per kg	(1) Wood (price, availability)
(2) Cost of supply bottles	(2) Charcoal (price, availability)
(3) Cost of cooking-heating	(3) Kerosene (price, availability)
equipment	(4) Coal (price, availability)
(4) LPG distribution networks	
(5) Per family income	
(6) Population growth in LPG	
supply areas	
(7) Social acceptability as home-	
cooking fuel	

(B) *Transportation*

(1) Cost per kg	(1) Cost of gasoline-driven cars
(2) Cost of vehicle conversion	(2) Cost of diesel-driven vehicles

(C) *Industry*

(1) Price and availability	(1) Price and availability of
(2) Fuel and feedstock demands	alternative fuels and feedstocks

Table 7.7 (continued)

DETERMINANTS	MAJOR POTENTIAL SUBSTITUTES
(IV) KEROSENE AND JET FUELS	
(A) Residential and Commercial Sectors	
(1) Price	(1) Cost and availability of other cooking/heating fuels (LPG, wood, charcoal, coal, etc.)
(2) Cost of kerosene lamps and/or cooking/heating equipment	(2) Availability and cost of electricity (mainly for lighting)
(3) Per family income	
(4) Population growth	
(5) Social acceptability as a cooking fuel	
(B) Transportation	
(1) Aviation demands	
(2) Price relative to diesel and gasoline for possible dilution of these primary fuels	
(C) Chemical Feedstocks	
(1) Demand for chemical intermediates (plastic, fertilizers, ethylene, etc.)	(1) Cost and availability of alternative feedstocks (e.g., natural gas, LPG)
	(2) Price and availability of imported chemical intermediates
(V) FUEL OIL	
(A) Electric power generation	
(1) Price	(1) Natural gas (price, availability, systems costs)
(2) Demand for electricity	(2) Coal (price, availability, systems cost)
(3) Generating systems-expansion plans (if based on fuel oil)	(3) Hydro (availability, systems cost)
(B) Industrial Uses	
(1) Price	(1) Natural gas, coal, LPG, wood, charcoal, straw, other biomass fuels (price, availability, systems costs)
(2) Energy input requirements for various industrial processes	(2) Potentials for energy conservation
(3) Levels of industrial outputs by industry or process	
(4) Needs for self-owned generating plants (permanent or backup service)	
(C) Commercial and Domestic (for heating purposes)	
(1) Price	(1) Total system costs of using alternative fuels
(2) Degree days below reasonable comfort range	(2) Potentials for energy conservation

mately, force out-migration because long-term human survival without access to a minimum of cooking fuel is not possible because the majority of our food stuffs have to be cooked or fried to become digestible (grains, meats, beans and pulses).

In spite of the wide recognition of the pervasiveness of this problem few reliable statistics of existing demand/supply relationships are available. The major source of information has been the FAO, which publishes annual statistics of forest products[81] as well as crop and livestock production.[82] The latter makes it possible to estimate the approximate available quantities of crop residues and animal dung. Based on these and a few other, country-specific data sources, Hughart has prepared 1976/77 and 1990 estimates of country by country per capita demands and supplies for fuel woods, crop residues, and animal dung. These projections are a highly useful aggregate analysis of the problem in a worldwide context.[83] However, the estimated data are no more than a first approximation. The FAO bases its data on published government statistics which often are quite unreliable. In some cases, for example, they include only production data from managed forests. Furthermore, the FAO data do not disaggregate industrial and domestic fuelwood and charcoal use. In addition, many of the published FAO data are subject to considerable doubt as to their accuracy. As Hughart indicates, FAO data for Peru, for example, indicate consumption levels of only about one third of those of a specific household consumption survey, while for Bangladesh, FAO figures are many times higher than the results of a special study undertaken for the Bangladesh Energy Study. He concludes that "this diagnostic procedure is at best a very crude tool for assessing the traditional fuel situation. A more reliable and meaningful examination of this situation would require a much more extensive research effort."

These research efforts have to be not only country- but also region-specific. Transportation costs severely restrict the utilization of firewood beyond a narrow radius. According to a 1975 Indian survey by R. Mathur, cited by Hughart, 70 to 100% of fuel requirements will be obtained from forests that are no more than 10 km from a village; beyond a distance of 15 km, fuelwood use virtually stops and is replaced by crop residues, dung, or charcoal. Of course, in regions where the latter substitutes are in equally short supplies, distances traveled to gather fuel may be somewhat longer. However, for physical reasons they rarely exceed one or two days on foot or animal. In a number of countries,

---

[81]FAO, *Yearbook of Forest Products*, Rome, annual.

[82]FAO, *Production Yearbook*, Rome, annual.

[83]A series of 13 maps identifying the most susceptible regions have been published in K. Shapiro, et al., p. 23 ff.

Table 7.8. Major Demand Determinants for Traditional Fuels (Wood, Charcoal, Crop Residues, Dung).

A. Residential and Commercial Sectors
 1. Local/regional population
 2. Availability as a common property resource
 3. Market price
 4. Family income and income distribution
 5. Degree-days below reasonable comfort range
 6. Availability and price of other fuels including appliance costs (coal, LPG, kerosene, etc.)
 7. Knowledge, availability, and costs of efficient stoves
 8. Social customs (will largely affect alternative fuel choices)
B. Industrial and Agroindustrial Sectors
 1. Needs for crop drying and kiln fuels (e.g., tobacco, bricks, ceramics, etc.)
 2. Needs for process steam and in-plant electricity (e.g., bagasse for sugar mills, kiln-drying of lumber, process heat of pulp and paper mills, etc.)
 3. Availability as a residual of production (e.g., bagasse, wood chips and shavings, coconut husks)
 4. Market prices
 5. Costs and availability of alternative fuels

gathering the necessary fuel for cooking consumes as much as one-quarter of the total available working hours of a family, while in some regions in Upper Volta and Niger fuel purchases require 20–30% of an average working family income. Because of the high cost of transportation, demand and cost studies have to be undertaken on a regional or subregional basis. Ecological conditions often vary drastically from one region to the next. In the Andes countries, for example, average biomass (i.e., wood) production on a per capita basis is more than ample. However, most of it is concentrated in the unpopulated tropical flatlands of the interior while in the populated highlands and the desertlike coast severe shortages are the rule.

While local availability will be a major factor in the consumption of fire-wood, another factor that will affect demand are climate.[84] Hughart reports that consumption in highlands is as much as four times greater than in tropical lowlands because of the need for fuel for heating in addition to cooking. The use of properly designed cooking stoves instead of open fires may reduce fuel needs by up to 50%.[85] Higher local income levels will result in the substitution

[84] In Thailand's better endowed regions average annual consumption is estimated at about 1 ton per capita. Similar estimates are made for other countries where local supply is not a constraint.
[85] Specific experimental results of net efficiencies of 23–28% compared to 3–5% for open fires are reported in FAO, *Improvement of Fuelwood Cooking Stoves and Economy in Fuelwood Consumption*, Report No. 1315, Rome 1961, p. 12.

by other fuels, such as kerosene, LPG or even electricity. Coal may be used in some regions where it is abundant, as in the central regions of Colombia where many small, family-owned coal mines are operating. A list of the likely demand determinants is shown in Table 7.8. Detailed studies on a regional basis are needed to evaluate their validity and quantitative effects.

8
Issues in Investment Planning

Energy demand management is closely related to and dependent upon energy supply management, which in turn depends on investment and investment planning. Any discussion of energy policies and demand management would be incomplete without a review of these interrelationships. Although a complete analysis of energy investment planning cannot be undertaken within a single chapter of a book devoted to demand management, we will present in this chapter a summary of the principal issues in investment planning that affect the demand side.

8.1 INTRODUCTION

There are two major avenues through which energy supply and, by implication, investment affect demand: availability and quality. If energy is not available, potential demand cannot be satisfied. But even if it is available, the quality characteristics of a given energy supply will determine its usefulness and productivity. Six specific characteristics of energy supply that will directly or indirectly affect demand can be identified:

(1) Availability
(2) Timeliness of availability
(3) Quality
(4) Absolute supply price
(5) Relative supply price
(6) Availability and costs of energy-use dependent facilities and equipment

Availability is, of course, fundamental for satisfying potential demand. But to make specific energy resources available at given locations when needed often requires complex and capital-intensive delivery systems. Before such systems are installed, therefore, assurances are needed that their overall benefits will exceed their costs. Only after they have been installed and are connected to the premises of users can they satisfy potential demand. Electric power grids and natural gas distribution networks are prime examples. Because of the high initial costs of such systems, many areas, even in the developed and industrialized countries, do not have access to certain forms of energy. For example, it is generally not realized that in the United States, the country with the most highly developed natural gas distribution system in the world, natural gas availability is limited to only about 55% of all households.[1] In developing countries, access to modern forms of energy supplies is more restricted. Electricity, for example, which is available to between 90 and 95% of all households in the industrialized nations, is usually not widely available. In Thailand, a middle-income developing country, only about 28% of the population has access to electricity,[2] while in Bangladesh, one of the world's poorest countries, the percentage is only about 3%.[3] What we can say, then, is that supply often creates demand or, more accurately, that new supply could satisfy potential or underlying demand that was not met before because of a lack of supply.

Once a new energy resource has been made available to potential users, this can affect energy demand in three different ways, especially if the new source is cheaper and more convenient to use. First, total energy use in terms of effective consumption of all types of energy is likely to increase over and above what it would have been in the absence of the new source. Second, demand for previously used energy resources is likely to decline. And third, total energy use in terms of physical energy input units (as measured in kcal or btu) may increase or decrease, depending on the use-efficiency of the new source relative to that of the previously used ones.[4]

One example of new energy systems creating their own demand, is the deliberate development of low cost electricity sources to attract power-intensive industries. The aluminum smelting industry of Canada, the second largest in the world, is a typical case. However, similar arguments may be applied to energy systems supplying more ubiquitous markets, such as the expansion of electricity or natural gas distribution systems into new areas, the development of LPG or LNG distribution facilities, the introduction of coal or lignite briquettes, and the marketing of new charcoal to replace dwindling supplies of traditional biomass-based fuels.

Figure 8.1 illustrates, as an example, the likely effects of introducing network-based electricity to an area that had no such services previously available. To simplify the exposition, the analysis is presented in comparative static terms. In practice, the effects of introducing electricity are likely to be felt over a number of years, as potential users make the necessary investments in electricity-using appliances over a period of time to take full advantage of the new

---

[1]National Gas Survey, U.S. Department of Energy, 1979.
[2]National Energy Administration, Thailand, 1979, 1979 data.
[3]Gunter Schramm, *Bangladesh Energy Sector Study,* Asian Development Bank, Regional Energy Survey, February 1981, 1978/79 data.
[4]e.g., use of high-efficiency gas or electric stoves instead of inefficient wood-burning ones.

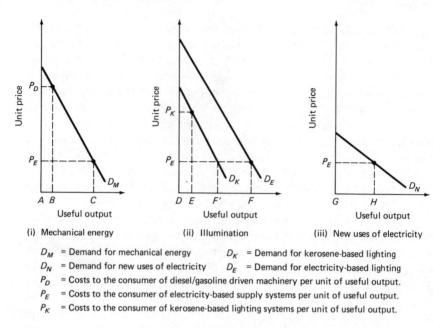

D_M = Demand for mechanical energy D_K = Demand for kerosene-based lighting
D_N = Demand for new uses of electricity D_E = Demand for electricity-based lighting
P_D = Costs to the consumer of diesel/gasoline driven machinery per unit of useful output.
P_E = Costs to the consumer of electricity-based supply systems per unit of useful output.
P_K = Costs to the consumer of kerosene-based lighting systems per unit of useful output.

Figure 8.1. Effects of electrification on overall energy demand.

energy source. The three graphs in the figure represent three distinct types of energy use.

First, Figure 8.1(i) indicates the demand for an output such as mechanical energy for pumping, where the new electricity supply can substitute for another form of energy that was previously used, with no significant change in the quality of the useful output. The horizontal axis is in units of useful output, e.g., gallons of water pumped, while the vertical axis shows the effective total cost to the customer (including fuel as well as the annuitized and prorated appliance capital costs), per unit of useful output. Since the overall costs of mechanical energy from network-based electricity supply are usually much lower than the costs of operating diesel or gasoline consuming motors, nonelectricity consumers will shift to electricity use, while their demand will also increase from AB to AC, due to the reduction in effective costs. Other examples in this category include kerosene refrigerators, diesel autogenerators of electricity, and so on.

The second graph, Figure 8.1(ii) shows the demand for a service such as illumination or lighting, where electricity will displace another energy source (say kerosene), but with a significant improvement in the quality of the final output. The higher quality of output results in the demand curve shifting out from D_K to D_E. If there was no such shift, demand would have increased from

DE to DF' because of the lower price of electricity, but with the demand shift, the overall need for lighting would be DF. (See Chapter 13 for a quantitative example of a quality induced shift in the demand curve).

The third category of demand shown in Figure 8.1(iii) arises from new uses of electricity that were previously not thought of or considered infeasible because of technical problems or prohibitively high costs. Typical examples include small power tools, television, air conditioning, and so on. The consumption, GH, therefore, represents an entirely new or induced market for electricity.

The foregoing discussion illustrates how the introduction of a cheaper and higher quality energy source will stimulate demand in a variety of ways. In other words, controlling supply is also an effective tool for demand management.

Timeliness of energy supply is a more subtle but highly important precondition for satisfying demand effectively or curtailing demand. Both supply and demand may vary over time, often stochastically. If this is the case and demand at any point in time exceeds available supply, shortages will result, especially in the case of nonstorable forms of energy like electricity. Such temporary shortages usually are very costly in terms of losses in energy-dependent output or convenience. They are particularly high if the energy shortages are unexpected.[5] But even if such shortages are more or less expected, their costs can be substantial and considerably exceed the additional costs of increasing the supply capacity or reliability of the energy system.

The quality dimensions of a given energy supply consist of two main components: one is related to the reliability of supply which, in turn, is related to the timeliness of supply, as discussed above. The second relates to the actual physical characteristics of the supply source. Electric current that varies in terms of voltage, natural gas that varies in terms of its constituent components, low-quality gasoline, diesel, and other petroleum products, or coal or other solid fuels that vary greatly in terms of contaminating admixtures or water-content all affect the efficient operation of an energy-source dependent system. Poor, or uncontrolled, variations in quality may make energy use expensive in terms of output losses and maintenance costs.

The absolute price of energy (i.e.), absolute with respect to the costs of all other nonenergy goods and services) affects the amount of energy that will be used by various types of activities. As practically all studies of this relationship between aggregate energy demand and costs of supply have shown, demand

---

[5]See Mohan Munasinghe and Mark Gellerson, "Economic Criteria for Optimizing Power System Reliability Levels," *The Bell Journal of Economics,* Vol. 10, Spring 1979, pp. 353–65; and Mohan Munasinghe "The Costs of Electric Power Shortages to Residential Consumers," *Journal of Consumer Research,* Vol. 6, March 1980, pp. 361–9.

elasticity, at least in the long-run, is not zero. In other words, capital and labor will be substituted for energy, or users will do with less energy.[6] These issues have been discussed in detail in Chapter 7.

The relative prices of alternative energy supplies will affect demand to the extent that these alternatives can be substituted for each other. Such substitutions, in response to price changes, may be almost instantaneous, if little or no changes are required in the appliances or equipment using the respective fuels,[7] or they may take decades, as in the case of substitutions from oil-fired to alternative electric generating systems.

Finally, the costs, availability, and quality of the energy-using equipment and appliances will affect demand, because producers and users will generally be more interested in overall systems costs than in the specific energy costs by themselves, at least in the long run. In electric power systems planning, for example, this systems cost approach is used as a matter of course, with supply cost comparison generally based on the delivered costs of electricity at the point of distribution, rather than on the initial costs of energy inputs.[8] The capital and recurrent costs of electricity using equipment must also be included to capture the final users' viewpoint.

Overall, then, it must be concluded that supply policies and, as an important part of such policies, investment policies for energy systems are of fundamental importance for the management of energy demand.

8.2   DEFINING THE OBJECTIVE FUNCTION

One of the complexities in energy investment planning is that energy resources are generally not demanded for their own sake, but are utilized to provide certain types of services such as transportation; mechanical, chemical, or heat energy for processing; and comfort through space conditioning. Users of energy, then, are primarily interested in maximizing the net benefits from (or

---

[6]See, for example, Robert S. Pindyck, *The Structure of World Energy Demand,* MIT Press, Cambridge, MA, 1979. Pindyck's estimate of the long-run industrial energy demand elasticity in industrial countries ranges between 0.7 and 0.8 (page 180) and for the residential sector as high as 1.1 (page 118). Estimates made by others vary substantially from these values, however, although all are significantly different from zero (see also Pindyck's tables 4.12 and 5.13). For a report on the most recent findings see: EMF 4 Working Group, "Aggregate Elasticity of Energy Demand," *The Energy Journal,* Vol. 2, April 1981, pp. 35–75.

[7]In Bangladesh, for example, kerosene and high-speed diesel fuels were traditionally priced exactly the same until July, 1980 when prices were raised at differential rates, with diesel then priced some 21% above kerosene. While the two fuels are not perfect substitutes a certain amount of kerosene can be blended into diesel fuel without drastically affecting engine performance. Within one month of the price change, kerosene sales increased by some 20%, while diesel sales fell by approximately the same quantity. *From:* Gunter Schramm, *Bangladesh Energy Sector Study,* Asian Development Bank, Manila, February 1981, page 62.

[8]See also the case studies on the power sectors of Thailand and Sri Lanka. Similar considerations apply in the choice of diesel versus gasoline-powered cars, for example.

minimizing the costs of) these desired outputs, rather than those of the energy component only. Hence, the objective function should focus on the optimization of the ultimate output or service rather than on that of the energy inputs only. This means that substitutions in the form of capital, labor, or alternative energy resources are to be taken explicitly into account.

For example, desired levels of space conditioning for a given structure can be achieved with large inputs of energy—be they gas, oil, coal, sunshine, or others; however, the use-efficiency of any of these energy inputs will depend on a number of complementary factors, such as the efficiency of the energy conversion equipment used or the amount of insulation provided. Clearly, trade-offs are possible between these various inputs of energy, energy equipment, and other factors in order to bring about a specified level of output (i.e., desired temperature and humidity levels).

Energy investment planning, therefore, should focus on total systems costs and benefits, rather than on physical energy use or energy costs and output values only (see also Chapter 6).

Two related criteria can be used in the economic evaluation of investment projects: maximization of net benefits or minimization of costs. Maximization of net benefits is the more general approach, which is to be used whenever the questions are how to optimize the use and development of a given energy resource, or whether the supply of a new energy resource can be justified (e.g., rural electrification, programs for biogas digesters, reforestation programs, extension of natural gas supplies into new service areas, etc.). This approach is particularly important for countries that are energy exporters and have to allocate their resources between domestic uses, direct exports, or energy-resource based process industries (e.g., chemicals based on hydrocarbon feedstocks, electroprocessing based on low-cost hydro, etc.). To make optimal allocation decisions in such situations requires the explicit determination of all benefits and costs over time.

By contrast, a cost minimization approach eliminates the need to measure the value of the benefits provided. This approach can be used if it is certain that a given level of energy services must be provided in similar quantities and qualities, whatever the supply source. The question then becomes simply one of selecting the lowest-cost source of supply. However, if there is any doubt at all whether a given supply of energy will be made available, or may only be made available in different quantities and qualities, a cost minimization approach will be inadequate to resolve the issue.[9] Typically, cost minimization is used in the case of electric power systems expansions.

The following discussion deals with some concepts that are important for the

---

[9]A cost minimization approach contains the implicit assumption that the demand for the particular increment of energy to be provided will not be affected by the differences in the supply price or quality of the various alternative energy sources under consideration.

evaluation of the economic consequences of alternative energy development decisions. A general benefit-cost approach must be employed, with the framework of analysis widened from the usual comparative project-evaluation to a comprehensive program approach which makes it possible to assess the relative merits of widely differing policy alternatives, alternatives that may consist of a series of government investments, private investments, of changes in laws or regulations, or of a combination of some or all of these measures.

Many of these alternatives have consequences that can lead to markedly different secondary development patterns on a regional, national, or sometimes even international basis. For example, complete deregulation of natural gas prices and natural gas uses, including imports, in the United States would have significantly different consequences on the utilization patterns of gas versus coal versus fuel oil than a continuation of the present policies, consisting of partial price-deregulations, controls over imports and import contract conditions, and gas allocation rules that require, among other things, the phasing-out of natural gas as a fuel for electric power plants. Policy changes would have direct effects on the pricing and utilization patterns of competing fuels; they would affect crude oil imports, refinery output mixes, and the fortunes of existing and potential coal mines in various regions of the United States and, perhaps, even overseas.[10] They could also affect the location of industry, and most definitely would have a profound effect on the feasibility of the 40 billion dollar Alaska natural gas pipeline.[11] Any valid comparison of the merits of potential alternatives, therefore, has to take into consideration the expected net benefits of such secondary effects. More often than not these will occur in the private rather than the public sector. What is needed, then, is an analytical framework that takes account of both.

Following usual benefit-cost practices the basic objective of the function developed here is the maximization of net benefits or income, which leads to maximum economic efficiency in resource allocation. Under certain restricting assumptions discussed earlier, net benefit maximization is equivalent to minimization of economic costs. Making efficiency, at least initially, the sole criteria for the choice among alternatives has obvious technical advantages. If we ignore, for the moment, the many troublesome problems that are common to any and all objective functions, we find that the choice of efficiency as the goal provides us with an unambiguous ranking order among alternatives.[12] Any one

---

[10]Changes in natural gas prices and use regulations could potentially affect the huge Exxon Carbon-coal mining development in northern Columbia, which is looking for markets for its thermal coal all along the Gulf and Atlantic sea coasts.

[11]See also the case study on the Alaska pipeline.

[12]Problems such as the evaluation of uncertainty and risk, of unpredictable changes in technology and relative price levels, of the valuation of intangibles and nonmarketed outputs, of positive or negative externalities or spillovers, and the troublesome issue of individual and collective time preferences, to mention but a few.

that yields more in terms of income will always appear to be preferable to one that yields less.

Obviously, since income is not an inferior good, it is generally desirable to have more instead of less. However, as every economic textbook proves beyond doubt, it is not really income that we ought to maximize but welfare, or total utility. The result is that an analysis which tells us how to maximize income does not necessarily tell us whether or not we are also maximizing welfare. Nevertheless, the choice of income maximization could be defended if the government would be willing to approach more equitable distribution patterns by redistributing income from the original beneficiaries to those whom the authorities feel to be more deserving. Another alternative is to use appropriate shadow prices as a weighting device. This issue is discussed below.

As long as we are dealing with the comparison of a few individual projects designed to provide similar outputs, we will often find that the distributional consequences of any one of them are roughly the same so that they can be safely disregarded. However, the specific purpose of the function developed here is to analyze not individual projects but whole development strategies, strategies that may be drastically dissimilar in their consequences. This means that we cannot bypass the issue of income distribution. Given this inescapable conclusion, does it still make sense to begin by analyzing the efficiency solution?

The answer is: Yes, it does. First of all, it is useful to know by itself which of the various alternatives will result in the largest increase in total net income to the national economy. Second, only if we do know how much income is obtainable can we make an assessment of the "costs" that various distributional or nonquantifiable objectives may have. The efficiency measure provides us with something of a yardstick that can be used to measure the consequences of the latter even if it cannot tell us what their real value is. In some cases we might find that these costs are unacceptably high. In others it might turn out that the proposed beneficiaries of a specific distributional objective may voluntarily opt for compensatory payments instead, if they find that such payments could make them better off. Knowing the economic losses associated with the realization of nonefficiency goals would greatly facilitate evaluations of potential trade-offs between the multidimensional objectives that usually form part and parcel of any comprehensive energy development program.

What we can conclude is that despite its shortcomings, an income maximization function serves a useful purpose for the evaluation of alternative development policies and investment decisions. What we must remember, however, is that finding the most efficient solution by itself does not answer the question: which of the various policies are preferable in terms of overall community welfare? It only provides us with a tool that can be used to calculate the necessary sacrifices in terms of total income that are needed in order to include other and possibly broader social objectives.

Ideally, the unweighted income maximization function should tell us which of the various alternative development strategies will result in the greatest net increase in real per capita or, perhaps, per family income for the relevant population as a whole. The relevant population consists of the present and future population living within the decisionmaker's jurisdiction. If the latter is represented by a national government, then the analysis should provide an estimate of the net increase in real national income. If it is a regional government, the relevant population is the regional one, and the income to be maximized should be the regional rather than the national one.[13]

The objective function has been formulated in terms of total net benefits, whereby net benefits represent the present-value excess of all social benefits over all social costs, and all required inputs have been treated as costs without an attempt at further disaggregation. A crucial aspect of such an objective function which has as its goal the maximization of net social benefits resulting from a series of interdependent public and private activities, is that it has to account for the opportunity costs of foregone alternatives in all sectors of the relevant economy. In addition, it has to account for the fact that energy users will try to maximize net income (or well-being) from energy-derived outputs. This requires explicit evaluation of the joint costs and benefits of the energy-using systems, rather than of energy supplies alone.

8.3   ECONOMIC COST-BENEFIT CRITERIA FOR PROJECT EVALUATION

Economic cost-benefit analysis of projects, which is the topic of this section, is typically one of the important analyses carried out in the critical appraisal stage—the point at which a final decision is made regarding the acceptance or rejection of a project. In addition to this economic test, we note that a number of other aspects including technical, institutional, and financial criteria also need to be considered in project appraisal.[14] We summarize next some criteria commonly used in the economic or cost-benefit test of a project.[15]

The most basic rule for accepting a project is that the net present value (NPV) of benefits is positive:

$$\text{NPV} = \sum_{t=0}^{T} (B_t - C_t)/(1 + r)^t$$

---

[13]However, these different accounting viewpoints also can lead to sharp differences in expected net benefits or costs from a common project. This has been well illustrated in the discussion of the Alaska gas pipeline project in Chapter XII.

[14]For a discussion of the broader issues involved in energy project evaluation, see Mohan Munasinghe, "Economic Analysis of Small Renewable Energy Projects and National Energy Policy," *Journal of Ambient Energy,* (forthcoming) 1982.

[15]For details, see Ezra J. Mishan, *Cost-Benefit Analysis,* Praeger Publishers, New York, 1976, Ch. 25–30.

where B_t and C_t are the benefits and costs in year t, r is the discount rate, and T is the time horizon.

As described later, B_t, C_t, and r are defined in economic terms and appropriately shadow priced. In particular, the shadow price of r is the accounting rate of interest (ARI).[16] If projects are to be compared or ranked, the one with the highest (and positive) NPV would be the preferred one, i.e., if $NPV^I >$ NPV^{II} (where NPV^i = net present value for project i), then project I is preferred to project II, provided also that the scale of the alternatives is roughly the same.[17]

The internal rate of return (IRR) is also used as a project criterion. It may be defined by:

$$\sum_{t=0}^{T} (B_t - C_t)/(1 + IRR)^t = 0$$

Thus, the IRR is the discount rate which reduces the NPV to zero. The project is acceptable if IRR > ARI, which in most normal cases implies NPV > 0 (i.e., ignoring cases in which multiple roots could occur).[18]

Another frequently used criterion is the benefit-cost ratio (BCR):

$$BCR = \left[\sum_{t=0}^{T} B_t/(1 + r)^t \right] \bigg/ \left[\sum_{t=0}^{T} C_t/(1 + r)^t \right]. \text{ If BCR}$$

$$> 1, \text{ then NPV} > 0$$

and the project is acceptable.

Each of these criteria has its strengths and weaknesses, but NPV is probably the most useful. In the case of energy projects, the benefits of two alternative technologies are often equal (i.e., they both serve the same need or demand). Then the comparison of alternatives is simplified. Thus: $NPV^I - NPV^{II} = \sum_{t=0}^{T} (C_t^{II} - C_t^I)/(1 + r)^t$; since the benefit streams cancel out. Therefore $\sum_{t=0}^{T} C_t^{II}/(1 + r)^t > \sum_{t=0}^{T} C_t^I/(1 + r)^t$; implies $NPV^I > NPV^{II}$. In other words

---

[16]See Annex to this chapter for details of ARI.

[17]More accurately, the scale of each of the projects under review must be altered so that, at the margin, the last increment of investment yields net benefits that are equal (and greater than zero) for all the projects.

[18]This occurs if the annual net benefit stream changes sign several times. Problems of interpretation occur if alternative projects have widely differing lifetimes, so that the discount rate plays a critical role.

the project which has the lower present value of costs is preferred. This is called the least-cost alternative (when benefits are equal). However, even after selecting the least-cost technology, it would still be necessary to ensure that the project would provide a positive NPV.

8.4   SHADOW PRICING OF COSTS AND BENEFITS

As outlined in Chapter 4, shadow prices are the imputed, "correct" economic or social values of inputs or outputs whose market values are either distorted, or for which market prices are not available. Formally, they are defined as the increase in welfare resulting from any marginal change in the availabilities of commodities or factors of production.[19] In other words, a shadow price represents the true economic or social opportunity costs of such inputs and outputs at the margin.

The objectives of shadow pricing may be either efficiency- or social-goal oriented. Efficiency-oriented shadow prices try to establish the actual economic values of inputs and outputs, while socially-oriented shadow prices take account of the fact that the income distribution between different societal groups or regions may be distorted in terms of overall social objectives. This may call for special adjustments in their values, usually by giving greater weight to benefits and costs accruing to the poor relative to the rich. In our analysis, we will place primary emphasis on efficiency shadow pricing (as in Chapter 4 and the first part of this chapter). But we will also make use of social shadow prices and social weights, mainly for determining subsidized energy prices and lifeline rates (Chapter 5).

Frequent causes of market price distortions are import duties, export taxes and subsidies, indirect taxes, price controls, imperfections in labor markets, and absolute shortages of input factors, such as skilled labor, foreign exchange, and investment capital.

Nonpriced inputs and outputs must be shadow-priced to reflect their economic opportunity costs. Major categories of such nonpriced inputs and outputs are public goods and externalities. Public goods are defined as those goods and services that are freely available to all without payments once they have been made available. Examples relevant to energy projects include transportation and navigation facilities, radiowaves, flood protection facilities, and police protection.[20]

Externalities are defined as beneficial or adverse effects imposed on others

---

[19]I. M. D. Little and J. A. Mirrlees, *Project Appraisal and Planning for Developing Countries,* Basic Books, New York, N.Y., 1974. See also L. Squire and H. G. van der Tak, *Economic Analysis of Projects,* World Bank, Johns Hopkins Press, Baltimore, 1975.
[20]e.g., for protecting atomic power plants against mass demonstrations and sabotage.

for which the originator of these effects cannot charge or be charged, as the case may be. Major energy-related examples are: air and water pollution, unaesthetic alterations of the environment, erosion and related flood damages and sedimentation, radiation, threats to health and safety, secondary effects of energy developments on surrounding communities (e.g., boom town effects), nonpriced recreational benefits, and improved access to or through project areas (see also Section 6.7).

Unfortunately, many of the environmental, aesthetic, and health and safety related externalities are not only difficult to measure in physical terms, but even more difficult to convert into monetary equivalents (i.e.. to measure the "willingness to pay" of the parties affected by the externalities").[21] Quite often, therefore, the approach taken by governments is to impose regulations and standards, expressed in physical measurements only, that try to eliminate the perceived external damages.[22] This approach, however, may often be highly inefficient, because no attempt is made to compare the costs of compliance with the real benefits provided (i.e., damages avoided).[23]

Social shadow prices have been developed in order to adjust market prices to reflect social ranking orders. In many countries, and particularly in many developing ones, the distribution of income is highly uneven, with a large number of people living at the margins of existence. Therefore, as mentioned earlier, the change in income or consumption of a poor man, attributable to a given project, is sometimes given a higher weight than the same change in consumption of a rich man. Similar adjustments are often made in favor of projects in lagging regions. Such regionally oriented, preferential developments are common throughout the world. They may consist of subsidies, special credits to

---

[21] For a discussion of these issues, see, for example, Joseph J. Seneca and Michael K. Taussig, *Environmental Economics*, Second Edition, Prentice-Hall, Englewood Cliffs, N.J., 1979, Chapter 3.

[22]This, for example, is the approach taken in most environmental legislation in the United States. See, for example, the 1970 Clean Air Act Amendments, the Federal Water Pollution Control Act Amendments of 1972, The Endangered Species Act of 1973, and the Toxic Substances Control Act of 1976.

[23]In the U.S., the Reagan Administration ordered the Environmental Protection Agency in 1981 to provide formal benefit and cost evaluations of any new, as well as existing, environmental regulations. A review of the many issues involved can be found in: Richard N. L. Andrews and Mary Jo Waits, *Environmental Values in Public Decisions, A Research Agenda,* School of Natural Resources, University of Michigan, Ann Arbor, April 1978. For an analysis of the effects of fixed standards on potential benefits and costs in project analysis, see: Gunter Schramm, "Accounting for Non-Economic Goals in Benefit-Cost Analysis," *Journal of Environmental Management,* (1973), 1, pp. 129–150. Much of the recent works in multiobjective planning and benefit-cost evaluations has centered in the water resources area. For a useful overview of the work in the field see Charles W. Howe, *Benefit-Cost Analysis for Water System Planning,* American Geophysical Union, Washington, D.C. 1971, and David C. Major, *Multiobjective Water Resources Planning,* American Geophysical Union, Washington, D.C., 1977.

regional economic activities, direct government investments, tax rebates and other means, all of which are examples of explicit or implicit shadow pricing.[24]

In the energy field social shadow pricing is practiced implicitly in many countries, usually through means such as preferential tariffs to poor consumers for electricity and gas (see Chapter 5),[25] through subsidized prices for household fuels such as kerosene,[26] or through mandatory allocation and pricing rules that give preference to specific classes of users, such as households as compared to commercial or industrial users.[27] A specific form of implicit shadow pricing is provided through uniform pricing rules for energy on a countrywide basis, regardless of differences in delivery costs. In some countries, such as Thailand, for example, the tariffs are even inverted, i.e., lower prices are charged in low-density, high distribution-cost areas than in high density, low distribution-cost ones (in the case of electricity.)[28]

Many of these shadow prices are implicit only and are usually determined more on the basis of political decisions than explicit and predetermined economic and social criteria. Nevertheless, for planning and investment purposes, formal rules and methodologies are required in order to evaluate the respective benefits and costs of alternative decisions.

Most attempts at applying economic shadow prices of marketed commodities have been made in developing countries. The reason is that in these countries market prices are far more subject to distortions due to regulations, high rates of un- and underemployment, indirect taxation, and absolute shortages of investment capital, foreign exchange, and skilled labor. These are generally of lesser importance in developed countries. In the latter, efforts at economic

---

[24]An important example of such regional, group-oriented preferential treatment in the United States is provided by the Reclamation Act of 1902, which accords highly favorable and preferential treatment to western irrigation farmers (such as zero interest rates and 50 year repayment periods). In Canada, for many years the Federal Government has operated a fiscal equalization scheme based on differences in provincial per capita income levels. It also has provided massive amounts of financial assistance to regions that are considered economically deprived, with such aid amounting to some 11% of total federal expenditures in the early 1970s. *From:* Gunter Schramm, ed., *Regional Poverty and Change,* Canadian Council on Rural Development, Ottawa, 1976, p. 3.

[25]This is usually done by setting low prices for the first few units of essential consumption and higher rates for subsequent units. Such inverted rate structures are common in many countries of the world and have also been introduced more recently in some of the industrialized nations. See, for example, the electric rate structure in Sri Lanka (Table 9.17).

[26]See, for example, the natural gas pricing structure in Bangladesh (Tables 11.7 and 11.10).

[27]In the United States, for example, the 1978 Natural Gas Act allowed substantial increases of the well-head prices of natural gas, but special provisions were included in the law to make industrial users absorb a very large share of the initial rise in prices. See also Walter A. Rosenbaum, *Energy Policies and Public Policy,* Congressional Quarterly Press, Washington D.C., 1981, p. 158. User-group price discrimination is also quite apparent in the Bangladesh data shown in Table 11.7.

[28]See the case study on Thailand's power tariffs.

shadow pricing more often are directed towards the evaluation of nonpriced commodities and environmental, health and safety effects.

One methodology of economic shadow pricing for developing countries generally follows the work of Little and Mirrlees, which was later elaborated and extended by Squire and van der Tak.[29] They developed a methodology that uses uncommitted public income in border prices as the unit of account, or numeraire, expressed in units of the local currency converted at the official rate of exchange.[30] Using border prices as the unit of account has the advantage that all C.I.F. import costs and all F.O.B. export earnings are already expressed in border prices (i.e., they have a shadow price conversion factor equal to one).

However, locally purchased inputs and outputs must be converted to border prices by means of appropriate conversion factors (C.F.). The basic objective is to establish the value of these local inputs and outputs in terms of their true economic opportunity costs. The question asked, therefore, is: "What would be the economic contribution of one additional unit of a given input or output if it were made available (expressed in terms of the numeraire selected)?" Some formal methodologies for deriving conversion factors for a sample of input and output factors (underemployed labor, capital goods and electric energy) are presented in the Annex to this Chapter.

While, formally, shadow prices are supposed to measure values at the margin, in practice data limitations make it necessary to use average values instead. Also, most shadow prices represent nationwide averages only and the level of aggregation is usually quite high. These limitations can have significant implications for their use in specific project evaluations.

The use of shadow prices in project analysis also requires the evaluation of likely changes in factor availabilities. If these are significantly affected, rates of utilization will change, and, as a consequence, the real opportunity costs of these factors will change as well. For those that are already in short supply (such as skilled labor, managerial personnel, or other specific inputs such as transport facilities) shadow prices must be used that indicate their full, foregone economic value in their best alternative uses.[31]

The shadow pricing of nonpriced public goods and of externalities due to environmental, health, and safety effects, is usually based on two broad approaches. One is that of establishing the "willingness to pay" of users or affected parties to either utilize the goods and services or avoid the damages

---

[29]I. M. D. Little and J. A. Mirrlees, *Project Appraisal and Planning for Developing Countries;* L. Squire and H. G. van der Tak, *Economic Analysis of Projects.*
[30]See the annex to this Chapter for details of this numeraire.
[31]For an analysis of these problems in the context of agricultural and water resource development policies see: Gunter Schramm, "Human and Institutional Factors," *Natural Resources Journal,* Vol. XVI, October 1976, pp. 923–937.

imposed. This approach is usually used to measure the value of consumption of goods or services. The other approach is to try to measure objectively the value of these goods, services, or damages in terms of reductions of production costs (e.g., through lower transportation costs from better highways), or in terms of losses of outputs from damages caused (e.g., peeling paint, or loss of earnings from sickness). Attempts at measuring "willingness to pay" have been made through direct survey research methodologies, or indirect measures, such as the travel-cost method used frequently in the evaluation of benefits from nonpriced or underpriced outdoor recreational activities.[32] Within the confines of this volume it is not possible to present and analyze these oftentimes ingenious and sophisticated methodologies. The reader is advised to consult some of the references cited in the footnote below.[33]

8.5   INVESTMENT-DEMAND MANAGEMENT INTERACTIONS

In this section we analyze investment-demand management interactions. Two examples are given which show how investment policies affect the ultimate level of energy consumption, mediated through energy prices and the availability of supply. The first case is appropriate to large integrated commercial energy delivery systems such as electricity and natural gas. In the second example, the implementation of a small decentralized renewable energy technology is the means of effectively meeting and controlling energy demand.

Electric Power

The economic evaluation of an electric power project consists of three basic steps: (a) demand forecast, (b) least cost investment program, and (c) cost-benefit analysis. The investment decision also interacts with the output pricing policy, which in turn affects the demand as summarized in Figure 8.2.

The load-demand forecast is made by systematically taking into account all factors that might have quantitative relationships with future electricity consumption such as price, income and economic growth, population growth,

---

[32]See Marion Clawson and Jack Knetsch, *The Economics of Outdoor Recreation,* Johns Hopkins Press, Baltimore, 1967.

[33]For a sampling of the very large literature on the subject see, for example, Lewis Perl and Frederick Dunbar, "Cost-Effectiveness and Cost-Benefit Analysis of Air Quality Regulations," *The American Economic Review. Papers and Proceedings,* March 1982; Alan Randall and George Tolley, "Revealing Willingness to Pay; The Effects of Experimental Formats, "paper presented at the *94th Annual Meeting, The American Economics Association,* Washington, D.C., December 27–30, 1981; John V. Krutilla and Anthony C. Fisher, *The Economics of Natural Environments,* Johns Hopkins Press, Baltimore, 1975; E. P. Seskin and M. H. Peskin, *An Analysis of the Benefits and Costs of Water Pollution Control,* The Urban Institute, Washington, D.C., 1975.

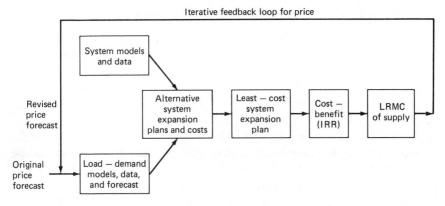

Figure 8.2.   Interaction of investment decision, price and demand forecast.

energy substitution, etc., as well as judgemental factors including effects of energy conservation programs. The assumptions made regarding the future evolution of electricity price is especially critical. Several alternative long-run investment programs (or system expansion plans) are then identified to meet the demand growth, subject to various constraints including minimum reliability of supply standards, environmental and safety criteria, and so on.[34] The least cost or cheapest alternative (in present discounted value terms and using shadow prices) is selected as the optimal program.

However, as discussed in Chapter 4, economic theory indicates that the benefits (to society) of electricity consumption will be maximized if the output prices are set equal to the marginal costs of supply based on the least-cost plan. In practice, prices will be adjusted to reflect not only long-run marginal costs (LRMC) of supply, but also other financial, social, and political criteria. Whatever future prices result, they must be compared to the prices assumed in making the original demand forecast. If there is inconsistency then the demand forecast may have to be adjusted and the investment plan reviewed again.

The above considerations indicate that investment and pricing decisions are closely interrelated and supply policies are inextricably linked to demand management.[35]

---

[34]Ideally, the reliability level also should be optimized. For details, see Mohan Munasinghe, "A New Approach to Power System Planning," *IEEE Trans. PAS,* Vol. PAS-99, May–June 1980, pp. 1198–1209, also available as Reprint No. 147, World Bank, Washington, D.C.; and Mohan Munasinghe and Mark Gellerson, "Economic Criteria for Optimizing Power System Reliability Levels," *The Bell J. of Econ.,* Vol. 10, Spring 1979, pp. 353–65, also available as reprint No. 112, World Bank, Washington, D.C.

[35]For details, see Mohan Munasinghe, "Optimal Electricity Supply: Reliability, Pricing, and System Planning," *Energy Economics,* Vol. 3, July 1981, pp. 140–52.

Solar Photovoltaic (PV) Energy for Agricultural Pumping

The economic evaluation of a decentralized new renewable technology such as the photovoltaic (PV) irrigation pumping system discussed below, and the decision to implement this technology illustrates the close interaction of the economic and financial analyses in controlling or meeting demand.

We assume that the agricultural area under consideration is so remote that electrification via the main power grid would be prohibitively expensive. The principal competitor of the PV system in this case is the diesel pump. Since the comparison is between two discrete alternatives, we need to only look at a single representative farm. But as discussed below, both economic and financial calculations need to be carried out, but carefully separated.

Consider a typical 1 hectare farm that requires about 20 m³ per day of water with a head of 5 m. Since the benefits are identical in both cases, we first seek to find the least-cost alternative by comparing diesel and PV system costs. The basic data for a simplified hypothetical example are summarized in Table 8.1.[36]

Investment Decision.   Let us compare the present value of costs (PVC) of the two alternatives over a 10 year period in (economic) shadow prices (Border rupees, BRs.).

$$\textit{Solar:} \ \mathrm{PVC}_{SE} = 60,000 + \sum_{t=0}^{9} 1000/(1.1)^t = \mathrm{BRs.} \ 66,760$$

$$\textit{Diesel:} \ \mathrm{PVC}_{DE} = 36,000 + 36,000/(1.1)^5 + \sum_{t=0}^{9} [1,050/(1.1)^t + 1,200$$
$$\times (1.03/1.10)^t]^t = \mathrm{BRs.} \ 74,540^{37}$$

On the basis of the above results, the solar PV system is the least cost alternative for irrigation pumping in this area.

Next, assume that due to the increased irrigation, the shadow priced value of the farmer's annual output of grain increases from BRs. 10,000 to BRs. 20,500 per hectare, based on the export price of grain.[38] The present value of benefits for the 10 year period is given by

$$\mathrm{PVB}_E = \sum_{t=0}^{9} (20,500 - 10,000)/(1.1)^t = \mathrm{BRs.} \ 70,970$$

---

[36]This example serves only to illustrate the principles involved. In actual practice, a range of parameters (e.g., different farm sizes, insolation rates, cropping patterns, system costs, etc.) would have to be considered.

[37]The compounding effect of the real increase in fuel prices of 3 percent per annum will partially offset the effect of discounting at the opportunity cost of capital of 10 percent.

[38]This increase in output value is net of any changes in costs of other inputs such as fertilizer and labor.

Table 8.1. Basic Data for Solar PV and Diesel Pumping Systems.

	SOLAR PV		DIESEL	
	MARKET PRICES (DOM. Rs.)	SHADOW PRICES (BORD. Rs.)[1]	MARKET PRICES (DOM.RS.)	SHADOW PRICES (BORD. Rs.)[1]
Initial Cost	66,000	60,000	28,800	36,000[3]
Annual Maintenance Costs	1,100	1,000[2]	1,500	1,050[4]
Annual Fuel Costs	—	—	840	1,200[5]
Lifetime	10 years		5 years	
Discount Rate (%)	15[6]	10[7]	15[6]	10[7]
Annual Inflation Rate (%)	10	—	10	—

[1]All foreign costs converted into border rupees (BRs.) at official exchange rate (OER) of US$1 = BRs. 20.
[2]All foreign exchange costs with 10% import duty.
[3]Import subsidy of 20% on diesel pumps provided to farmers.
[4]Conversion factor = 0.7 based on spare parts and labor.
[5]Subsidy of 33% provided on diesel fuel imported at US$0.3 per liter. International fuel price assumed to rise at 3% per annum in real terms.
[6]Bank borrowing rate for farm loans.
[7]Opportunity cost of capital used as proxy for ARI.

Since NPV = PVB_E − PVC_{SE} = BRs. 4,210, there is a positive net benefit to the country by installing solar PV pumps in farms. These results are summarized in Table 8.2. Therefore, from a national viewpoint, the government is justified in taking a policy decision to encourage the use of solar PV systems by farmers, i.e., to stimulate demand and encourage this form of energy consumption.

Implementing the Investment Policy.   However, farmers will make their decisions on irrigation pumping on the basis of private financial costs and benefits. Therefore, let us analyse the present value of costs of the solar and diesel alternatives in financial terms, using market prices (Domestic rupees, DRs.):

$$Solar: PVC_{SF} = 66,000 + \sum_{t=0}^{9} 1100 \times (1.10/1.15)^t = DRs. 75,080^{39}$$

$$Diesel: PVC_{DF} = 28,800 + 28,800 \times (1.10/1.15)^5$$
$$+ \sum_{t=0}^{9} [1500 \times (1.10/1.15)^t + 840 \times (1.03$$
$$\times 1.10/1.15)^t] = DRs. 72,110^{40}$$

[39]In the financial calculation, the compounding effects of the 10 percent annual inflation rate will partially offset the effect of discounting at the market interest rate of 15 percent.
[40]The compounding effects of the 3 percent per annum increase in real fuel prices is included.

Table 8.2. Economic and Financial Tests for Investment Decision. (All amounts are in present value terms)

ITEM	A NATIONAL VIEWPOINT: SHADOW PRICES (BRs.)	B PRIVATE VIEWPOINT (INITIAL): MARKET PRICES BEFORE POLICY CHANGES (DRs.)	C PRIVATE VIEWPOINT (INTERIM): MARKET PRICES AFTER FIRST POLICY CHANGE (DRs.)	D PRIVATE VIEWPOINT (FINAL): MARKET PRICES AFTER SECOND POLICY CHANGE (DRs.)
Solar PV Costs	$PVC_{SE} = 66,760$	$PVC_{SF} = 75,080$	$PVC_{SF} = 75,080$	$PVC_{SF} = 75,080$
Diesel Costs	$PVC_{DE} = 74,540$	$PVC_{DF} = 72,110$	$PVC_{DF} = 85,070$	$PVC_{DF} = 85,070$
Irrigation Benefits	$PVB_E = 70,970$	$PVB_F = 73,670$	$PVB_F = 73,670$	$PVB_F = 78,000$

CONCLUSIONS

Condition		Consequence
A. *Shadow Priced Values:*		
$PVC_{SE} < PVC_{DE}$	—	Solar PV is the economically preferred least-cost technology.
$PVB_E > PVC_{SE}$	—	Investment in solar PV is economically justified.
B. *Market Priced Values Before Policy Change:*		
$PVC_{SF} > PVC_{DF}$	—	Farmers will prefer diesel to solar PV pumps, financially.
C. *Market Priced Values After First Policy Change:*		
$PVC_{SF} < PVC_{DF}$	—	Farmers will prefer solar PV to diesel pumps, financially.
$PVB_F < PVC_{SF}$	—	Farmers will find solar PV pumps financially unprofitable to install.
D. *Market Priced Values After Second Policy Change:*		
$PVC_{SF} < PVC_{DF}$	—	Farmers will prefer solar PV to diesel pumps, financially.
$PVB_F > PVC_{SF}$	—	Farmers will find solar PV pumps financially profitable to install.

Clearly, since $PVC_{SF} > PVC_{DF}$, the average farmer who wished to acquire a pumping system, would prefer to purchase a diesel pump rather than a solar PV system. Because the market price signals are distorted the private individual's decision does not coincide with the national policy decision based on shadow prices (as summarized in Table 8.2). Therefore, the government must adjust the market prices or adopt other policies that will make the solar PV option attractive to farmers.

First, the authorities could remove the 10 percent import duty on solar PV systems and spare parts. Then $PVC_{SF} = DRs. 68,250$, and the solar PV pump is the least cost alternative in both financial and economic terms—but suppose that this measure would also require the reduction of import duties on many similar photovoltaic and electric components, which the government is unwilling to consider.

A second policy alternative is to raise the price of diesel fuel until PVC_{DF} exceeds PVC_{SF}—but let us assume that there is a strong antiinflation lobby that can block increases in the diesel fuel price on the grounds that this would raise transport costs too much.

A third approach might be for the government to legislate that farmers could no longer buy diesel pumps. This nonprice policy option is cumbersome and has its own host of implementation difficulties. A fourth possibility is for the government to provide low interest agricultural loans or credits to buy solar pumps.

Finally the government could remove the import subsidy on diesel motors. This increases PVC_{DF} to DRs. 85,070, and since this is greater than PVC_{SF}, the farmers will voluntarily prefer solar PV pumps to diesel pumps. Let us assume that this policy alternative is selected for implementation (see Table 8.2).

But we still have to consider the financial benefits aspect. The irrigation program increases the market value of farm output from DRs. 8,500 to DRs. 17,425 based on the guaranteed government purchase price for grain (which is 85 percent of the world market or export price). The present value of financial benefits for 10 years of output is:

$$PVB_F = \sum_{t=0}^{9} (17,425 - 8,500) \times (1.10/1.05)^t = DRs. 73,670$$

Since PVB_F is less than the value $PVC_{SF} = DRs. 75,080$, the average farmer will be financially worse off if he adopts the (least cost) solar PV technology.

The government must therefore make a further change in policy. For example, it could raise the domestic grain price to 90 percent of the world market price.[41] Then the new $PVB_F = 78,000 > PVC_{SF}$ and farmers will voluntarily begin using the solar PV systems (see Table 8.2).

Thus, this case study shows that demand management by implementation of a given investment decision (based on shadow prices) may require a related series of wide-ranging policy decisions by the government.

ANNEX TO CHAPTER 8
SELECTED TOPICS IN SHADOW PRICING

In this annex, we discuss in somewhat greater detail several topics in shadow pricing referred to already in Chapter 8, including numeraire, application of shadow prices, and conversion factors.

NUMERAIRE

To derive a consistent set of economic prices for goods and services a common yardstick or numeraire to measure value is necessary, as illustrated by a simple example.

If one wishes to compare avocados with coconuts, the equivalent units might be either one avocado for one coconut, or one kilo of avocados for one kilo of coconuts. In the first instance, the common yardstick is one fruit; in the second, it is the unit of weight. Clearly, if the weights of the two types of fruits are different, the result of the comparison will depend on the numeraire used.

With a numeraire of economic value the situation is more complicated, because the same nominal unit of currency may have a different value depending on the economic circumstances in which it is used. For example, a rupee's worth of a certain good purchased in a duty free shop is likely to be more than the physical quantity of the same good obtained for one rupee from a retail store, after import duties and taxes have been levied. Therefore, it is possible to distinguish intuitively between the border-priced rupee, which is used in international markets free of import tariffs, and a domestic-priced rupee, which is used in the domestic market subject to various distortions. A more sophisticated example of the value differences of a currency unit in various uses arises in countries where the aggregate investment for future economic growth is considered inadequate. In these instances, a rupee of savings that could be invested to increase the level of future consumption may be considered more valuable than a rupee devoted to current consumption.

The choice of the numeraire, like the choice of a currency unit, should not influence the economic criteria for decisionmaking except in relation to magnitude, provided the same consistent framework and assumptions are used in the analysis. For example, only one difference exists between a study using cents as units and one using rupees (where the rupee is defined as one hundred cents). In the study using cents all monetary quantities will be numerically one hundred times as large as in the one using

---

[41]They may, however, have to also provide an equivalent subsidy to urban grain consumers to keep down their cost of living.

rupees. Therefore, a numeraire may be selected purely on the basis of convenience of application.

A most appropriate numeraire in many instances is a unit of uncommitted public income at border prices. Essentially, this unit is the same as freely disposable foreign exchange available to the government, but expressed in terms of units of local currency converted at the official exchange rate.[1] The discussion in the next section is developed in relation to this particular yardstick of value. The border-priced numeraire is particularly relevant for the foreign exchange scarce developing countries. It represents the set of opportunities available to a country to purchase goods and services on the international market.

APPLYING SHADOW PRICES

The estimation and use of shadow prices is facilitated by dividing economic resources into tradeable and nontradeable items. The values of directly imported or exported goods and services are already known in border prices, that is, their foreign exchange costs converted at the official exchange rate. Locally purchased items whose values are known only in terms of domestic market prices, however, must be converted to border prices, by multiplying the former prices by appropriate conversion factors (CF). Therefore, tradeables and nontradeables are treated differently.

For those tradeables with infinite elasticities—of world supply for imports and of world demand for exports—the cost, insurance, and freight (C.I.F.) border price for imports and the free-on-board (F.O.B.) border price for exports may be used with a suitable adjustment for the marketing margin. If the relevant elasticities are finite, then the change in import costs or export revenues, as well as any shifts in other domestic consumption or production levels or in income transfers, should be considered (see section on Conversion Factors for details). The free trade assumption is not required to justify the use of border prices since domestic price distortions are, in effect, adjusted by netting out all taxes, duties, and subsidies.

To clarify this point, imagine a household in which a child is given an allowance of twenty pesos a month as pocket money. The youngster may purchase a bag of lollipops from the grocery store at a price of two pesos. Since the parents want to discourage consumption of sweets, however, they impose a fine of one peso on each bag of lollipops. The fine is exactly like an import duty, and the child must surrender three pesos for every bag of candy (valued at its domestic price, inside the household). From the family's perspective however, the total external payment for the item is only two pesos,

---

[1] Ian M.D. Little and James A. Mirrlees, *Project Appraisal and Planning for Developing Countries*, Chap. 9; and Lyn Squire and Herman G. van der Tak, *Economic Analysis of Projects*, Chap. 3. A numeraire based on private consumption is advocated in: Partha Dasgupta, Amartya Sen, and Stephen Marglin, *Guidelines for Project Evaluation* (UNIDO), United Nations, New York, 1972. For a disscussion of the different assumptions used in various approaches to shadow pricing, see *Social and Economic Dimensions of Project Evaluation*, Harry Schwartz and Richard Berney (eds.), Inter-American Development Bank, Washington, D.C., 1977; and Deepak Lal, *Methods of Project Analysis*, World Bank Staff Occasional Papers No. 16, World Bank, Washington, D.C., 1974.

because the one peso fine is a net transfer within the household. Therefore, the true economic cost, or shadow price, of the bag of lollipops (valued at its border price) is two pesos, when the impact of the fine on the distribution of income between parent and child is ignored.

Two types of nontradeable economic resources are discussed. A nontradeable is conventionally defined as a commodity whose domestic supply price lies between the F.O.B. and C.I.F. prices for export and import, respectively. Items that are not traded at the margin because of prohibitive trade barriers, such as bans or rigid quotas, are also included within this category. If the increased demand for a given nontradeable good or service is met by the expansion of domestic supply or imports, the associated border-priced marginal social cost (MSC) of this increased supply is the relevant resource cost. If decreased consumption of other domestic or foreign users results, the border-priced marginal social benefit (MSB) of this foregone domestic consumption or of reduced export earnings would be a more appropriate measure of social costs.

The socially optimal level of total consumption for the given input (Q_{opt}) would lie at the point where the curves of MSC and MSB intersect. Price and other distortions lead to nonoptimal levels of consumption $Q \neq Q_{opt}$ characterized by differences between MSB and MSC. More generally, if both effects are present, a weighted average of MSC and MSB should be used. The MSB would tend to dominate in a short-run, supply constrained situation; the MSC would be more important in the longer run, when expansion of output is possible.

The MSC of nontradeable goods and services from many sectors can be determined through appropriate decomposition. For example, suppose one rupee's worth, in domestic prices, of the output of the domestic construction sector may be broken down successively into components. This would include capital, labor, and materials, which are valued at rupees $a_1, a_2, \ldots, a_n$—in border prices. Since the conversion factor of any good is defined as the ratio of the border price to the domestic price, the construction conversion factor equals $\sum_{i=1}^{n} a_i$.

The standard conversion factor (SCF) may be used with nontradeables that are not important enough to merit individual attention or lack sufficient data. The SCF is equal to the official exchange rate (OER) divided by the more familiar shadow exchange rate (SER), appropriately defined. Converting domestic priced values into border price equivalents by applying the SCF to the value is conceptually the inverse of the traditional practice of multiplying foreign currency costs by the SER (instead of the OER), to convert to the domestic price equivalent. The standard conversion factor may be approximated by the ratio of the official exchange rate to the free trade exchange rate (FTER), when the country is moving toward a freer trade regime:[2]

$$\text{SCF} = \frac{\text{OER}}{\text{FTER}} = \frac{eX + nM}{eX(1 - t_x) + nM(1 + t_m)}$$

---

[2] For a more detailed discussion of alternative interpretations of the SCF and SER, see Lyn Squire and Herman G. van der Tak, *Economic Analysis of Projects*; and Bela Balassa, "Estimating the Shadow Price of Foreign Exchange in Project Appraisal," *Oxford Economic Papers*, Vol. 26, July 1974, pp. 147–68.

where X = F.O.B. value of exports, M = C.I.F. value of imports, e = elasticity of export supply, n = elasticity of import demand, t_x = average tax rate on exports (negative for subsidy), and t_m = average tax rate on imports.

The most important tradeable inputs used in the energy sector are capital goods and petroleum-based fuels. Some countries may have other fuels available, such as natural gas or coal deposits. If no clear-cut export market exists for these indigenous energy resources, then they cannot be treated like tradeables. In addition, if there is no alternative use for the fuels, an appropriate economic value is the MSC of production or of extracting gas or coal, plus a markup for the discounted value of future consumption foregone, or "user cost" (see Chapter 4 for details). If another high value use exists for this fuel, the opportunity cost of not using the resource in the alternative use should be considered the economic value of the fuel. The most important nontradeable primary factor inputs are labor and land, the next subjects for discussion.

Consider a typical case of unskilled labor in a labor surplus country—for example, rural workers employed for dam construction. The foregone output of workers used in the electric power sector is the dominant component of the shadow wage rate (SWR). Complications arise because the original rural income earned may not reflect the marginal product of agricultural labor and, furthermore, for every new job created, more than one rural worker may give up former employment.[3] Allowance must also be made for seasonal activities such as harvesting, and overhead costs like transport expenses. The foregoing may be represented by the following equation for the efficiency shadow wage rate (ESWR):

$$ESWR = a \cdot m + c \cdot u,$$

where m and u are the foregone marginal output and overhead costs of labor in domestic prices, and a and c are corresponding conversion factors to convert these values into border prices.

Consider the effect of these changes on consumption patterns. Suppose a worker receives a wage W_n in a new job, and that the income foregone is W_o, both in domestic prices; note that W_n may not necessarily be equal to the marginal product foregone m. It could be assumed, quite plausibly, that low-income workers consume the entire increase in income ($W_n - W_o$). Then this increase in consumption will result in a resource cost to the economy of $b(W_n - W_o)$. The increased consumption also provides a benefit given by $w (w_n - w_o)$, where w represents the MSB, in border prices, of increasing domestic-priced private sector consumption by one unit. Therefore,

$$SWR = a \cdot m + c \cdot u + (b - w)(W_n - W_o)$$

The letter b represents the MSC to the economy, resulting from the use of the increased income. For example, if all the new income is consumed, then b is the relevant consumption conversion factor or resource cost (in units of the numeraire) of making available to consumers one unit worth (in domestic prices) of the marginal

---

[3]See John R. Harris and Michael P. Todaro, "Migration, Unemployment, and Development," *American Economic Review*, Vol. 60, March 1970, pp. 126–42.

basket of goods they would purchase. In this case $b = g \cdot CF_i$, where g is the proportion or share of the i th good in the marginal consumption basket, and CF_i is the appropriate conversion factor.

The corresponding MSB of increased consumption may be decomposed further: $w = d/v$, where $1/v$ is the value (in units of the numeraire) of a one-unit increase in domestic-priced consumption accruing to someone at the average level of consumption $\bar{c}$. Therefore, v may be roughly thought of as the premium attached to public savings, compared to "average" private consumption. Under certain simplifying assumptions, $b = 1/v$. If $MU(c)$ denotes the marginal utility of consumption at some level c, then $d = MU(c)/MU(\bar{c})$. Assuming that the marginal utility of consumption is diminishing, d would be greater than unity for "poor" consumers with $c < \bar{c}$, and vice versa.

A simple form of marginal utility function which could be assumed is $MU(c) = c^{-n}$.

$$\text{Thus, } d = MU(c)/MU(\bar{c}) = (\bar{c}/c)^n$$

Making the further assumption that the distribution parameter $n = 1$, gives:

$$d = \bar{c}/c = \bar{i}/i$$

where $\bar{i}/i$ is the ratio of net incomes, which may be used as a proxy for the corresponding consumption ratio.

The consumption term $(b - w)$ in the expression for SWR disappears if at the margin (a) society is indifferent to the distribution for income—or consumption—so that everyone's consumption has equivalent value; and (b) private consumption is considered to be as socially valuable as uncommitted public savings, that is, the numeraire.

The appropriate shadow value placed on land depends on its location. Usually, the market price of urban land is a good indicator of its economic value in domestic prices, and the application of an appropriate conversion factor, such as the SCF, to this domestic price will yield the border-priced cost of urban land inputs. Rural land that can be used in agriculture may be valued at its opportunity cost, the net benefit of foregone agricultural output. The marginal social cost of other rural land is usually assumed to be negligible, unless there is a specific reason to the contrary. Examples might be the flooding of virgin jungle because of a hydroelectric dam that would involve the loss of valuable timber, or spoilage of a recreational area that has commercial potential.

The shadow price of capital is usually reflected in the discount rate or accounting rate of interest (ARI), which is defined as the rate of decline in the value of the numeraire over time. Although there has been much discussion concerning the choice of an appropriate discount rate,[4] in practice the opportunity cost of capital (OCC) may be

[4]See, for example, Stephen Marglin, "The Social Rate of Discount and the Optimal Rate of Investment," *Quarterly Journal of Economics*, 1963; Ezra J. Mishan, *Cost-Benefit Analysis*, Chap. 31–34; and Arnold C. Harberger, *Project Evaluation*, Macmillan and Co., London, 1972.

used as a proxy for the ARI, in the pure efficiency price regime. The OCC is defined as the expected value of the annual stream of consumption, in border prices net of replacement, which is yielded by the investment of one unit of public income at the margin. A simple formula for ARI, which also includes consumption effects, is given by:

$$\text{ARI} = \text{OCC} \left[s + (1 - s)w/b \right],$$

where s is the fraction of the yield from the original investment that will be saved and reinvested.

Usually the estimation of shadow prices on a rigorous basis is a long and complex task. Therefore, the energy sector analyst is best advised to use whatever shadow prices have already been calculated. Alternatively, the analyst would estimate a few important items such as the standard conversion factor, opportunity cost of capital, and shadow wage rate. When the data are not precise enough, sensitivity studies may be made over a range of values of such key national parameters.

CONVERSION FACTORS (CF)

For a given commodity, the conversion factor may be defined as the ratio of its border price (in terms of the chosen foreign exchange numeraire) to its domestic market price. As an example, we analyze below the general case of an input into the energy sector that is also consumed by nonenergy sector users. The input is supplied through imports as well as domestic production.

TS is the total supply curve consisting of the sum of WS and DS which are the world and domestic supply curves (see Figure 8.A.1). DD is the domestic demand curve of nonenergy sector users, while TD is the total domestic demand when the requirements of the energy sector PD are included. The domestic market price of the input is driven up from P to $P + dP$ due to the effect of the increased demand PD. The corresponding world market C.I.F. (import) prices are smaller by a factor $1/(1 + t_m)$ where t_m is the rate of import duty on the input. The level of consumption due to nonenergy sector users only (i.e., without the power sector demand), would be $Q = D$, of which M would be supplied through imports and S from domestic production. The total domestic consumption with PD included is $Q = dQ$, which is composed of nonenergy sector demand $D - dD$ and energy sector demand $(dQ + dD)$. In this situation, the respective quantities $M + dM$ and $S + dS$ are imported and domestically supplied.[5]

Let us consider the marginal social cost (MSC) of meeting the energy sector demand. Firstly, there is the increased expenditure of foreign exchange (i.e., already in border prices) for extra imports:

$$d\text{FX} = \text{M} \cdot d\text{P} \cdot (1 - e_w)/(1 + t_m)$$

---

[5]For convenience, we develop the formula in terms of increments dP, dQ, etc.

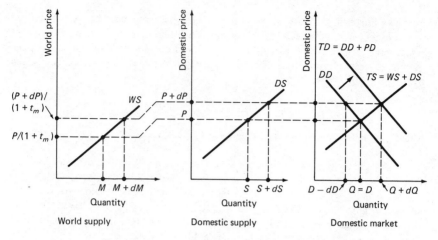

Figure 8.A.1.   Conversion factor for an importable.

where $e = P \cdot dM/M \cdot dP$ is the price elasticity of world supply. Secondly, we consider the increased resource cost of additional domestic supply:

$$dRC = f \cdot (S \cdot dP \cdot e_d)$$

where $e_e = P \cdot dS/S \cdot dP$ is the price elasticity of domestic supply, and f is the factor which converts the cost of supply from domestic to border prices. Next, the decrease in the benefit of consumption of nonenergy sector consumers should be considered as a cost:

$$dBD = w_c \cdot (D \cdot dP \cdot n_d)$$

where $n_d = P \cdot dD/D \cdot dP$ is the relevant price elasticity of demand (absolute value), and w_c is the social weight attached to domestic consumption at the relevant consumption level. Fourthly, there is the effect due to the increase in income of nonenergy sector consumers:

$$dI_c = (b_c - w_c) \cdot [D \cdot dP \cdot (n_d - 1)]$$

where b_c is the shadow priced resource cost of increasing domestic consumption (of other goods and services) by one unit, and once again w_c is the corresponding benefit. Finally, we evaluate the impact of increased income accruing to domestic producers:

$$dI_p = (b_p - w_p) \cdot (S \cdot dP)$$

where b_p and w_p represent the shadow priced resource cost and marginal social benefit (MSB) respectively of a one-unit increase in the income of the producers.

The total cost of the input to the energy sector in shadow prices is

$$dBP = dFX + dRC + dBD + dI_c + dI_p$$

The corresponding cost of the input in domestic prices is:

$$dDP = P \cdot (dD + dQ) = dP \cdot (D \cdot n_d + S \cdot e_d + M \cdot e_w)$$

Therefore, the expression for the conversion factor may be written:

$$CF_M = \frac{dBP}{dDP}$$
$$= \frac{M[(1 + e_w)/(1 + t_m)] + S(fe_d + b_P - w_p) + D(b_c n_d + w_c - b_c)}{Me_w + Se_d + Dr_d}$$

Often, this expression may be considerably simplified:
Tradeable Input
Case I. Small country assumption, i.e., $e_w \to \infty$:

$$CF_M = \frac{1}{1 + t_m}$$

Nontradeable Input
Case IIa. No imports and perfectly elastic supply, i.e., $M = 0$ (thus $S = D$) and
$e_d \to \infty$:
or Case IIb. No imports, perfectly inelastic demand and no income effects, i.e.,

$$M = 0, n_d = 0, w_p = b_p \text{ and } w_c = b_c:$$
$$CF_2 = f$$

Case IIIa. No imports and perfectly inelastic demand, i.e.,

$$M = 0, n_d = 0, w_p = b_p \text{ and } w_c = b_c:$$

or Case IIIb. No imports, perfectly inelastic supply, and no income effects, i.e.,

$$M = 0, e_d = 0, w_p - b_p \text{ and } w_c = b_c:$$
$$CF_3 = b_c$$

We note that the corresponding conversion factor for an energy sector input which is domestically produced for export as well as for nonenergy sector domestic consumption may be derived analogously:

$$CF_x = \frac{X[(n_w - 1)/(1 - t_x)] + S(fe_d + b_p - w_p) + D(b_c n_d + w_c - b_c)}{Xn_w + Se_d + Dn_d}$$

Where X represents exports, t_x is the tax rate on exports, n_w is the price elasticity of world demand, and the other symbols are as defined earlier.

Part B
Case Studies

In the following chapters, a number of energy problems are analyzed to illustrate the application of the theoretical principles presented in the first part of the book. Practical solutions to these problems are discussed, with particular emphasis on demand management issues. Energy investment and supply-side issues are also examined, to the extent that they interact with demand management policies.

The topics have been selected to cover a variety of energy supply subsectors and uses. The countries studied differ significantly in their degree of economic development, ranging from low income nations such as Bangladesh and Sri Lanka, through the middle income countries Brazil, Costa Rica, and Thailand, to a high income industrialized nation, the United States. The chosen countries also differ widely in terms of size, geographic location and climate, energy resource endowments, and usage patterns.

9

Sri Lanka: Energy Management Issues in a Low Income, Oil Importing Country

The energy situation in a typical low income developing country, Sri Lanka, is examined in this chapter, highlighting, in particular, demand management issues. The first section below contains an introduction to the Sri Lanka energy sector and a brief overview of the principal energy problems the nation faces. In Section 9.2, the management of electricity demand is analyzed, with particular emphasis on power tariff policy. Finally, the main issues involved in choosing among several different energy sources, including kerosene, traditional fuels, and liquified petroleum gas (LPG), for domestic cooking, are examined in Section 9.3.

9.1. BACKGROUND AND OVERVIEW OF THE ENERGY SECTOR

Sri Lanka is a tropical island off the southern tip of India with a surface area of about 25,000 square miles, a population of 14.9 million, and an annual per capita gross national product (GNP) of US$240 in 1980. The total energy consumption in the same year was about 3.7 million tons of oil equivalent (toe), consisting of 60 percent traditional fuels, 27% petroleum products, and 13% electricity.

Woodfuel use has grown little during the past decade, except for a small increase in the last few years, due to the growth of industrial consumption. Commercial energy consumption has also been roughly constant during the period 1970–77, with a decline in oil use being counterbalanced by increasing electricity consumption. Since 1978, however, changes in government policies have resulted in a spurt in economic growth and an associated increase in commercial energy demand of about 7% per annum.

Table 9.1 provides a summary breakdown of energy consumption in Sri Lanka during 1979. Households consumed over half of the total energy used,

Table 9.1. Sri Lanka Energy Consumption Breakdown in 1979 (10^3 toe).

CONSUMER CATEGORY	ENERGY CONSUMPTION							
	TOTAL		WOODFUEL		PETROLEUM PRODUCTS		ELECTRICITY[1]	
	10^3 toe	PER-CENT	10^3 toe	PER-CENT	10^3 toe	PER-CENT	10^3 toe	PER-CENT
Domestic	1890	55	1600	73	205	23	85	23
Industrial	936	27	500	23	237	27	199	55
Transport	420	12	—	—	420	48	—	—
Commercial	170	5	100	4	1	—	69	19
Agriculture and Other	27	1	—	—	17	2	10	3
TOTAL	3443	100	2200	100	880	100	363	100
Losses (percent of gross supply)	5		not applicable		10		15	
Gross Supply 10^3 toe	3605		2200		978[2]		427	
Gross Supply Share of total	100		61		27		12	

Source: Sri Lanka Government data and authors' estimates.
[1]Assuming that 1 toe is equivalent to 3570 kWh, based on a 30% conversion efficiency for average thermal generation.
[2]Excluded reexports of petroleum products.

while industry, transport, and commercial accounted for 27, 12, and 5%, respectively. Analysis by energy source indicates that 61, 27, and 12% of the gross energy supplied consisted of woodfuel, oil, and electricity respectively; almost identical to the 1980 percentages. The household sector used 73 percent of the traditional fuel, and the rest was consumed mainly by industries. Almost half of total oil consumption was for transportation, while residential and industrial users each accounted for about one quarter. In the other commercial energy subsector, electricity, industry consumed well over half the total consumption while the shares of households and commercial users were 23 and 19 percent respectively.

Sri Lanka's forests, a major renewable energy resource, have declined from 7.4 million acres (or about 45 percent of the total land area) in 1955, to about 4 million acres in 1980, due mainly to conversion to agricultural use. Heavy woodfuel use has also contributed to the extensive deforestation, since this source supplies about 85 percent of household energy needs. Thus, while the natural regenerative capacity of forest cover is estimated at one million tons (or 0.4 million toe) per year, woodfuel consumption has increased f.om about

3 to about 4.5 million tons per year between 1955 and 1980, with demand growth curtailed in recent years, mainly due to supply constraints.

Hydroelectric power is the second major indigenous energy resource in Sri Lanka. Out of a total estimated potential of over 2000 MW and 6500 GWh (2.2 million t.o.e.) to 369 megawatts (MW) of hydro capacity supplying 1600 gigawatt-hours (GWh) has already been developed. By 1987, another 572 MW of capacity supplying 1500 GWh will have been harnessed along the Mahaweli River. In the 1980 to 82 period however, a combination of rapid growth of demand unforeseen by power planners, and inadequate rainfall in the catchment areas of existing hydro schemes had resulted in power shortages of 8 hours a day for several months at a time. To meet this short-term deficit, the government has already installed 120 MW of gas turbines, and is expected to add a further 100 MW of diesel generating plant shortly.

Other known indigenous energy sources are relatively small. Several oil companies have signed exploration contracts with the government to drill off the northwest shore, but the prospect of a major find is not very great. The eastern part of the island shows some limited geothermal potential in the form of hot springs, although no systematic evaluation of this energy source has been attempted. Minor deposits of monazite sands (made up of about 10% Thorium and 0.1% Uranium Oxide) have also been identified. Renewable energy sources including biomass, solar, wind, and minihydropower offer limited short-term possibilities, and could also make a significant contribution in the longer term, if a coordinated program of development is adopted.

Oil imports make up the gap between total demand and indigenous energy supply. Since the oil price increase of 1973, petroleum imports have become an increasingly significant factor in the balance-of-payments. Between 1973 and 1978, the oil import bill grew rather slowly from about 10 to 18% of nonpetroleum export earnings, mainly because of good export performance, reduced economic growth, and strict import controls. Since 1978, however, the most recent round of international oil price increases and rapid acceleration of economic growth and imports following liberalization of the economy have exarcerbated the situation. Thus, in 1980, net petroleum imports amounted to 1.2 million toe or 35% of the value of nonoil exports, and this oil import burden is likely to continue to increase over the medium term.[1]

Principal Issues.

The chief problems and issues in the Sri Lanka energy sector have arisen from the three underlying factors discussed earlier, that is, the short-run electric

---

[1] In 1980, 1.96 million toe of petroleum products was imported, of which 0.74 toe was reexported in the form of refined products such as naptha, fuel oil, and bunkers.

power deficits, the increasing oil import bill, and the steady depletion of forest reserves.

There are five important issues to be dealt with in formulating a short-term energy strategy to support the economic development effort up to about 1985. First, the energy sector institutional framework must be rationalized to improve the integration, coordination, and consistency of energy policymaking and implementation. Unfortunately, political and bureaucratic constraints limit the extent of the reforms that can be realistically applied to the existing framework that consists of many energy institutions with overlapping responsibilities that fall under different ministries.

Therefore, as an initial step, a basic two-tier structure could be adopted. The first layer, consisting of the Ministry of Power and Energy (MPE) and the newly formed Authority for Natural Resources, Energy, and Science (ANRES), would share the responsibility for overall integrated planning and policymaking. In the second tier, line agencies involved with energy supply such as the Ceylon Electricity Board (CEB), Ceylon Petroleum Corporation (CPC), Mahaweli Development Authority (MDA), Forestry Department (FD), and Colombo Gas and Water Company (CGWC), as well as energy users like the Central Transport Board (CTB), Ceylon General Railways (CGR), Greater Colombo Economic Commission (GCEC), local authorities, and others, would continue to implement policies, undertake more decentralized planning, and carry out daily activities and operations. A good two way flow of information between the two tiers would be essential for the successful functioning of the system. The MPE and ANRES should act through the President, to overcome interministerial and interagency conflicts.

The second major short-run issue concerns pricing policies in the petroleum products subsector. In general, the pricing of most refinery outputs is consistent with international price levels. The principal divergences were in the retail prices of gasoline and kerosene. By early 1981, there was a premium of over 70% on gasoline prices (about Rupees 45 or US$2.50 per imperial gallon), justified mainly on the income distributional argument that private motor car owners are relatively better off. The surplus derived from this markup was partially offset by a 33 percent subsidy on the price of kerosene (about Rupees 17.50 or nearly US$1 per imperial gallon), because the latter was considered to be the poor man's fuel.

This situation was, however, a major improvement on the pre-September 1979 period, when the price of kerosene was about one sixth the 1981 price. At that time, it was estimated that less than one third of the total kerosene sold was used by low income households, the intended beneficiaries of the subsidy. Better off households accounted for about three fourths of the remaining kerosene consumption, while industrial and commercial users shared the rest. To reduce the massive budgetary subsidy of almost Rupees 400 million, the government started a phased program of kerosene price increases. This was coupled with a kerosene stamp scheme (KSS) of Rupees 9.50 per family per month to enable low income families to meet their basic needs for kerosene. More generally, further payments were provided to needy households under the stamp scheme, to subsidize the consumption of a package of basic commodities including foodstuffs as well as kerosene. As a result, kerosene sales declined by about 25 percent from 1978 to 1980, in contrast to a 18 percent increase between 1976 and 1978. Since the government already had the administrative machinery in place to identify target low income families, the implementation costs of the program were relatively small compared to the savings in the subsidy and reduction in "leakages" of cheap kerosene for industrial and commercial use.[2]

Elimination of the remaining kerosene subsidy by raising the price to economically efficient levels would be desirable. First, it would further reduce the gap between kerosene and gasoline prices, thus diminishing the incentive to blend the two fuels for use in gasoline engines. Second, it would permit the price of diesel (which is in short supply) to be raised relative to the heavier grades of oil, and thus encourage the consumption of the latter which is in excess supply. If the subsidy on kerosene is not eliminated, then an increase in diesel prices could lead to substantial substitution of kerosene for diesel.

In 1981, losses accounted for about 70 percent of the gas pumped into the piped supply system serving the capital city of Colombo. The remedying of this situation, preferably by closing down the pipeline and strengthening the bottled LPG distribution system in Colombo, could do much to ease demand on electricity, liquid fuels, and woodfuel. (See also Section 9.3 for an analysis of the relative demand for these types of energy for domestic cooking.)

The third issue revolves around the promising prospects for establishing a viable program of energy demand management and conservation. In the electricity subsector, relatively simple measures such as reducing power factors by installing capacitative correction, and increasing the efficiency of air conditioning and electric motors among large industrial and commercial customers, would make it possible to save up to 5% of total consumption. Similarly, about 15% of oil use could be conserved with straightforward improvements of boilers and combustion practices, substitution of the lighter fractions by heavier grades of refined oil, and increased efficiency of heat exchangers used in the drying of tree crops. These conservation efforts would be greatly enhanced by appropriate price signals—in particular, by increasing and restructuring electricity tariffs (as described in Section 9.2), and adjusting the price differential between the lighter and heavier fuel oils (as discussed above).

---

[2]For a more detailed discussion of the merits of commodity subsidies to meet the basic needs of the poor, see Chapter 5.

The more efficient use of woodfuel (especially in domestic cooking) could result in a saving of about one third of the consumption in this category. This traditional fuel problem is discussed below as a long-term energy issue, and also in the last section of this chapter.

The fourth issue concerns the management of the temporary electricity shortages, including the need to more systematically analyze the short- and medium-term power investment program. As mentioned earlier, the CEB is installing several hundred megawatts of thermal plant (gas turbines and diesels) to meet the power deficit, since the hydroelectric schemes under development cannot be completed early enough. While the thermal additions would not have been the optimal supply response if the sudden shortfall had been correctly anticipated, this solution is nevertheless preferable to accepting the economic losses associated with widespread power shortages. However, the impact of these decisions on the medium-term least-cost power generation expansion program needs to be studied. The transmission and distribution networks also need to be optimized, especially since system losses are increasing. Finally, a study of the economic costs incurred by consumers due to power shortages should be carried out to establish a ranking of electricity consumers for the purposes of load shedding depending on user sensitivity to outages—if and when this contingency is likely to arise.

The fifth and final major short-run issue arises from the necessity to revise existing electricity tariffs. This problem is analyzed in some detail in the next section.

Beyond the essentially short-run energy problems discussed above, Sri Lanka also faces several critical medium- and long-term difficulties. First, the overall management and performance of the national economy will be strongly affected by the energy policies adopted, although the direct contribution of commercial energy production to the gross domestic product (GDP) is small (the 1980 share of commercial energy in GDP was only about 2%). This is true especially because of the large capital requirements of the energy sector and the substantial foreign exchange needs for importing oil during the next decade. For example, during the 1981–85 period, about 30% of public sector investments were allocated to electric power alone, and this trend is likely to continue. Thus, pricing policy, domestic resource mobilization, and the monitoring of possible inflationary effects due to such massive public sector investments will be important issues. At the same time, the net oil import bill is likely to triple between 1980 and 1985, and account for over 50% of the value of nonpetroleum export earnings. The adverse effect of oil imports on the balance-of-payments will continue to trouble the economy for at least this decade.

The second medium-term issue arises from the fact that the demand for wood (both as a source of timber and fuel) has exceeded the natural regenerative capability of forests for several decades. With 95% of households relying on woodfuel, the remaining forest cover will be completely denuded in about 25 years, if present trends continue. Roughly 6 million tons per year of additional woodfuel will be required in the short-run, implying the need for reforestation of about 0.7 million acres of land (with a five-year rotation period), plus a further 2% annual growth in replanted acreage to meet increasing demand.

The development of a viable program for the development of nonconventional energy sources is the third major medium- to long-term issue. Although there is an existing program, it could be better coordinated and strengthened in all its aspects, that is, research, development, demonstration and pilot projects, diffusion, and commercialization. Specific areas that appear to be promising include better use of agricultural wastes such as paddy husks; reactivation of minihydro sites for operating tea estates; solar heating installations for industrial, commercial, and residential use; solar drying of crops like tea, coconut, and tobacco; development and diffusion of more efficient woodfuel and charcoal stoves; and supplying electricity to remote areas using medium-sized wind generators (i.e., unit sizes up to 5 MW).

In the remainder of this chapter, we will examine several specific demand management problems and possible practical solutions, within the broad context of the energy sector issues mentioned above.

9.2 ELECTRICITY PRICING

Following the basic objectives and theory underlying energy pricing policy as described in Chapters 4 and 5, electricity pricing is normally carried out in two stages. First, the strict long-run marginal costs (LRMC) that satisfy the economic efficiency objective are estimated, as described below. Second, this strict LRMC is adjusted to arrive at an appropriate realistic tariff schedule which satisfies other constraints such as social-lifeline pricing considerations, utility financial requirements, simplicity of metering and billing, and so on.

Sector Background

The Ceylon Electricity Board (CEB), a statutory corporation established in 1969, has responsibility for the generation, transmission, and distribution of electricity. CEB supplies power direct to consumers and also sells in bulk to local authorities who retail to their own consumers. The Ministry of Power and Energy is responsible for supervision of CEB's policies.

CEB's present installed capacity is 499 MW, all interconnected in one system. The transmission and distribution system comprises 569 miles of 132 kV line with 17 substations, 214 miles of 66 kV lines with 9 substations, about 3,250 miles of 33 kV, and about 750 miles of 11 kV lines. The system control and load dispatching center is at Kolonnawa, and all important plants and switching centers can communicate by means of power line carrier.

CEB's generation expansion program (Table 9.2) to the end of this decade shows additions that will bring the installed hydro capacity to over 1000 MW, with a firm energy capacity of about 3,200 GWh. In the first four years of the next decade, additions to hydro capacity of over 900 MW and 2,000 GWh are proposed. To complement the hydro, several hundred megawatts of thermal plant will be needed between 1985 and 1994.

The total energy generated by CEB's power stations in 1980 was 1,668

Table 9.2. CEB Generating Facilities.

EXISTING	TYPE	YEAR OF COMMISSIONING	INSTALLED CAPACITY, MW
Old Laksapana	Hydro	1950/58	50
Inginiyagala	Hydro	1950	10
Uda Walawe	Hydro	1968	6
Winalasurendra (Norton Bridge)	Hydro	1965	50
Polpitiya (Maskeliya Oya Stage I)	Hydro	1969	75
New Laksapana (Maskeliya Oya Stage II)	Hydro	1974	100
Ukuwela (Polgolla)	Hydro	1976	38
Bowatenne	Hydro	1981	40
			369
Kelanitissa (Grand Pass)	Steam	1962	50
Kelanitissa	Gas Turbines	1980/81	60
Pettah	Diesels	1954	6
Chunnakam	Diesels	1954	14
			499
Under Construction			
Kelanitissa	Gas Turbines	1981/82	60
Canyon	Hydro	1982/83	60
Victoria	Hydro	1984	210
Kotmale	Hydro	1985/87	201
Colombo	Diesels	1984	70–100
Future Hydro			
Randenigala	Hydro ⎫		122
Rantembe	Hydro ⎬ 1988/90		48
Broadlands	Hydro ⎭		30
Samanalawewa	Hydro ⎫		240
Upper Kotmale	Hydro ⎪		120
Kukule	Hydro ⎪		250
Ratnapura	Hydro ⎪ 1990/94		35
Uma Oya	Hydro ⎪		150
Bing Hamala	Hydro ⎪		110
Jasmin	Hydro ⎭		40

GWh, about 89% of it hydro. CEB supplied about 203,000 consumers, including 218 local authorities who distributed electricity to another 210,000 consumers, making the total number of consumers about 413,000. Of these, about 302,000 were residential. This suggests that at present only about one household in eight has an electricity connection. Rural electrification has been extended to over 2,000 of a total of 25,000 villages, but electricity consumption per head was less than 100 kWh in 1980. The majority of households use firewood for cooking and kerosene for lighting.

The Bowatenne hydro station was commissioned in 1981 and Canyon is scheduled to be commissioned in 1982. Victoria and Kotmale will come on line in 1984 and 1985. In an attempt to prevent capacity and energy shortages before enough new hydro becomes available, three 20 MW gas turbine units were installed in 1980/81 and three more in early 1982. Further generation will be needed in addition to the six gas turbines, to supply the forecast energy needed in 1983, 84, and 85. CEB will also order between 70 MW and 100 MW of diesel generators for installation by 1984.

Demand for Electricity

Growth in consumption of electrical energy over the period 1961 to 1980 has averaged 9.5% p.a., ranging from 17.8% in 1966 to 3.3% in 1974. From 1972 to 1977 there was a period of weaker growth, but it appears that a rate higher than the long-term trend is now being established.

The short- and long-term annual growth rates for each class of consumer are summarized in Table 9.3. Energy sales since 1961 are analyzed. Table 9.4 shows electric energy sales by user category.

CEB prepares an annual short term energy sales forecast from forecasts of the consumption of each class of consumer within each of its geographical divisions. In so doing, historical trends are considered for each of the 23 areas which together make up the divisions. Due weight is given to known prospective loads such as housing schemes, industrial parks, commercial areas, individual industries and commercial properties, and rural electrification schemes. For longer term forecasting, correlation and regression analyses are used. All consumers are divided into three categories, and growth rates are related to gross domestic product, to value added in mining and manufacturing and export processing, and to population. To accommodate deviations from the trend, upper and lower limit estimates are made.

Based on this data the following forecast was prepared:

Year	1981	1982	1983	1984	1985	1986	1987	1988	1989	1990
Generation (GWh)	2112	2354	2584	2884	3313	3648	3958	4308	4691	5108
Peak (MW)	447	498	546	610	698	769	834	908	989	1077

Table 9.3. Annual Growth in Energy Demand.

	DOMESTIC %	INDUSTRIAL %	COMMERCIAL %	LOCAL AUTHORITIES %	TOTAL %
1961–80[1]	8.8	11.2	7.8	8.3	9.5
1961–65	3.6	14.5	3.8	7.4	8.7
1965–70	6.3	18.3	6.3	12.0	13.0
1970–75	6.3	8.8	6.8	5.7	7.8
1975–80	18.0	4.4	14.4	8.0	8.4
1977–78	11.9	14.0	7.4	9.0	11.6
1978–79	29.0	6.2	27.7	7.2	11.7
1979–80	30.0	3.0	14.8	17.0	11.4

[1]1980 sales are as estimated without load shedding.

Table 9.4. Energy Sales by Category.

	1961 %	1966 %	1971 %	1976 %	1978 %	1979 %	1980 %
Domestic	15.5	11.3	8.9	9.5	10.3	11.9	13.8
Industrial	33.3	43.6	51.7	51.4	51.0	49.1	44.9
Commercial	21.7	17.0	12.9	13.9	13.7	15.8	16.1
Local Authorities	29.5	28.1	26.5	25.2	25.0	23.2	25.2
	100.0	100.0	100.0	100.0	100.0	100.0	100.0

A typical expected load duration curve (LDC) for the CEB system in 1984 is shown in Figure 9.1. The LDC shows the number of hours in the year during which the level of MW demand in the system will equal or exceed a given value. The figure also gives the loading order of different generating plants, i.e., the machines with the lowest operating costs are put into operation earliest. We note, however, that until the Victoria hydroplant is commissioned in 1984, gas turbines will need to be run during both peak and off-peak periods, to meet generation requirements.

Calculating Strict LRMC[3]

The strict LRMC of supply for the CEB system is derived in this section. The calculations are carried out in constant 1981/82 Rupees using an official exchange rate of US$1 = Rupees 20. Strict LRMC is first estimated in border prices using appropriate conversion factors applied to market prices (see Chap-

[3]For further details of the theory and methodology, see: Mohan Munasinghe and Jeremy Warford, *Electricity Pricing,* Johns Hopkins Univ. Press, Baltimore, 1982.

ters 4 and 8). Then the border priced results are transformed by dividing by the standard conversion factor (SCF), to yield the final shadow priced strict LRMC in domestic prices (Standard Conversion Factor (SCF) = 0.9).

Strict LRMC may be defined broadly as the incremental cost of optimum adjustments in the system expansion plan and system operations attributable to an incremental demand increase that is sustained into the future. However, LRMC must be evaluated within a disaggregated framework. This structuring of LRMC is based chiefly on technical grounds and may include: differentiation of marginal costs by time of day, voltage level, geographic area, season of the year, and so on. The degree of structuring and sophistication of the LRMC calculation depends on data constraints and the usefulness of the results given the practical problems of computing and applying a complex tariff. For example, the LRMC of supplying each individual consumer at each moment of time can, in theory, be estimated; however, this might not lead to a tariff structure that could be applied in practice.

Three broad categories of marginal costs may be identified for purposes of the LRMC calculations: 1) capacity costs, 2) energy costs, and 3) consumer costs. Marginal capacity costs are basically the investment costs of generation,

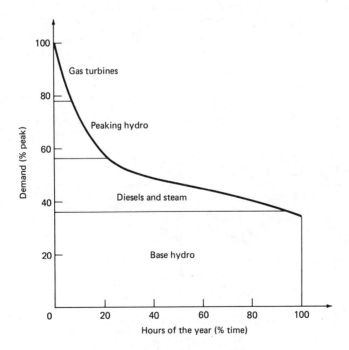

Figure 9.1.   Annual lead duration curve (1984).

transmission, and distribution facilities associated with supplying additional kWh. Marginal energy costs are the fuel and operating costs of providing additional kWh. Marginal customer costs are the incremental costs directly attributed to consumers including costs of hookup, metering, and billing. Wherever appropriate, these elements of LRMC must be structured by time of day, voltage level, and so on, as mentioned earlier.

The first step in structuring is the selection of appropriate pricing periods. By examining the system load duration curves, it is possible to determine periods during which demand presses on capacity, e.g., at a particular time of day, or in a given season of the year. Because the CEB system does not exhibit marked seasonability of demand, we choose only two rating periods by time-of-day, i.e., peak and off-peak. Other aspects of structuring will be introduced later during the analysis.

Marginal Capacity Costs

Consider Figure 9.2 which shows the typical annual load duration curve (LDC) for the system ABEF in the starting year 0, as well as the two rating periods: peak and off-peak. As demand grows over time, the LDC increases in magnitude, and the resultant forecast of peak demand is given by the curve D in Figure 9.3, starting from the initial value MW_0. The LRMC of capacity may be determined by asking the following question: What is the change in system capacity costs ΔC associated with sustained increment ΔD in the long

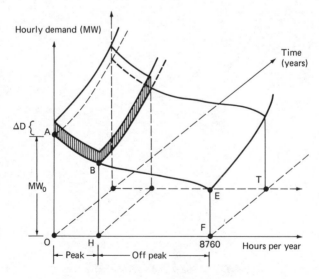

Figure 9.2. Typical annual load duration curve (LDC).

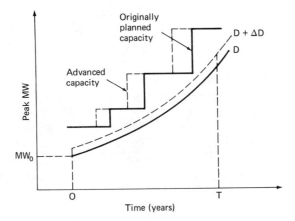

Figure 9.3.   Forecast of peak power demand.

run peak demand (as shown by the shaded area of Figure 9.2 and the broken line $D + \Delta D$ in Figure 9.3). Consequently, the LRMC of generation would be $\left(\dfrac{\Delta C}{\Delta D}\right)$, where the increment of demand ΔD is marginal *both* in time, and in terms of MW.[4]

In an optimally planned system, the change in the expansion program to meet a new incremental load Δ would normally consist of advancing the commissioning date of future plant or inserting new units such as gas turbines or peaking hydro plants (see Figure 9.3). If system planning is carried out using a computerized model, it is relatively easy to determine the change in capacity costs ΔC by simulating the expansion path and system operation, with and without the demand increment ΔD.[5] Even if such a computer model is unavailable, as in the CEB case, it is possible to use simple considerations to derive marginal capacity costs, as described below.

Generation Capacity Cost

The CEB power system utilizes a combination of hydroelectric units and gas turbines to generate electricity. The peak demand, which typically consists of

[4]In theory, ΔD can be either positive or negative, and generally the ratio $(\Delta C/\Delta D)$ will vary with the sign as well as the magnitude of D. If many such values of $(\Delta C/\Delta D)$ are computed, it is possible to average them to obtain LRMC. However, the easiest procedure would be to consider only a single representative positive increment of demand.
[5]If a more sophisticated tariff structure having more rating periods is used, then the LRMC in any rating period may be estimated by simulating the computerized system expansion model with a sustained load increment added to the LDC during that period, i.e., just as in the case of the peak period analysis.

a four hour period from about 6 P.M. to 10 P.M., will be met over the next 6 years by operating both gas turbine and hydroelectric units. Although the plant mix will change after 1984, on average approximately 50% of peak period demand will be supplied by gas turbines and the balance by peaking hydro units during the 1982–87 period. Therefore generation costs are calculated on the basis of this mix of the two types of peaking plants.

The generation capacity costs for the hydro plant and gas turbine are summarized in Tables 9.5, 9.6, and 9.7. The weighted capacity cost per KW installed in 1981/82 border prices is Rs. 4464. This measures the long-run marginal cost of meeting a KW increment in peak demand at the busbars (including the reserve margin).

Table 9.5. Generation Costs for Typical Gas Turbine Plant.
(1981/82 Rupees — Border Prices)

Total cost for 60 MW	=	Rs. 233×10^6
Unit cost	=	Rs. 3883 per kW
Adjustment for 20% reserve	=	Rs. 3883×1.2 = Rs. 4660 per kW
Annuitize at 11% for 20 years	=	Rs. 4660×0.126 = Rs. 587/kW/year

Table 9.6. Generation Costs for Typical Peaking Hydroelectric Plant (Canyon: Phase II).
(1981/82 Rupees — Border Prices)

Total cost for 30 MW	=	Rs. 231×10^6
Unit cost	=	Rs. 7700 per kW
Adjustment for 10% reserve	=	Rs. 7700×1.1 = Rs. 8470 per kW
Annuitize at 11% for 45 years	=	Rs. 8470×0.111 = Rs. 940/kW/year[1]

[1]For comparative purposes the costs of capacity (excluding storage costs) based on Victoria and Kotmale (410 MW) is about Rs. 980/kW/year.

Table 9.7. Generation Capacity Costs.
(Rs./kW/year — 1981/82 Border Prices)

UNIT	CAPACITY COST	WEIGHTED CAPACITY COST[1]
Hydro	940	470
Gas turbine	587	293
Weighted total	—	763

[1]The relative weights are .5 for hydro and .5 for gas turbine.

Transmission and Distribution Capacity Costs

Next, the LRMC of transmission and distribution (T&D) are calculated. Generally, all T&D investment costs are allocated to incremental capacity since the designs of these facilities are determined principally by the peak kW that they carry rather than the kWh. The concept of structuring by voltage level may be introduced at this stage. Consider three supply voltage categories: high, medium, and low (HV, MV, LV). Consumers at each voltage level are charged only upstream costs. Therefore, capacity costs at each supply voltage must be identified.

The planned investment program for HV transmission capacity (132/220kV) for 1982–87 is summarized in Table 9.10 and projected HV demand is derived in Tables 9.8 and 9.9. Using the results presented in these tables, and the average incremental cost approach,[6] the incremental HV capacity cost can be calculated as the ratio of discounted investment to incremental demand: $AIC_H = 1256 \times 10^3/232 = $ Rs. 5414 per kW. Based on a similar procedure, the average incremental MV capacity cost is: $AIC_M = 795 \times 10^3/224.9 = $ Rs. 3535 per kW. Similarly, the average incremental LV capacity costs: $AIC = 186.1 \times 10^3/131.6 = $ Rs. 1414 per kW.

Strict LRMC of Capacity at HV, MV, and LV

The capacity costs at the HV, MV and LV levels calculated earlier are consolidated in Table 9.11.

Marginal Energy Costs

The marginal energy costs during the peak and off-peak periods are estimated by looking at the typical pattern of generation plant operation during the next

---

[6]Suppose that in year i, ΔMW_i and I_i are the increase in demand served (relative to the previous year) and the investment cost, respectively. Then the AIC of capacity is given by:

$$AIC = \left[\sum_{i=0}^{T} I_i/(1 + r)^i \right] \bigg/ \left[\sum_{i=L}^{T+L} \Delta MW_i/(1 + r)^i \right]$$

where r is the discount rate (e.g., the opportunity cost of capital) and T is the planning horizon (e.g., > 10 years). We note that in the AIC method the actual additional increments of demand are considered as they occur, rather than the hypothetical fixed demand increment ΔD used (more rigorously) in calculating generation LRMC. However, because there is no problem of plant mix with T&D investments, AIC and the hypothetical increment method will yield similar results, while AIC is usually much easier to calculate using readily available planning data. The lag period L recognizes that the benefits of new plant are realized some years after the investment. In this study $L = 2, 1$, and 0, for investments at the HV transmission, MV distribution, and LV distribution levels, respectively.

Table 9.8. System Peak Demand at HV/MV/LV/1981–1990.

YEAR	GEN. GWh	SYSTEM PEAK AT GEN. MW (1)	STATION USE AND TRANS. LOSS (8%) (2)	SYS. PEAK AT MV (3) = (1) – (2)	HV CONS. PEAK DEM. (4)	MV LOSSES (9.8%) (5) = [(3) – (4)] × .098	SYSTEM PEAK AT MV (6) = (3) – (4) – (5)	MV CONS. PEAK DEMAND (7)	LV LOSSES (9.8%) (8) = [(6) – (7)] × .098	SYSTEM PEAK AT LV (9) = (6) – (7) – (8)
1981	2112	447	35.8	411.2	4.9	39.8	366.5	126.2	23.5	216.7
1982	2354	498	39.8	458.2	6.8	44.2	407.2	137.8	26.4	243.0
1983	2584	546	43.7	502.3	8.1	48.4	445.8	143.4	29.6	272.8
1984	2884	610	48.8	561.2	10.9	53.9	496.4	157.9	33.2	305.3
1985	3313	698	55.8	642.2	28.6	60.1	553.5	177.6	36.8	339.1
1986	3648	769	61.5	707.5	32.1	66.2	609.2	195.5	40.5	373.2
1987	3958	834	66.7	767.3	34.9	71.8	660.6	211.0	44.1	405.5
1988	4308	908	72.6	835.4	36.3	78.3	720.8	228.5	48.2	444.1
1989	4691	989	79.1	909.9	37.6	85.5	786.8	252.2	52.4	482.2
1990	5108	1077	86.2	990.8	39.0	93.3	858.5	272.3	57.4	528.8

Table 9.9. Summary of Projected Peak Demand at HV, MV, and LV. 1981–1990

YEAR	SYSTEM PEAK AT GENERAT.	SYSTEM PEAK AT HV	ΔHV	DISCOUN. ΔHV	SYSTEM PEAK AT MV	ΔMV	DISCOUN. ΔMV	SYSTEM PEAK AT LV	ΔLV	DISCOUNT. ΔLV	DISCOUNT FACTOR (AT 11%)
1981	447	411.2	—	—	366.5	—	—	216.7	—	—	1
1982	498	458.2	47.0	42.3	407.2	40.7	36.7	243.0	26.3	23.7	.901
1983	546	502.3	44.4	36.1	445.8	38.6	31.3	272.8	29.8	24.2	.812
1984	610	561.2	58.4	42.7	496.4	40.6	29.7	305.3	32.5	23.8	.731
1985	698	642.2	81.0	53.4	553.6	57.2	37.7	339.1	33.8	22.3	.659
1986	769	707.5	65.3	38.8	609.2	55.6	33.0	373.2	34.1	20.3	.594
1987	834	767.3	59.8	32.0	660.6	51.4	27.5	405.5	32.3	17.3	.535
1988	908	835.4	68.1	32.8	720.8	60.2	29.0	441.1	35.6	17.2	.482
1989	989	909.9	74.5	32.3	786.8	66.0	28.6	482.2	41.1	17.8	.434
1990	1077	990.8	80.9	31.6	858.5	71.7	28.0	528.8	46.6	18.2	.391

NOTE: Discount rate used is 11%.

Results

1. Σ Discounted ΔHV (1984–89) = 232 MW
2. Σ Discounted ΔMV (1982–88) = 224.9 MW
3. Σ Discounted ΔLV (1982–87) = 131.6 MW

Table 9.10. 220/132 kV (HV) Transmission
Investment Costs Rs. 10^6 (1981/82).

YEAR	FOREIGN COST B.P.	LOCAL COST D.P.	LOCAL COST B.P.	TOTAL LOST B.P.	DISCOUNT FACTOR (AT 11%)	DISCOUNTED TOTAL COST
1981	—	—	—	—	1	—
1982	347	68	59	406	.901	366
1983	567	92	80	647	.812	525
1984	104	71	62	166	.731	121
1985	70	16	14	84	.659	55
1986	105	45	39	144	.594	86
1987	140	60	52	192	.535	103
					Total (1982–87) =	1256

Note:
1. Costs for 1986 and 87 are estimated while those up to 1985 inclusive are based on the current investment program
2. For local costs BP = D.P. × 0.874, based on the following analysis:

Cost Component	Weight	Conversion Factor	Weighted Average
Weighted Average			
Materials	0.3	1.0	0.3
Skilled labor	0.28	1.0	0.28
Unskilled Labor	0.42	0.7	0.294
		Local Costs Conversion Factor =	0.874

3. Foreign costs are given without customs duty and are therefore at B.P.

few years. As explained earlier, thermal plant will tend to dominate until the longer gestation hydro projects come on stream after about 1984/85. Therefore, incremental energy costs up to 1984 are based essentially on the costs of generating additional kWh during the peak and off-peak periods, using gas turbines.

After 1984, marginal peak period energy generation will still be based on gas turbines, but off-peak energy costs will be based, on average, on a 50:50

Table 9.11. Capacity Costs.

VOLTAGE LEVEL	CAPACITY COST/ kW (BORDER PRICES)	ANNUITIZED CAPACITY COST, RS/kW/YEAR BORDER PRICES
1. Generation weighted total	6565	763[1]
2. HV	5414	606[2]
3. MV	3535	407[2]
4. LV	1414	168[2]

[1]From Table 3.2.
[2]The lifetimes for HV, MV, and LV plant are taken as 40, 30, and 25 years, respectively. With an assumed opportunity cost of capital of 11%, the resulting annuitizing factors are: 0.112, 0.115, and 0.119, respectively.

Table 9.12. Marginal Energy Costs at the Generators.

PLANT TYPE	FUEL USE (LITERS/ kWh)	FUEL COST (RS./LITER)	INCREMENTAL COSTS (RS./kWh)	
			BORDER PRICES	DOMESTIC PRICES[3]
1. Gas turbine	0.384	5.03[1]	1.93	2.15
2. Diesel	0.25	3.65[2]	0.91	1.01

	STORAGE COSTS[4] (RS. × 10[6]/YEAR)	ENERGY STORED (GWh)	INCREMENTAL COST (RS./kWh): DOMESTIC PRICES
3. Hydro (based on Victoria and Kotmale)	262	1118	0.23

[1]Equivalent to $40 per barrel (high speed diesel).
[2]Equivalent to $29 per barrel (furnace oil).
[3]Domestic Price = Border Price/0.9.
[4]Annuitized at 11% over 50 years: annuity factor = 0.111.

mix of new diesel and hydro plant. The relevant peak and off-peak incremental kWh costs at the generators are shown in Tables 9.12 and 9.13 together with the assumptions used. These costs have to be increased by corresponding peak and off-peak loss coefficients in the transmission and distribution networks to obtain the marginal energy costs at the HV, MV, and LV levels, as described later.

Summary of Strict LRMC

The results obtained in Tables 9.11 and 9.12 are combined with the estimate of losses and expenditures on O&M, to determine strict long run marginal capacity costs at various voltage levels shown in Tables 9.14 (border prices) and 9.15 (domestic prices). The underlined values in each table indicate LRMC of peak period capacity in Rs./kW/year at each voltage level. Corresponding incremental energy costs (domestic prices) are summarized in Table 9.16.

Analysis of Existing Tariff

The existing tariff structure is compared with LRMC in Table 9.17. Current basic electricity prices in all consumer categories are significantly below LRMC, with the greatest discrepancy being in the low voltage (LV) class, especially for domestic consumers. For example, households are charged Rs. 0.35 to 0.70 per kWh, whereas the average LRMC is about Rs. 2.50 per kWh.

Table 9.13. Long Run Marginal Energy Costs: Peak and Off-Peak (Domestic Prices).

A. *Peak Period Energy Cost* = Rs. 2.15 per kWh (Gas Turbines—from Table 9.12)

B. *Off-Peak Period*

YEAR	1982	1983	1984	1985	1986	1987	1988	1989	1990	TOTAL
Increment generated (kWh)	1	1	1	1	1	1	1	1	1	—
Discounted increment[1]	0.901	0.812	0.731	0.659	0.593	0.535	0.482	0.434	0.391	5.54
Generation cost (Rs./kWh)[2]	2.15	2.15	2.15	0.62	0.62	0.62	0.62	0.62	0.62	—
Discounted cost (Rs./kWh)[1]	1.94	1.74	1.57	0.41	0.37	0.33	0.30	0.27	0.24	7.17

Weighted Off-Peak Energy Cost (1982–90) = 7.17/5.54 = Rs. 1.29 per kWh

[1] At 11% discount rate.
[2] Based on gas turbines (1982–84) and 50% diesel; 50% hydro thereafter.

Table 9.14. LRMC of Capacity by Voltage Levels (Border Prices).

VOLTAGE LEVEL	GENERATION			HV (220kV, 132kV)			MV(33kv, 11kV)			LV (400V)		
	CAPACITY COST (Rs./kW/yr)	O&M + A&G	TOTAL	CAPACITY COST (Rs./kW/yr)	O&M + A&G	CUMULATIVE TOTAL	CAPACITY COST (Rs./kW/yr)	O&M + A&G	CUMULATIVE TOTAL	CAPACITY COST (Rs./kW/yr)	O&M + A&G	CUMULATIVE TOTAL
Generation	763	98	*861* 936									
HV				606	108	*1650* 1829						
MV							407	248	*2484* 2754			
LV										168	99	*3021*

Note: 1. The O&M plus A&G costs as a % of capacity costs are worked on the basis of 1.5% for generation, 2% for HV, 7% for MV, and 7% for LV.
2. Peak time losses are worked out on the basis of 8% for HV, 9.8% for MV, and 9.8% for LV. Thus generation costs at HV level are: 861/(1 − 0.08) = 936; generation plus HV costs at MV level are: 1650/(1 − 0.098) = 1829; and so on.

311

Table 9.15. LRMC of Capacity by
Voltage Levels (Domestic Prices).

LEVEL	LRMC[1] (RS./ kW/yr)
Generation	957
HV	1833
MV	2760
LV	3357

[1]Domestic Price = Border Price (from Table 9.14)/ 0.9.

Table 9.16. LRMC of Energy by
Voltage Levels (Domestic Prices) (Rs./
kWh)

	PEAK[1]	OFF-PEAK[2]
Generation	2.15	1.29
HV	2.34	1.36
MV	2.59	1.44
LV	2.87	1.54

[1]Peak period losses (of incoming energy) at HV, MV, and LV are 8%, 9.8%, and 9.8%, respectively. Therefore HV cost = 2.15/(1 − 0.08) = 2.34, and so on.
[2]Off-peak period losses at HV, MV, and LV are 5%, 6%, and 6%, respectively (of incoming energy). Therefore HV cost = 1.29/(1 − 0.05) = 1.36, and so on.

During the last two years, the addition of a fuel surcharge has effectively doubled the basic price, but this is not a satisfactory solution. First, domestic consumption below 200 kWh is exempt from the fuel escalation clause, so that only about 16% of households are subject to the surcharge. Second, the surcharge on kWh also distorts the relationship between energy and demand charges for medium and high voltage (MW and HV) consumers as described below.

The structure of existing tariffs is distorted in at least two important ways. First, the ratio of LV to MV (or HV) charges is much lower than the corresponding value using LRMC, indicating that LV consumers are subsidized relative to higher voltage customers. Second, the capacity charges paid by MV and HV consumers are about one tenth of the LRMC, thus making the ratio of existing energy to capacity charges very low relative to the comparable ratio derived from LRMC. The imposition of the fuel surcharge (which applies only to the kWh tariff) exacerbates this distortion. As a result, there is no effective

Table 9.17.   Existing Electricity Tariffs and Estimated Long Run Marginal Cost (LRMC).

CATEGORY	TARIFFS			LRMC		
		ENERGY CHARGE: (RS/kWh)			ENERGY COST (RS./kWh)	
	CAPACITY CHARGE (RS./kVA/mo)	NORMAL	INCL. 135% FUEL ADJ. SURCHARGE	CAPACITY COST RS/kW/mo.	OFF PEAK	PEAK
High Voltage	20	0.60	1.40	153	1.36	2.34
Medium Voltage						
General	20	0.60	1.41 ⎫	230	1.44	2.59
Local Authority	18	0.35	0.82 ⎬			
Low Voltage						
General						
— over 50 kVA	22	0.60	1.41 ⎫			
— under 50 kVA	nil	0.60	1.41			
Local authority	20	0.35	0.82	280	1.54	2.87
Domestic				approx. Rs. 2.50 per		
— up to 50 kWh	nil	0.35	0.35 ⎬	average kWh		
— 50–350 kWh	nil	0.45	0.75			
— over 350 kWh	nil	0.70	1.65 ⎭			

price signal to peak period consumers regarding the high costs of supplying additional kW, and inadequate incentives for shifting away from the peak to the off-peak period.

In summary, the low average level of prices encourages wasteful consumption and reduces CEB's revenues, necessitating government equity contributions to meet sector financial needs, while distortions in the price structure relative to LRMC lead to undesirable patterns of electricity use.

9.3.   THE CHOICE OF A DOMESTIC COOKING FUEL

In this section, we analyze the economic and policy issues underlying the correct choice of a domestic cooking fuel. We focus upon urban areas because in the immediate future the choices available there are substantial, and extending these same choices to rural areas would involve considerable cost of networks such as LPG and electricity. Furthermore, the low incomes in rural areas would preclude high-capital-cost options unless substantial subsidies were to be provided by the government. Our calculations are limited to five options: wood in open fires and in closed stoves, kerosene in wick-type stoves and in pressure-type stoves, and LPG. For LPG a range of values is given for lack of precise technical parameters. We do not consider charcoal because of lack of data, and we exclude piped gas and electricity because they are more limited as options for upper-income households only.

Introduction

As described in Chapter 6, the demand for energy products should be analyzed in the context of the need for the final useful service provided by this energy when used in conjunction with other economic resources.[7] Thus, for example, households use energy to cook their meals; they produce the requisite heat by combining capital equipment (e.g., a stove), fuel, and some labor. The total cost of producing the desired output will depend upon technical parameters such as the efficiency of conversion, and energy input value of the fuel; and upon economic parameters such as initial capital charges, fuel costs, and labor costs. To produce enough heat to cook a meal a number of different modes and fuel combinations can be used—wood with open and closed stoves, coal, charcoal, kerosene in wick or pressure stoves, LPG, piped gas, or electricity. Since each of these combinations involves different values of the technical and economic parameters noted, the overall costs will vary.

In an equilibrium, disregarding personal preferences for one or the other type of cooking method, the private marginal cost of effective cooking heat should equate across these different methods. In practice, this will not happen for two principal reasons. First, different cooking methods have other characteristics besides technical efficiency and economic cost, e.g., in Sri Lanka wood is traditionally preferred to charcoal;[8] kerosene is messy and has an unpleasant smell; LPG is considered dangerous; and so on. Secondly, the imperfections of the capital markets—particularly for the small amounts of cooking equipment involved—mean that at a given time the capital unit owned (such as a stove) cannot be sold for its economic value, and the switch over to a lower marginal-cost method will only occur with some time lag. Furthermore, even if private marginal costs were to be equated, these could differ substantially from marginal economic efficiency costs.

This case study indicates that the private cost of wood understates its economic opportunity cost (because of the high shadow cost of deforestation). Thus a private market equilibrium of cooking modes wherein the private marginal cost of cooking with wood equals the private marginal costs using other means is socially undesirable to the extent that the marginal efficiency cost of wood cooking is likely to be higher than with other methods, unless other competitive fuels are also subsidized.

In the following discussion, we provide an illustration of this approach applied to the case of cooking fuels in urban areas of Sri Lanka—the capital city of Colombo in particular. Technical and economic parameters for such an analysis are not well established or easy to estimate. Therefore, the results are

---

[7]See, for example, Kelvin Lancaster, "A New Approach to Consumer Theory," *J. Pol. Econ.,* Vol. 74, 1966, pp. 135–145.

[8]While in Thailand the opposite is the case.

presented for illustrative purposes only and are subject to a certain margin of error. However, given the critical situation of rising petroleum prices, potential deforestation and the midterm exhaustion of major hydro sites, it is necessary to focus on the allocative efficiency of alternative energy forms by some form of analysis as described here. What we present below can be elaborated if more exact values for the parameters are found. However, even with the approximations used in our analysis, the calculations indicate general policy directions. They also identify areas that require further analysis.

The Marginal Private Costs of Urban Domestic Cooking[9]

Marginal cost calculations are usually made relative to some natural quantitative unit, which for the case of energy may be a kilowatt-hour, or btu, or gigajoule (GJ). Here, however, we shall first compute the marginal annual private cost (using market prices) of providing cooking heat for an average family of six people. A family or household is the natural consumption unit for this product. We assume that the demand for the output, i.e., cooked meals remains constant. As prices change, a family will not cook more (or less) but rather switch from one mode to another. With the exception of open fires, the sunk-cost aspect of capital-equipment means that the decision is decidedly long term, and therefore the marginal-cost perspective of the decisionmaker is likely to be with respect to the total cost for a fixed period of time. In estimating these total costs, we exclude any labor involved for two reasons. First, the labor used in generating cooking heat as such—distinct from the labor used in the culinary process itself—is insignificant except in the case of wood. Second, even with wood the labor cost is very low because of the low shadow price of labor. This is even more true when the labor is done by members of the family for whom alternative opportunities are not significant because it does not interfere with other work or income earning opportunities at the times it is undertaken.

The amount of useful energy output for a year's cooking can be estimated at about 5.42 GJ.[10] Thus, the annual marginal cost is given by:

$$MCA_i = \frac{5.42/e_i}{EV_i} \cdot P_i + aK_i$$

where e_i is the energy efficiency of the mode i, EV_i is the energy input content in GJ per unit of fuel type i, P_i is the price of fuel i, a is the annuitization factor, and K_i is the cost of the capital stock required in mode i.

---

[9]All calculations are in constant (1979) Sri Lanka rupees (Rs.). For the period during which data was gathered US$1 = Rs. 16 (approx.)

[10]Based on estimates of firewood used by an average family of six in the *Consumer Finances Surveys* (1953, 1963, 1973), and an energy content of 14GJ/metric ton.

Table 9.18. Capital Costs of Cooking Equipment (Rupees).

| | CAPITAL COST | | ANNUITIZED COST[1] | |
COOKING FUEL AND MODE	MARKET PRICES (DRS.)	SHADOW PRICES[2] (BRS.)	MARKET PRICES (DRS.)	SHADOW PRICES (BRS.)
1. *Wood*				
Open Fire	0	0	0	0
Closed Stove	100	90	14.6	15.3
2. *Kerosene*				
One Burner				
—Wick Stove	200	180	27.2	30.6
—Pressure Stove	300	270	40.8	45.9
3. *LPG*				
One Burner Stove				
and Cylinder	700	630	95.2	107.1

[1]Annuity factors are 0.136 and 0.170, based on real discount rates of 6 and 11 percent, respectively, in market and shadow prices, over 10 years.
[2]Shadow Price = Market Price × Conversion Factor (CF). CF = 0.9 (see Chapter 8).

Approximate capital costs (K_i) for the various modes are shown in Table 9.18; they range from zero for an open fire (a few stones) to Rs. 700 for a simple one-burner LPG stove plus the cylinder. For the kerosene and LPG options, the most rudimentary one-burner commercial stoves are taken as the basis for the computations. Larger two-burner models of better quality can be as much as Rs. 600–700 for kerosene, and over Rs. 1,000 for LPG. On the other hand, rudimentary handcrafted equipment can probably be made for much lower costs, perhaps as low as Rs. 200.

Annual capital-use cost would be derived by applying the appropriate annuitization factor to the capital cost. The resultant annual private capital charges shown in Table 9.18 are about Rs. 14 for wood stoves, Rs. 27 to 41 for kerosene stoves, and Rs. 95 for LPG equipment. Corresponding (shadow pricing) efficiency costs are also shown in the table. As expected, the capital cost element is highest in the case of LPG and lowest in the case of wood. The various modes ranked in ascending order of capital cost are: open fires, wood stoves, kerosene wick stoves, kerosene pressure stoves, and LPG stoves.

The fuel cost component, at 1979 private market prices, changes this ordering considerably as can be seen in Table 9.19, under the "Fuel Cost" column. Kerosene has the lowest fuel cost, LPG is next,[11] while wood is most expensive

[11]Lack of data on LPG precluded comparable computations of efficiency and energy content. We took instead an estimate of annual usage by a "customer" as shown in a Colombo Gas Works feasibility study. The study gave two values (100 and 150 kg), which we use as low and high estimates.

Table 9.19. Marginal Private Cost of Cooking Heat in Urban Areas Per Family/Per Annum in Market Prices (Domesic Rupees)
(required output = 5.42 GJ)

FUEL AND MODE	EFFICIENCY	INPUT IN GJ	INPUT VALUE IN GJ	UNITS REQUIRED	UNIT PRICE	FUEL COST	CAPITAL COST	TOTAL COST AT MARKET PRICES
WOOD								
Open Fire	.08	67.75	15/MT[1]	4.52 MT	264/MT	1193.3	0	1193
Closed Stove	.15	36.13	15/MT	2.41 MT	264/MT	636.2	13.6	650
KEROSENE								
Wick Stove	.43	12.6	.157/IG[2]	80.3 IG	2.48/IG	280	27.2	307
Pressure Stove	.51	10.6	.157/IG	67.6 IG	3.48/IG	240	40.8	281
LPG								
High Requirement	—	—	—	150 kg	2.6/kg	390	95.2	485
Low Requirement	—	—	—	100 kg	2.6/kg	260	95.2	385

[1]MT = metric ton.
[2]IG = imperial gallon.
Source: see text.

regardless of the type of converter used. It can be seen that fuel costs dominate annualized capital costs. Thus the very low capital charges for wood cooking are not enough to outweigh the high fuel cost. The overall marginal cost (last column of the table) for open fires is decidedly higher than for the other modes, while even a closed stove is more expensive than LPG if the latter's "high" fuel requirements are taken. Both wood and LPG have marginal costs higher than kerosene stoves of either type. Even if the capital costs of kerosene stoves were in fact substantially greater—say Rs. 500 instead of Rs. 200–300—this conclusion would still hold, as the overall marginal cost for kerosene would be in the range Rs. 315–380, still below all other options.

For the private decisionmaker—the household or family unit—it seems clear that in terms of overall cost, kerosene was the best option, given the heavily subsidized 1979 kerosene prices. However, rough data on usage patterns in 1980 suggest that a very large number of urban households, perhaps 60 percent or even more, continued to use wood, while only about 35 percent used kerosene, with the rest using gas (mostly bottled LPG, and some piped) and electricity.

Why was there such a high use of wood given that its costs were nearly twice that of kerosene, and why was there such a low use of LPG given that it was only slightly more expensive than kerosene, but far less costly than wood? Various explanations can be put forward. The high price of wood was a recent phenomenon with prices doubling in the preceding two or three years. Kerosene prices, on the other hand had been steady at Rs. 3.48 per imperial gallon.[12] Adjustment usually involves time lags—first, for recognition of increased fuel costs; second, to decide to accumulate savings for the needed capital investments. Therefore, the process of substitution had not really started by 1979. Further, we have noted earlier that traditional preferences favor wood for its "taste" qualities. As a result the equilibrium is perhaps one in which wood does have a somewhat higher value in use. As to LPG, it is even less well known to low-income people than kerosene. (The latter is used widely for lighting, and is considered risky.) It may also be considered to be inconvenient, given the required refilling—say Rs. 50–100 per year for taxi fares to take bottles to a depot. However, these added transport costs are not high enough to make total LPG usage cost approach those of wood. Further, it could be argued that the use of only one bottle is inconvenient, as refilling might be necessary at inconvenient times. To avoid this, a user would have to invest in an additional bottle. At a cost of Rs. 450 per bottle, this would mean an additional Rs. 75 per annum in costs.

These two factors, therefore, could add a perceived cost of Rs. 125–175 for

---

[12]They remained so until 1980, but have risen sharply since then.

LPG use. This would make its total cost approach the costs of closed wood stoves. Because it is arguable whether these cost items are essential, we have not included them in our total, but it is important to recognize such "inconvenience" costs associated with LPG in trying to understand the use patterns despite the differences in marginal usage costs. Such "convenience" factors could result in an equilibrium in terms of consumer satisfaction that differs from observed marginal monetary costs.

Another factor that is likely to greatly reduce the adaptation rate of kerosene and LPG cookers is that of capital costs. While the applied real rate of discount of 6% and 11% is probably realistic, it does not reflect the need (and frequent inability) of families to accumulate the necessary funds to make the investment in the new appliance.

The Marginal Efficiency Costs of Urban Domestic Cooking

In the preceding discussion, comparative costs of alternative cooking methods were based on 1979 market prices. However, as described in Chapter 6, these marginal private costs are relevant only for analyzing and understanding how private individuals make their choice among alternative cooking methods. These choices must be taken into account in the design of any policy that tries to affect the use of different fuels. However, the policies themselves must be based on social optimization objectives.

Therefore, the cost comparison of alternative methods must also be done by using social opportunity costs, or shadow prices. For example, wood is in short supply and its cutting is causing deforestation; its market price therefore understates the economic opportunity cost. Equally, the 1979 kerosene prices of Rs. 3.48/gallon understated the opportunity cost, because the market price was kept low by the government subsidy policy. LPG, on the other hand, was in excess supply in Sri Lanka. Therefore, it had a low opportunity cost at the refinery.

Although it is not possible to specify with precision the opportunity costs for these three fuels, particularly for firewood, it is reasonable to approximate these in ranges to permit analysis of the relative marginal costs in question. Since the divergence of private and efficiency costs of firewood is not well established, we calculate first the efficiency costs assuming the market price of wood is a correct reflection of marginal efficiency cost but converted to border prices using the conversion factor of 0.9 as shown in Table 9.20 (open fire). Kerosene's opportunity cost is reasonably well established on the border or import-price basis of about Rs. 9 per imperial gallon.

LPG costs are less clear, though they were certainly below the Rs. 2.6/kg 1979 market price charged consumers. Operating costs for a proposed expansion of the Gas Works sales were about Rs. 1.3 million for 5.5 million kilo-

Table 9.20. Marginal Efficiency Costs of Cooking Per Family Per Annum in
Shadow Prices (Border rupees)

FUEL AND MODE	CAPITAL COST	UNITS REQUIRED	UNIT VALUE[3]	FUEL COSTS[1]	TOTAL COST AT SHADOW PRICES
WOOD					
Open Fire	0	4.52 MT[2]	237.6 MT	1074	1074
Closed Stove	15.3	2.41 MT	237.6 MT	585	600
KEROSENE					
Wick	30.6	80.3 IG[3]	9 IG	722.7	753
Pressure	45.9	67.6 IG	9 IG	608.4	654
LPG					
High	202.7	150 kg[4]	2 kg	300	503
Low	202.7 (107.1)	100 kg	2 kg	200	403

[1]Economic oppurtunity costs are assumed to be Rs. 9/IG for kerosene and Rs. 2/kg for LPG.
[2]MT = metric ton.
[3]IG = imperial gallon.
[4]kg = kilogram.

grams of LPG, or about Rs. 0.25/kg. To this must be added the ex-refinery economic opportunity costs of LPG. There was estimated to be about half the prevailing price of Rs. 1.4/kg. This implied a total efficiency cost of about Rs. 1/kg. To bias our estimate against LPG, and to reflect the fact that any substantial increase in the use of LPG would change the excess supply situation of Ceylon Petroleum Corporation (CPC) refinery,[13] we take Rs. 2/kg as the shadow value of LPG fuel in border prices.

In addition to the shadow cost of LPG, estimated above, and the private investment costs for stoves and bottles other facilities are required. These consist primarily of additional storage tanks, land, and some distribution equipment. Their costs were estimated in a recent feasibility study for the Colombo Gas Works at Rs. 22.5 million for 40,000 customers, in border prices, or Rs. 562.5 per household, equivalent to annual capital charges of about Rs. 95.6. These must be added to total capital costs of this option.

The results summarized in Table 9-20 show the open fire method to be the least economic alternative, followed by the kerosene options. The most economic method is LPG using either the high or low case. In contrast to the analysis in market prices, kerosene is no longer the least-cost option. This difference between market prices and social opportunity costs results in an inefficient allocation of resources through overutilization of kerosene. The incorrect

[13]Until 1980, Sri Lanka's refinery was operated to meet the needs of high-demand middle distillates, resulting in surpluses of certain other products; as a result, more than half of the potentially available LPG was flared.

pricing of fuels also overstates the cost of LPG usage in market prices. This also leads to some inefficiency—albeit less significant than for kerosene—and less LPG is likely to be used than a social optimum might dictate. The cost advantage of LPG and other fuels over wood would become even more pronounced if the real opportunity cost of firewood were to exceed its market price.

In conclusion, we find that the shadow pricing of kerosene and LPG alone, while using the market price for wood, shows the open fire to be the highest-cost option, followed by kerosene. LPG is the low-cost option. If the real economic costs of wood were higher than its market price this would enhance the cost advantage of LPG. However, if the wood price rise were to be substantial, kerosene would eventually become the lowest cost option at the margin, given the limited availability of LPG. The results suggest a role for kerosene as an economical cooking fuel at some future time, following an initial period of expanded use for LPG. The use of firewood should be encouraged only in the context of improved stoves, together with increased reforestation efforts to reduce its shadow value.

Conclusions and Policy Implications

The calculations presented here are only partial ones because they exclude some cooking methods. They also contain a number of uncertainties pertaining to technical energy-value contents and to prices. Nevertheless, they are robust enough to permit some tentative conclusions.

1. LPG, which is essentially disregarded in energy policy discussion in Sri Lanka because of its limited availability, may have an important role to play in easing the fuels problem in urban areas. Its marginal efficiency costs are probably lower in cooking than those of competing fuels—wood and kerosene. The importance of a proposed, medium-term project of bottled, LPG expansion will be to provide an initial demonstration to consumers, in the nature of propaganda and education about the use of gas for domestic cooking. In addition, such a project would provide some dynamism to the presently moribund LPG supply enterprise.

2. Because of the likelihood of a continued increase in deforestation problems and the consequent increasing efficiency costs of wood, pressures on potentially available LPG supplies will lead to sharply rising marginal costs of the latter, once the existing supply from the domestic refinery is fully appropriated. This will turn kerosene once again into an economical alternative. Thus, despite the apparent inefficiency of kerosene use now, policy options should be kept open for a flexible refocusing upon its promotion as a way to save scarce firewood. This is even more important for rural areas because in the latter LPG distribution costs would be too high. Furthermore, a basic rural network for kerosene marketing is already in place because of its widespread use as a lighting fuel.

3. The far greater thermal efficiency of closed stoves relative to open fires (recall we have overestimated the efficiency of open fires) makes this option potentially important both in view of the rising opportunity cost of firewood and because of the potential for further increases in world oil prices. It also suggests that the even more efficient process of charcoaling and use of closed stoves should be investigated further as potentially viable supply sources of additional (or substitute) cooking energy.

A number of important conclusions relevant to the analysis of energy use patterns in developing countries can be drawn from this specific case study. The first and most important one is that user choices among alternative energy sources will not only be determined by the financial costs of each alternative fuel delivered to the point of use and compared on a net heat content basis. Equally or more important will be the auxiliary costs to the user of converting to and utilizing each type of fuel. These will consist of the costs of labor, capital, and (sometimes) land related to receiving, unloading, storing, handling, ease of control in use and potential risks, and system maintenance. For households, capital availability to purchase applicances is likely to be a major constraint. Reliability of supply and transportation and uniformity of quality will also be important considerations. Habits may significantly affect use, particularly in situations in which the potential choice lies between a known, customarily-used energy material and new alternatives of unknown characteristics. Resistance to change may be formidable, and lack of knowledge and experience is perhaps one of the most difficult barriers to overcome in changing energy use patterns in developing countries. Only if all of these factors are systematically and carefully taken into consideration will it be possible to bring about desirable changes.

An important issue is that market prices to users frequently do not reflect the true national economic costs of a specific energy use. Appropriate shadow pricing is necessary to determine the difference, if any, and either changes in market prices or regulations affecting alternative choices may be necessary to bring about a realistic consumer choice pattern that also reflects real economic opportunity costs.

For governments, the fact of the rapidly rising relative costs of energy should act as a strong stimulant to evaluate systematically existing and projected future energy use patterns on a disaggregate basis and relate them to existing or potentially available sources of supply, evaluate the economic opportunity costs of use of the latter in a dynamic setting of changing demand and supply conditions, and attempt to bring about desirable changes in future energy utilization patterns. Such promotional activities, however, must take full account of the many related technical, economic, and social-psychological factors that affect and influence energy uses and fuel choices.

10

Thailand: Energy Management Issues in a Middle Income, Oil Importing Country

10.1 INTRODUCTION

Thailand is one of the few rapidly growing developing countries of the world. Its average annual GDP growth rate was in excess of 7.5% during the last twenty years. The increases in world petroleum prices seriously threaten this enviable performance. This is so because until now the country was almost completely dependent on imported crude oil or refined products to meet its petroleum products requirements. Petroleum accounted for almost 75% of total energy consumption in 1979. However, the country has a number of indigenous energy resources that can be developed to reduce this dependence on imported fuels: Among them are substantial deposits of natural gas, lignite, hydro, and significant, but rapidly dwindling, forest resources.

In the following case study some of the issues related to this heavy energy import dependence and the policies designed to reduce it are discussed. As will be seen, pricing policies are likely to play a prominent role. First, existing electricity tariffs are analyzed and compared with long-run marginal costs. This is followed by analyses of the economic costs of natural gas, of petroleum products, and of lignite. A special evaluation highlights the divergence of private and net economic costs in the pricing of liquid petroleum gases (LPG). An analysis of the fuel choices of the tobacco curing industry provides proof that energy users are more interested in total energy systems costs and benefits than specific fuel costs. A final section sketches-in some of the major changes that have occurred on the energy supply side in Thailand since these studies were completed.[1]

10.2 ENERGY AND THE ECONOMY

Thailand has a population of about 46 million people (mid-1979) and an area of 542,373 square kilometers. The largest concentration of people occurs in the

---

[1]The studies presented here were based on information available in 1979.

Greater Bangkok area, which has a population of nearly 5 million (December 1978 estimate). Roughly 70% of the labor force is employed in agriculture. Agriculture contributed 26% of GDP in 1979, although its relative share has been declining fairly steadily. Per capita GNP is roughly US$490 (1978 estimate), making Thailand one of the "lower middle-income group" nations of the world.[2]

The adult literacy ratio was estimated to be 84% in 1975. Regional differences in development are significant: Most of the manufacturing industry and much of agriculture (especially rice, cassava, and sugarcane) are concentrated in the central part of Thailand, around Bangkok and in the plains to the immediate north. The far northern and northeastern parts of the country are relatively poor.

The Thai economy has been undergoing a fairly steady structural transformation. The growth of cash cropping was followed by the development of an increasingly important manufacturing sector. For example, value added in manufacturing increased from 19.4% of GDP in 1975 to 22.0% by 1979. Future annual average growth rates are projected at close to 10%.

Parallel to the general structural transformation of the economy, there have been certain shifts in the relative importance of the major industrial groups. There has been a decline in the relative share of value added of the basic agroindustries—food processing, beverages, and tobacco—and a growth in the importance of textiles, rubber products, petroleum products and coal, and electrical machinery.

Thailand's recent economic performance has not been as good as the longer-term trends suggest. The Governor of the Bank of Thailand has ranked the following three items as the country's major economic problems for the 1980s: (1) the balance of payments deficit, (2) inflation, and (3) the need to mobilize considerable financial resources (both foreign and domestic) in order to sustain economic growth.

The deficit of imports over exports increased from $1.2 billion dollars in 1977 to an estimated $3.3 billion in 1981. A key problem in the balance of payments has been the increase in oil prices. The ratio between petroleum imports and total exports fluctuated around 30% between 1975 and 1979; it was projected to rise to 44.6% in 1980 and 44.8% in 1981.[3] Thus, oil imports were expected to become a more serious drain on scarce foreign exchange in the short run, until natural gas becomes available in 1982; however, after 1984, petroleum

[2]Bangkok population estimate is from Europa Publications, *The Far East and Australasia 1979/80*. Other data is from Asian Development Bank, "Thailand Country Program Paper, 1981–1983," 1980.
[3]Based on data in Government of Thailand "1980 IMF Consultation."

imports are projected to begin rising again. Petroleum consumption was forecast to total 13.9 million toe in 1981, 12.8 million toe in 1984, 14.5 million toe
in 1987, and 15.3 million toe in 1990.

The country's energy situation is subject to a number of important trends
and factors that require careful evaluation in the light of several potential,
alternative policy options. Throughout the 1970s imported petroleum was the
main source of its commercial energy supplies. In 1979 it accounted for 74.3%
of total, and some 95.5% of total commercial energy consumption.

However, some major changes are expected to take place soon. Offshore natural gas is under development now and will reach Bangkok by the end of 1981.
It will largely replace fuel oil now used in producing electricity. Furthermore,
large lignite deposits are under development. Towards the latter half of the
1980s, therefore, most of the rapidly growing electric power demands will be
supplied from gas and lignite-fired thermal plants, and to a lesser extent, by
hydro.

Gas will also play an increasingly important role in the household, industrial,
and transportation sectors. Proposed natural gas based extraction plants will
produce increasing amounts of LPG for household use, as a chemical feedstock,
as a transportation fuel replacing gasoline, and for export.

Total investments in energy supply systems planned for the 1980s are estimated at US$6.8 billion (in 1978 dollars), an amount about equal to the coun-
try's total gross domestic investments in 1978.

Table 10.1 details past trends in energy consumption. The four major categories of primary energy sources—lignite, petroleum, hydroelectricity, and
noncommercial—contributed the following shares:

	1970 %	1975 %	1979 %
Lignite	1.9	1.9	2.2
Petroleum	65.7	70.4	74.3
Hydro[4]	2.0	2.7	1.3
Noncommercial	30.4	25.1	22.2

It can be seen that petroleum consumption has become progressively more
important while noncommercial energy was declining in relative importance.
Energy consumption has increased an average of 5.8% per annum since 1973,
with considerable fluctuations in this trend. Petroleum consumption has
increased at about 6% per annum over this period (1977–79), and lignite,

---

[4]Includes net primary electricity imports, which are less than 5% of the total given for "Hydro."

Table 1O.1. Internal Energy Consumption by Energy Source. (thousands of toe).

	1965	1970	1973	% CHANGE	1974	% CHANGE	1975	% CHANGE	1976	% CHANGE	1977	% CHANGE	1978	% CHANGE	1979
LIGNITE	45	147	137	57.66	216	0.46	217	17.05	254	53.94	117	46.15	171	101.75	345
PETROLEUM	2429	5061	8368	-2.91	8125	-0.13	8115	13.17	9192	12.18	10321	9.32	11283	4.16	11753
LPG	5	54	91	12.08	102	20.58	123	14.63	141	9.21	154	13.63	175	17.71	206
Gasoline[1]	328	789	1251	7.11	1340	9.77	1471	11.35	1638	11.17	1821	5.71	1925	0.62	1937
Kerosene	60	116	188	15.42	217	-14.19	186	42.47	265	-3.02	257	7.01	239	28.03	306
Jet Fuel	546	296	796	-19.23	643	16.95	752	2.39	770	-10.65	688	2.90	708	10.73	784
Gas/Diesel Oil	998	2083	2943	-4.18	2820	-2.95	2737	17.06	3204	8.86	3488	7.45	3748	12.35	4211
Fuel Oil	379	1331	2478	-2.22	2423	4.41	2530	10.43	2794	20.68	3372	12.69	3800	-0.11	3796
Crude Oil[2]	113	392	621	-6.61	580	-45.52	316	20.25	380	42.36	541	27.17	688	-25.49	513
PRIMARY Electricity[3]	72	152	175	28.57	225	36.0	306	6.53	326	-9.21	296	-30.41	206	-1.46	203[3]
NONCOMMERCIAL[4]	N.A.	2343	2624	4.80	2750	5.27	2895	5.76	3062	5.78	3239	6.42	3447	1.22	3510
TOTAL	N.A.	7703	11304	0.10	11316	1.91	11533	11.28	12834	9.73	14083	7.27	15107	4.66	15811

Source: National Energy Administration of Thailand.
Notes: [1]Includes motor gasoline, aviation gasoline, and some naphtha.
[2]For internal use by refineries.
[3]Includes net imports of electricity and hydro, except for 1979 which is only hydro. Net imports ranged from −4000 toe in 1971 to 16,000 toe in 1978.
[4]Includes firewood, charcoal, bagasse, and paddy husks.

Table 10.2. Percent Distribution of Petroleum
Product Consumption.

	1970	1973	1975	1979
LPG	1.1	1.1	1.5	1.8
Gasoline[1]	15.6	14.9	18.1	16.5
Kerosene	2.3	2.2	2.3	2.6
Jet Fuel	5.8	9.5	9.3	6.7
Gas/Diesel Oil	41.2	35.2	33.7	35.8
Fuel Oil	26.3	29.6	31.2	32.3
Crude Oil[2]	7.7	7.4	3.9	4.4

[1]Includes motor gasoline, aviation gasoline, and some naptha.
[2]For internal use by refineries.

petroleum, and noncommercial energy consumption have each increased at approximately 5% per annum.

Lignite is produced domestically from three deposits: Mae Moh, Bang Pu Dum, and Li. The Mae Moh and Ban Pu Dum deposits are being mined for power generation and the Li deposits for tobacco curing. Some coal is also imported. A major expansion is underway at Mae Moh whose estimated reserves amount to as much as 650 million tons, enough to support mine-mouth electric generating plants with capacities up to 2000 MW. Petroleum consumption increased 10-fold between 1960 and 1979. Roughly 95% of all commercial energy is petroleum based. The growth in the relative importance of petroleum vis-a-vis other commercial fuels has slowed somewhat since the 1973 oil price increases. In 1974 and 1975, the total amount of petroleum energy consumed actually declined. The shares of the major petroleum products in total petroleum energy consumption are shown in Table 10.2.

The products which dominate the petroleum mix are gasoline, diesel, and fuel oils. The fuel oil is used principally for power generation and the gasoline and diesel principally for transportation. Gasoline and fuel oil consumption each grew at an average rate of 7.6% per annum between 1973 and 1979, and gas/diesel oil at 6.4% per annum. The fastest growth among the petroleum products has been for LPG: 14.6% per annum. This LPG growth is due to its low price relative to competing fuels. Overall, however, in 1979, LPG still only accounted for 1.8% of total petroleum product consumption.

Electricity consumption has risen quite rapidly over time. Table 10.3 shows absolute and percentage increases in peak and energy generation. Between 1970 and 1979 the former grew at average annual rates of 13.0% and the latter at 14.6%. The annual system load factor gradually increased from 62% to an average of about 67%.[5]

---

[5]The 1970 load factor of almost 71% was influenced by peak load restrictions.

Table 10.3. EGAT Generating Statistics, 1970-79.

| FISCAL YEAR | PEAK GENERATION | | ENERGY GENERATION | | ANNUAL LOAD FACTOR (%) |
	MW	% INCREASE	GWH	% INCREASE	
1970	748.35	17.28	4,095.31	21.62	62.47
1971	872.70	16.62	4,792.88	17.03	62.69
1972	1,028.80	17.89	5,711.15	19.16	63.37
1973	1,199.30	16.57	6,872.84	20.34	65.42
1974	1,256.30	4.75	7,258.62	5.61	65.96
1975	1,406.60	11.96	8,211.57	13.13	66.64
1976	1,652.10	17.45	9,414.48	14.65	65.05
1977	1,873.40	13.40	10,950.62	16.32	66.73
1978	2,100.60	12.13	12,371.67	12.98	67.23
1979	2,255.00	7.35	13,964.56	12.88	70.69

Source: Electricity Generating Authority of Thailand (EGAT).

Table 10.4 indicates the sectoral composition of total energy consumption in Thailand for 1971 and 1977. The only significant change is the rising share of industry, and apparent declining share of all the other categories which include households. However, the latter's predominant consumption consists of traditional fuels, for which no reliable statistics exist. The statistical trends shown in Table 10.4, therefore, might well be an error.

The major user of petroleum products in recent years was the transport sector followed by electricity (and water) and manufacturing. Annual average growth was most pronounced in the electricity sector with 15.2%, followed by the commercial and household sectors with 9.8%, and agriculture with 8.2%.

Table 10.4. Total Energy Consumption by Sector-Percentage Shares.[1]

SECTOR	1971	1977
Energy[2]	10.9	12.3
Industry	16.8	20.8
Transportation	23.2	24.5
Agriculture	6.7	6.5
Other[3]	42.4	35.9

[1]Excluding nonenergy uses.
[2]Energy consumption in energy sector is net of own use, losses, and energy lost in thermodynamic conversion in power stations and refineries.
[3]Includes commerce, government, households, and miscellaneous.
Source: Coopers and Lybrand Associates "Survey of Energy Utilization. Volume 1, Main Report," Washington, D.C. (March 1980).

Electricity consumption is dominated by industry with 64.1% followed by the household sector with 20.2% and commerce with 14.8%.

Little can be said about energy consumption in sectors other than industry and transport. Because of the elimination of taxes on LPG, there has been an interfuel substitution within the transport sector. Taxicabs, in particular, are being increasingly fueled by LPG. While each of the domestic refineries has some capacity to produce LPG, much of it must be imported. LPG imports rose from 3.5×10^6 liters in 1977 to 76.1×10^6 liters in 1979.[6] This still represented less than 5% of the total petroleum product import bill, but preliminary observation indicated that the LPG conversion rate had, if anything, accelerated in 1980. There is also an interfuel substitution occuring between gasoline and diesel, due to the availability of inexpensive diesel engines for small vehicles. Thailand has a relatively large domestic automobile assembly industry which may enable it to be fairly responsive to change in domestic fuel costs.

10.3   ELECTRICITY PRICING[7]

As in the case of the Sri Lanka case study, we use a two stage procedure to determine electricity tariffs. (See also Chapters 4 and 5.) In the first stage, the strict long-run marginal costs (LRMC) of supply are estimated to meet the economic efficiency objective. The second step consists of adjusting strict LRMC to yield the final tariff structure which satisfies all the other goals of pricing policy described in Chapter 4.

Sector Background and Supply

The power sector of Thailand consists of 3 principal state enterprises classified by their functions. The Electricity Generating Authority of Thailand (EGAT) is responsible for electricity generation and transmission. The Metropolitan Electricity Authority (MEA) is responsible for electricity distribution in the capital city of Bangkok and the surrounding metropolitan area which has a population of 6 million. The Provincial Electricity Authority (PEA) distributes electricity to the rest of the country which has a population of approximately 38 million. Other government agencies which influence the operations of the three main enterprises include the Ministry of Finance, Office of the National Economic Social Development Board, the Budget Bureau, National Energy Administration, and the Power Policy Committee. The Power Policy Committee composed of a chairman (Deputy Prime Minister) and members repre-

---

[6]National Energy Administration, "Oil and Thailand 1978–1979."

[7]This study was carried out in 1979, on the basis of information available at the time. While the final numerical results would be different if more recent data was used, the broad policy conclusions are still valid for the early 1980s.

senting the 3 electric utilities and some of the government agencies described above play an active role in reviewing and recommending to the cabinet all future electric power development policy and tariff policy.

EGAT's electric power system consists of 4 geographic regions. The central, northeastern, and northern regions are interconnected by 230 kV, 115 kV, and 69 kV transmission lines. The southern region is weakly linked to the central region. A new transmission line is under construction and is expected to be in service by 1980.

MEA and PEA buy most of their electricity from EGAT, but PEA still owns and operates about 300 small diesel plants which are used to serve areas not yet within the reach of EGAT's transmission grids. These isolated stations serve many remote village areas.

As of 1978, the total installed generating capacity of EGAT's system was approximately 2,742 MW. It consisted of 7 hydro (909 MW), 3 oil-fired (1568 MW), 2 lignite-fired (66 MW), 5 gas turbine (165 MW), and 5 diesel plants (35 MW), as shown in Table 10.5. In the MEA's distribution system, the primary and secondary networks are mostly of overhead radial type at 12 kV and 22 kV for primary distribution, and 400/230 volts for secondary distribution. The PEA's subtransmission and distribution networks are also predominantly overhead radials energized at 11 to 33 kV (primary) and 400/230 volts (secondary). In 1977, the total energy supply was 11,048.2 GWh. EGAT's share of generation amounted to 10,777.3 GWh or about 97.55% of the total supply, the remainder was supplied by isolated diesel plants of PEA, and some energy was purchased from Laos. Shares of energy generation by the various types of generating plants of EGAT were as follows:

Hydro	30.6%
Oil-fired	65.5%
Lignite-fired	2.7%
Diesel	0.4%
Gas turbine	0.8%
	100.0%

EGAT's power development plan has been formulated to supply adequate power and energy to meet the load forecast. For the period 1978–1986, the major plant additions will consist of 1,013 MW of hydro plants and 1,667 MW of thermal plants consisting of lignite, oil, natural gas, and/or nuclear plants, as well as 240 MW of gas turbines. (See Table 10.6 for details).[8]

[8]This expansion program is likely to be modified after 1984/5 given the much larger off-shore gas reserves recently discovered and the expectation of substantial increases in proved coal reserves. The impact on LRMC and pricing, however, will be relatively small.

Table 10.5. Thailand: EGAT's Existing Electricity Generating Plants As of December 1977.

TYPE	PLANT	LOCATION (REGION)	INSTALLED CAPACITY			COMMISSIONING DATE
			UNIT SIZE	NUM-BER	TOTAL (MW)	
Hydro	Bhumibol	4	70	6	420	May 1964–Aug. 1969
	Sirikit	4	125	3	375	Jan. 1974–July 1974
	Ubolratana	2	8	3	25	Feb. 1966–June 1968
	Sirindhorn	2	12	2	24	Nov. 1971
	Chulabhorn	2	20	2	40	Nov. 1973
	Nam pung	2	3	2	6	Oct. 1965
	Kang Krachan	1	19	1	19	Aug. 1974
	Subtotal			19	909 (33%)	
Oil-fired steam	North Bangkok	1	75	2	150	Mar. 1961–June 1963
		1	87.5	1	87.5	Nov. 1968
	South Bangkok	1	200	2	400	Nov. 1970–Nov. 1972
		1	300	3	900	July 1974–Nov. 1977
	Surat Thani	3	30	1	30	Feb. 1973
	Subtotal			9	1,567.5 (57%)	
Lignite-fired steam	Mae Moh	4	6.25	1	6.25	Nov. 1960
	Krabi	3	20	3	60	June 1964–June 1968
	Subtotal			4	66.25 (3%)	
Gas turbine	South Bangkok	1	15	3	45	April 1969–Jan. 1970
	Nakon Ratchasima	2	15	1	15	June 1968
	Udon Thani	2	15	1	15	June 1969
	Hat Tai	3	15	3	45	Aug. 1971–Nov. 1977
	Surat Thani	3	15	3	45	N.A.
	Subtotal			11	165 (6%)	
Diesel				24	34.6 (1%)	
	Total			67	2,742.35 (100%)	

Source: EGAT.

MEA's power distribution development program for the period 1978–1986 was formulated for the expansion of the subtransmission and distribution system services to the outskirts of the existing service area, system reinforcement, and expansion of the primary voltage (12–24 kV) and low voltage (220–380 V) systems.

PEA's power distribution development plan has been based on a short-range and a long-range plan and includes the rural electrification program for the

Table 1O.6. Future Power Plants on EGAT System to Meet Forecast Demand to 1990.

PLANT	FUEL TYPE	INSTALLED CAP.	DEPENDABLE CAP.	COMMISSIONING DATE	ENERGY (GWH)	
					AVERAGE	FIRM
Mae Moh No. 2	Lignite	75 MW	71.3 MW	Jan. 1979	490.0	—
Srinagarind No. 13	Hydro	360 MW	358.5 MW	Sept. 79–March 80	1,178.7	888.1
Bang Pakong—Gas Turbine—I	Oil/then Gas	240 MW		Oct. 1980		
Berge Thermal Plant	Oil/then Gas	75 MW	71.3 MW	Jan. 1981	525.0	—
Bang Pakong—Gas Turbine—II	Oil/then Gas	240 MW		April 1981		
Mae Moh	Lignite	75 MW	71.3 MW	July 1981	490.0	—
Bhumibol	Hydro	90 MW		August 1981		
Pattani	Hydro	72 MW	53.5 MW	Oct. 1981	208.8	116.8
Bang Pakong Combined Cycle (I)	Gas	360 MW (120 MW increment waste heat boiler)	342.0 MW	April 1982	1,890.0	—
Lower Ouae Yai	Hydro	38 MW	38.0 MW	August 1982	170.5	140.0
Bang Pakong Combined Cycle (II)	Gas	360 MW (120 MW increment waste heat boiler)	342.0 MW	Oct. 1982	1,890.0	—
Bang Pakong Unit 1	Gas	550 MW	522.5 MW	July 1983	3,850.0	—
Mae Moh	Lignite	150 MW	142.5 MW	Jan. 1984	985.0	—
Khao Laem	Hydro	300 MW	241.1 MW	March 1984	765.3	457.6
Mae Moh	Lignite	150 MW	142.5 MW	July 1984	985.0	—
Bang Pakong Unit 2	Gas	550 MW	522.5 MW	August 1984	3,850.0	—
Lang Suan	Hydro	135 MW	135.0 MW	Oct. 1984	333.9	225.6
Srinagarind	Hydro	360 MW	360.0 MW	Oct. 1985		
Chiew Larn	Hydro	120 MW	118.6 MW	Jan. 1986	563.6	499.2
Mae Moh	Lignite	300 MW	285.0 MW	Sept. 1986	1,840.0	—
Combined Cycle	Oil/Gas	180 MW	171.0 MW	Oct. 1986	945.0	—
Gas Turbines Retired	Diesel Oil (−)	−165 MW		Oct. 1986		
Mae Moh	Lignite	300 MW	285.0 MW	March 1987	1,540.0	—
Mae Moh	Lignite	300 MW	285.0 MW	Sept. 1987	1,840.0	—
Upper Ouae Yai	Hyrdo	560 MW	527.8 MW	Oct. 1987	1,109.5	963.6
Thermal Capacity	Lignite/Gas Oil/Nuclear/ Coal?	900 MW	855.0 MW	April 1989	6,400.0	—
Combined-cycle	Oil/Gas	180 MW	171.0 MW	July 1990	945.0	—

Source: EGAT

period 1978–1986. The investment program consists of extending the networks, extending and revamping the PEA-owned diesel plants, and electrifying some 12,000 villages.

The Demand for Electricity

The average annual growth rate of consumption in Thailand during the past 10 years was 17%, with a high of 25% in 1968/69 and a low of 5.3% in 1973/1974, the year of the energy crisis and energy conservation measures. For the MEA area, the average growth rate was 15% with a high of 19% in 1972/1973 and a low of 1.7% in 1973/1974. In the PEA area, the average growth was 24% over the decade of 1967–1976. By the end of 1977, the total electric energy consumption in Thailand was about 9,640.36 GWh, with a peak load of 1,909 MW. The relative energy consumption of major customer groups is shown in Table 10.7. Demand forecasts are based on the analysis of historic growth and on the known future industrial projects in the MEA and PEA area. The results of this forecast were checked and revised through a comparison with an independent forecast that used a correlation analysis of historic consumption and socioeconomic indicators such as GNP, population, and price.

For the period of 1978–1986, the projected growth rate of energy consumption for all categories in the MEA area is nearly 8% per annum, while that in the PEA area, which includes the rapidly expanding rural electrification load, is over 14%. Demand by EGAT's direct customers, who are large industrial establishments, is expected to grow by 5%. Total EGAT sales as shown in Table 10.8 are the sum of the estimated requirements of MEA, PEA, and its own, direct customers. The electric energy and demand requirements of the whole power sector in Thailand for the period 1978–1986 are estimated to grow at an average rate of 10 and 11% per annum, respectively.

Table 10.7. Thailand. Electric Energy Consumption by Major User Group as a Percentage of Total Consumption (1977).

| | PERCENT | | | |
	MEA	PEA	EGAT	TOTAL
Residential	11.16	7.61	—	18.77%
Small business	8.80	5.12	—	13.92%
Medium and Large Business	42.09	17.97	4.75	64.81%
Others	0.42[1]	1.33[2]	—	1.75%
Street Lighting	0.34	0.41	—	0.75%
	62.81	32.44	4.75	100%

Remarks: [1]Off-peak rate consumers
[2]Water works, agricultural pumping, and temporary consumers

Table 10.8. Total EGAT Requirements.

YEAR	PEAK-GENERATION		ENERGY GENERATION		LOAD FACTOR %
	MW	% INCREASE	10^6 KWH	% INCREASE	
1978	2,098.0	11.99	12,135.0	11.60	66.03
1979	2,343.0	11.68	13,542.0	11.59	65.98
1980	2,614.0	11.57	15,056.0	11.18	65.75
1981	2,948.0	12.78	16,748.0	11.24	64.85
1982	3,274.0	11.06	18,631.0	11.24	64.96
1983	3,636.0	11.06	20,533.0	10.21	64.47
1984	3,966.0	9.08	22,361.0	8.90	64.36
1985	4,314.0	8.77	24,265.0	8.51	64.21
1986	4,667.0	8.18	26,243.0	8.15	64.19
78–86	—	10.51 p.a.	—	10.12 p.a.	—

PEA is presently engaged in a rapid expansion of its rural distribution network. As a consequence, it is expected to increase its share of total system peak capacity requirements from 41% in 1978 to 57% by 1990, and in energy requirements from 36% to 50%. This increasing share of PEA demands will result in a decline in the overall system load factor from 66% in 1978 to 64% in 1990, inspite of a slight increase in MEA's load factor.

Calculating Strict LRMC

As described in greater detail in Section 9.2, we define strict LRMC as the incremental cost of optimal adjustments in the system expansion plan and system operations attributable to a sustained incremental demand increase. The LRMC analysis recognizes three principal categories of costs: capacity or KW charges, energy or kWh charges, and consumer charges, with further possible structuring or differentiation of costs by voltage level, time-of-use, geographic area, and so on. Calculations have been carried out in constant 1979 Baht using shadow prices where appropriate, and an official exchange rate of US$1 = Baht (B) 20.4. Only an outline of the main steps and summarized results are presented here, to avoid duplicating the very similar computations described in the Sri Lanka case study (Chapter 9).

First, appropriate peak and off-peak periods must be determined by examining the typical system load duration curve shown in Figure 10.1. The period during which demand presses on capacity appears to be from about 1900–2130 hours. However, given the magnitude of other peaks adjacent to the principal peak, it would be more prudent to define the peak period over the broader interval 0830 to 2300 hours, with the off-peak period occurring during the remainder of the day. If the duration of the peak period is defined too narrowly,

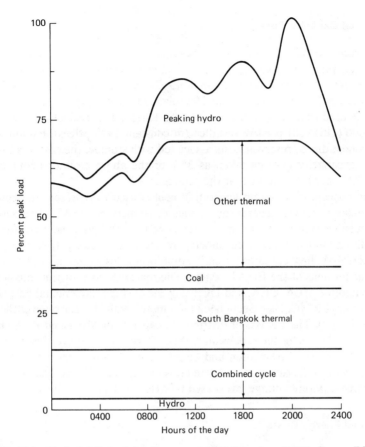

Figure 10.1. Typical daily load duration curve for EGAT system. (1984). Source: EGAT.

peak-load pricing is likely to cause a shift in the peak to the off-peak period. Seasonal variations are not significant enough to warrant further analysis at this stage.

For EGAT's present system, the base load is generally provided by the oil-fired thermal plants. The small hydro and lignite plants are operated more or less continually. The Sirikit hydro plant with 375 MW capacity operates at a constant intermediate level during the daytime, while the Bhumibol hydro plant with 420 MW capacity provides for the remainder of the load. The spinning reserve during the peak is provided by the unused capacity of these two main hydro plants. The future pattern of operation will be similar, with the loading order of generation units consisting of base hydro, a variety of thermal plants, and, finally, peaking hydro, as shown in Figure 10.1.

Marginal Capacity Cost

Following the analysis described in Section 9.2, in the case of the EGAT system, the optimal response to a peak period increment would be the addition of the low plant factor (peaking) hydro units. Therefore, the costs of the 300 MW Khao Laem hydro project (about 29% planned plant factor in a normal year) may be used as a representative case. The Khao Laem project will cost about B19,000 (US$950) per kW installed (in constant 1979 prices). Excluding the component due to reservoir costs associated with storage, the cost of adding to peaking capacity may be taken as 35% of the above costs or about B6650 (US$332) per kW installed, at the generator.[9]

The concept of a system generating pool is used to represent the common transmission network shared by various consumers of EGAT. Two separate transmission systems with distinct costs, one for MEA and the other for PEA and direct industrial (DC) customers, are assumed to go out from the pool to the corresponding consumer area.[10] Appropriate losses are also allocated to various portions of the EGAT system. The marginal costs of transmission are computed for MEA, PEA, and DC using the average incremental cost (AIC) to estimate LRMC. (See Section 9.2 for a more detailed description of the AIC methodology). The LRMC of distribution capacity of MEA and PEA are the incremental costs for the expansion of their distribution systems. The reinforcement of their subtransmission and distribution systems to meet the expected load growth in the future is calculated (also using the AIC method) at the two and three different voltage levels used by MEA and PEA, respectively.

Marginal Energy Costs

The estimation of incremental energy costs is more complicated and requires the analysis of the typical future pattern of system growth and operation. As shown in Figure 10.1, all new thermal units will be run as base load plant with little variation in generation throughout the day. Hydro will be fully utilized to displace as much of the least efficient other thermal plant as possible, with peaking hydro units being used especially to meet the maximum demand between 1830 and 2130 hours. The small amounts of hydro used during the off-peak period 2300–0800 hours are required to meet technical and other constraints (e.g., maintaining minimum river flows for navigation, etc.)

Thus, the thermal plants are used to meet energy requirements in excess of

---

[9]This is comparable to the alternative of adding a peaking gas turbine unit at a capacity cost of about B7000 (US$350) per kW installed.

[10]An alternative approach that is not used here would be to aggregate all transmission costs and have a common transmission cost for all customers.

the total hydro supply. Clearly, any incremental energy, irrespective of when it is required, will ultimately necessitate increased thermal generation, and this situation will continue into the future. Given a mix of thermal plants similar to the units shown in Figure 10.1 as the source of additional long-run energy, this yields an LRMC of energy of B0.58 per kWh at the point of generation, during both peak and off-peak periods. These costs must be grossed up by relevant loss coefficients to yield the LRMC of energy at different voltage levels, times-of-day, and geographic areas.

Summary of Strict LRMC

The total LRMC of capacity and energy are summarized in Table 10.9. The data shown include the appropriate adjustments for losses, O&M costs, etc.

Analysis of Existing Tariffs

Although EGAT's transmission costs to PEA areas are higher than transmission costs to the MEA areas because of the longer distances and lower load densities, the wholesale tariff charged to PEA is much lower than the one charged to MEA. This is the result of government policy that tries to equalize the residential retail tariffs of PEA and MEA by permitting PEA a higher markup from bulk to retail tariff, thus compensating for PEA's higher distribution costs. In order to cover the reduced revenues due to the lower PEA bulk tariff and balance its revenues, EGAT had to increase the tariff charged to MEA and to its industrial direct customers.

Consumer-Related Costs

Incremental consumer-related costs are considered to be the cost of providing meters, reader meters, billing, and maintenance of meters. The estimation of the consumer-related costs is based on the records of the average cost of reading and billing for both MEA and PEA systems in 1976 and the cost of meters used by different types of consumers (annuitized over their lifetime). The consumer-related costs thus calculated are shown in Table 10.10.

In general, the consumers in PEA areas are effectively subsidized by the consumers in MEA areas, at least in relative terms.[11] (Table 10.11.) For example, in the case of residential consumers, the existing MEA average price per kWh is about 3% higher than the PEA rate (B943 vs. B917), while PEA marginal costs are 36% higher than for MEA (B1.62 vs. B1.19). The distortion of

---

[11]In relative terms because in real terms tariffs are below LRMC for all groups of customers.

Table 10.9. LRMC of Capacity and Energy (Domestic baht).

	CAPACITY COSTS (B PER KW)								ENERGY COST (B/KWH)	
	GENERATION		TRANSMISSION		GENERATION AND TRANSMISSION	GENERATION AND TRANSMISSION	SUBTRANSM. AND DISTRIB.	TOTAL CAP. COSTS (B/		
COST IN	TOTAL	ANNUAL	TOTAL	ANNUAL	ANNUAL COST[1]	MONTHLY COST	MONTHLY COST	KW/MONTH)	PEAK	OFF-PEAK
EGAT	7000[1]	767[2]							0.611[1]	0.598[1]
MEA		791[3]	888	98[2]	889	74.1	—	74.1	0.630[3]	0.610[10]
69kV						75.6[4]	4.6	80.2	0.643[4]	0.618[11]
12–24kV						76.4[5]	28.8	105.2	0.649[5]	0.622[12]
220–380V						85.2[6]	94.8	180.0	0.724[6]	0.678[13]
PEA&DC		843[7]	2060	227[2]	1070	89.2	—	89.2	0.671[7]	0.636[14]
22kV						92.0[8]	36.3	128.3	0.692[8]	0.649[15]
220–380V						107.5[9]	142.2	249.7	0.808[9]	0.723[16]

[1] Includes station use and HV losses (peak = 5%; off-peak = 3%).
[2] Annuity factor 0.106 (discount rate of 10% for 30 year lifetime) applied and annual O. M & A costs added (generation = B 25/kW/year; MEA Transmission = B4/kW/year; PEA Transmission = B 9/kW/year).
[3] Includes transmission peak losses = 3%.
[4] Includes peak losses = 2%.
[5] Includes peak losses = 3%.
[6] Includes peak losses = 13%.
[7] Includes transmission peak losses = 2%.
[8] Includes peak losses = 3%.
[9] Includes peak losses = 17%.
[10] Includes off-peak losses = 10%.
[11] Includes off-peak losses = 1.3%.
[12] Includes off-peak losses = 2%.
[13] Includes off-peak losses = 10%.
[14] Includes off-peak transmission losses = 6%.
[15] Includes off-peak losses = 2%.
[16] Includes off-peak losses = 12%.

Table 1O.1O. Consumer-related Costs.

TYPE OF CUSTOMER	TOTAL COST (B/MONTH)	
	MEA	PEA
Residential	17.31	15.97
Small Business		
—220/380 V	32.15	29.87
—12–24 kV	316.01	313.73
Large Business and Industrial		
—220/380 V	109.67	103.90
—12–24 kV	354.38	349.61
69 kV	354.38	

tariffs relative to real costs leads to incorrect decisions about electricity use in the two areas. Therefore, the case for maintaining the existing subsidy to PEA must rest on arguments such as regional development, stemming migration to Bangkok, and other sociopolitical reasons.

Lifeline Rates

The average monthly consumption of MEA and PEA residential customers was 200 kWh and 60 kWh, respectively, a ratio of 3.3/1. This shows that lower income households, which are concentrated in provincial areas, use considerably less electricity than higher income urban households. However, rural per capita incomes also are far lower than urban ones, the urban/rural ratio being about 2.6/1. In the past, increases in power tariffs have normally been across the board, with the result that their effects on lower income users were more severe than those on higher income users. The introduction of a strict marginal cost pricing approach would prevent a significant proportion of low income groups from obtaining even a minimum supply of electricity because of the higher distribution costs to them. This may call for the introduction of a new tariff schedule comprising two steps, with the initial one representing a relatively low lifeline rate, while the second would more closely reflect actual LRMC.

The setting up of a low subsidized block would require the definition of a "consumption threshold." According to government statistics about 46% of the people in Thailand are said to be poor. A determination of minimum basic electricity needs for poor households was made by considering the kWh consumed by electric appliances required to maintain a minimum level of comfort: (1) lighting, (2) cooking, (3) an electric fan, (4) a radio, and (5) an electric

Table 10.11. Comparison between Average Tariff Levels of
Existing and LRMC Based Tariffs.

| | AVERAGE TARIFF LEVEL (฿/KWH) | | LOAD FACTOR |
TYPE OF CONSUMERS	EXISTING[1]	LRMC BASED TARIFF[2]	(LF)[3] (%)
EGAT			
MEA	0.6057	0.759	73
PEA	0.4795	0.841	58
Industrial	0.7200	0.830	61
Special rate	0.6376	—	70
Off-Peak	0.6142	0.610	74
Over all	0.5680		
MEA			
Residential	0.9434	1.19 (220–380V)	49
Small Business	0.8933	1.19 (220–380V)	49
Medium Business	0.7663	0.923 (12–24kV)	50
Large Business	0.7006	0.778 (69 kV)	74
On-Off Peak	0.6939	—	—
Special Rate	0.6653	—	57
Over all	0.7898	NA	—
PEA			
Residential	0.917	1.62 (220–380V)	40
Small Business	0.975		50
General Business	0.841	1.01 (12–24kV)	52
Medium Business	0.791	1.01 (12–24 kV)	52
Large Business	0.700	0.914 (12–24kV)	72
Mining	0.849	NA	NA
Irrigation	0.501	NA	NA
Water Works	0.848	NA	NA
Over all	0.846	NA	—

[1]From revenues and energy sales in 1978.
[2]kW charge converted to kWh charge:

$$฿/kWh = \frac{฿/kW/months}{730 \times LF}.$$

[3]From load research.

iron. The kWh requirements for these minimum residential uses were estimated to be 40 kWh per month. The tariff level for this subsidized would have to be set on the basis of the average income of poor households. If the lifeline were implemented, 13% and 52% of the residential users in the MEA and PEA areas, respectively, would be affected.

Adjustments to Strict LRMC

Both the levels and structures of the strict LRMC values as calculated above are different from the existing tariffs. The present level of demand charges are much lower while the energy charges are higher. A set of suggested adjusted tariffs based on LRMC are shown in Tables 10.12 to 10.14. EGAT's LRMC was adjusted in order to avoid a overly large increase in the demand charge, while the energy charge was maintained at a level at least equal to the fuel costs of thermal plants. The adjusted tariffs of MEA and PEA at the retail level reflect social and other constraints as discussed earlier. The estimate of EGAT's revenue generation in a test year, using this tentative tariff, is about 35 percent higher compared to that generated by the existing tariff (about Baht 0.57 per kWh). Further adjustments could be made if the financial surpluses were judged either to be inadequate or excessive.

The introduction of such a revised tariff structure would have several advantages:

First, it would produce enough revenues to eliminate the current heavy sub-

Table 10.12. Suggested Adjusted Tariff Structure for EGAT.
EGAT's Adjusted Tariff.

TARIFF METHOD	TYPE OF CONSUMERS	ENERGY CHARGE (฿/KWH)	DEMAND CHARGE (฿/KW)
E1	MEA	0.5897	66
E2	PEA	0.6777	74
E3	Direct customers	0.6557	74

Table 10.13. Suggested Adjusted Tariff Structure for MEA.

TARIFF METHOD	TYPE OF CONSUMERS	DESCRIPTION	ENERGY CHARGE (฿/KWH)	DEMAND CHARGE (฿/KW)
M1	Residential	Lighting and power	First 5 kWh or less: 5.00฿ Next 35 kWh : 0.73 Rest : 1.24	—
M2	Small business	Demand less than 30 kW	M2.1 at 220–380 V: 1.24 M2.2 at 12–24 kV : 0.98	— —
M3	Large business	Demand over 30 kW	M3.1 at 220–380 V: 0.78 M3.2 at 12–24 kV : 0.64 M3.3 at 69 kV : 0.60	100 78 70
M4	Street Lighting		at 12–24 kV : 0.97	—

Table 10.14. Suggested Adjusted Tariff Structure for PEA.

TARIFF METHOD	TYPE OF CONSUMERS	DESCRIPTION	ENERGY CHARGE (฿/KWH)	DEMAND CHARGE (฿/KW)
P1	Residential	Lighting and power	First 5 kWh or less: 5.00฿ Next 25 kWh : 0.82 Rest : 1.70	—
P2	Small business	Demand less than 30 kW	P2.1 at low voltage : 1.45 P2.2 at high voltage: 1.04	— —
P3	Large Business	Demand over 30 kW	P3.1 at low voltage : 0.91 P3.2 at high voltage: 0.68	125 86
P4	Street Lighting		at low voltage : 1.04	—
P5	Irrigation		: 0.40	—

sidization of EGAT from general governmental funds[12] and to finance the expansion program of the three utilities. Second, it would create strong inducements to industrial users to reduce peak load demands because of the much higher demand charges. This would reduce EGAT's capacity requirements and lead to lower generating costs. Third, the use of the "lifeline" rate would enable poor consumers to cover their minimum needs of electricity. It probably would also induce more potential users in newly electrified, rural areas to convert to electricity use. This, in turn, would tend to reduce overall distribution costs.

10.4   NATURAL GAS COSTS AND PRICING

Some time ago natural gas was discovered in the Gulf of Thailand. Overall reserves in early 1981 were estimated at more than 7 trillion cubic feet (TCF). The gas was found in two separate deposits, one explored by Union Oil Company, the other by Texas Pacific. Several other potential deposits were indicated but not explored at that time.

The government has committed itself, through its fully-owned Natural Gas Organization of Thailand (NGOT), to build a gas pipeline from the nearest of those deposits owned by Union Oil to markets in and around Greater Bangkok. Negotiations with the owners of the second field, Texas Pacific, about delivery schedules, prices, and quantities, still have not been concluded by early 1982.

---

[12]These subsidies are provided in part through subsidized fuel oil prices.

Under the terms of Union Oil's contract with the government, field prices start at a level of US $1.30 per million btu for the first 75 million cubic feet per day (MMCFD) and decline to a minimum of $0.67 for the last incremental block of an overall maximum total of 250 MMCFD that is to be supplied under the contract. At a delivery rate of 250 MMCFD the average 1976 price amounts to US $1.05 per million btu. However, prices are subject to various price escalator clauses so that current prices are higher.[13]

The first delivery of natural gas from the Union field was expected to reach Bangkok by late 1981.[14] Projected gas utilization rates have been shown in Table 10.15. About 90% of the projected demand was expected from EGAT-owned electric power generating stations. These were the existing, oil-fired generating plants at South Bangkok with an installed, total capacity of 1300 MW, and the Bang Pakong generating units 60 miles east of Bangkok then under construction. The latter will reach a total capacity of 1,270 MW by 1983. Together, these plants were estimated to consume some 415 MMCFD by 1983.[15] An additional, 550 MW gas-fired thermal unit was planned for completion at Bang Pakong by 1984. This unit was projected to account for another 131 MMCFD of gas consumption, resulting in an overall demand of about 500 MMCFD of gas for electric power generation.

Projected industrial, commercial, and other natural gas uses, by comparison, were minor. They were projected to be no more than 29 MMCFD in 1981, rising to 84 MMCFD by 1995. However, industrial demand could be increased substantially if Thailand decides to proceed with a petrochemical complex based on natural gas as feedstock.

The Value of Natural Gas as Measured by Its Replacement Costs

The maximum economic value of the gas for use in the two EGAT powerplants and in most industrial uses is determined by the economic costs of the fuel oil that it replaces. For the time being, these uses cover some 95% of total projected gas utilization, so that the few other uses can be safely disregarded. However, for the second, 550 MW Bang Pakong thermal powerplant, planned for completion in 1984, the relevant value may well be determined by the delivered cost of electricity from an alternative, coal-fired powerplant at Mae Moh, because of the possibly lower costs of the latter compared to a fuel-oil-fired powerplant.

---

[13]In early 1982, wellhead prices had increased to $2.20/MCF. Because 40% of the gas price is tied to Singapore fuel oil spot prices, overall gas prices could conceivably decline if fuel oil prices continue their downward trend. Price data from: "Thailand, Economic Report," *Christian Science Monitor,* March 23, 1982, p. B.2.

[14]The pipeline was completed on time and delivered some 100 MMCFD by early 1982.

[15]At an average heat rate of 9800 btu/kWh net of station use and 1 million btu/MCF of gas.

Table 1O.15. Natural Gas Supply/Demand 1981–199O in MMSCFD (1OOO btu/c.f.).

DEMAND/ SUPPLY	YEAR	1981	1982	1983	1984	1985	1986	1987	1988	1989	1990
Electric sector demand		245	311	429	547	547	572	572	572	572	598
Industrial demand		26	29	32	35	39	41	44	46	47	48
Total gas demand		271	340	461	582	586	613	616	618	619	644
Total Union Oil and Texas Pacific gas supply		158	276	329	420	459	511	563	563	563	563
Supply/ demand (deficit)		(113)	(64)	(132)	(162)	(127)	(102)	(53)	(55)	(56)	(81)

Source: NGOT.

The economic costs of the displaced fuel oil consist of the sum of the C.I.F. costs of imported fuel oil, expressed in local currency at the prevailing rate of exchange, plus domestic storage and delivery costs, adjusted by the estimated current standard conversion factor of 0.93 to express them in equivalent border prices. Costs of imported fuel oil, in 1977, amounted to an average of ฿1.63 liter.[16] Average fuel oil costs to EGAT for the South Bangkok plant were somewhat lower, amounting to ฿1.58 in 1978. However, this was the result of the availability of lower-priced fuel oil from the price-controlled, domestic refineries. Delivery costs were quite small, amounting to between ฿0.02 to ฿0.04 or an average of, say, ฿0.03/liter. Expressed in 1977 prices, this represented a total economic cost of ฿1.66 per liter, or ฿0.432 per kWh of electricity produced.[17] By early 1979, actual costs had risen to about ฿1.75/liter, or ฿0.455 per kWh.

Imported fuel oil prices, rather than imported crude oil prices plus domestic refining costs were the relevant measure of the economic costs, because Thailand had been a net importer of fuel oil in recent years. These imports were rising rapidly. Without additional refinery capacity the availability of natural

---

[16]NEA, *Oil and Thailand* 1977, Bangkok, p. 17. Exchange rate US$1.00 = Baht 20.00.
[17]At an estimated consumption of 0.26 liters per kWh. *From:* NEA, *Power Tariff Study,* Bangkok, 1978, p. 32.

gas after 1981 was projected to reduce net imports of fuel oil for only one or two years. By 1984, total fuel oil demands are projected to exceed existing domestic refining capacity once more.

Because Thailand is a price taker with respect to petroleum imports, future rather than current relative fuel oil prices (i.e., after adjustment for general inflation) should be used in estimating the economic costs of fuel oil compared to natural gas. Such prices are difficult to predict. However, one important aspect of this unpredictability is the very aspect of uncertainty or risk related to the needs for importing a vital ingredient of national productive activity. This type of uncertainty is absent in the case of domestic natural gas supplies (or all other alternative domestic energy sources such as lignite, for example). Therefore a special risk premium ought to be added to the estimated economic costs of imported fuel oil. Such risks premiums are quite commonly paid, directly or indirectly, by main oil importing countries.[18] However, while the economic principle of applying such a risk premium is quite sound, establishing its appropriate magnitude is a rather difficult undertaking. In one way or another, its value should be related to the potential net losses the national economy might suffer if prices rise suddenly, or if supplies are interrupted or reduced. Here it is not possible to explore and discuss the many difficult issues that would have to be addressed in order to come up with such firmed-up estimates. However, a 10 to 15% premium does not appear to be unreasonable, given the experiences of 1973/74 and 1979/80. Including such a risk premium of 10%, the economic cost of imported fuel oil, on the basis of early 1979 prices, is estimated to be equal to ฿1.93/liter, or ฿0.501 per kWh of electricity.

The financial cost of imported fuel oil will be almost equivalent to the economic one (except for the risk premium) because of the use of the foreign exchange border price as the numeraire in the shadow pricing of imports, and the insignificance of domestic delivery costs.

It is possible that, beyond 1983, the appropriate marginal economic replacement costs of additional natural gas supplies used in the production of electricity may be represented by the prorated, appropriately shadow-priced costs of electricity produced by new coal-fired, mine-mouth power plants at Mae Moh.[19] It is estimated elsewhere that up to 2000 MW of thermal power gen-

---

[18]Examples are the elaborate and costly strategic oil reserve programs in the USA and various European countries; the willingness by the Canadian and Alberta governments to guarantee world market prices to tar sand projects, while domestic oil is price-controlled at lower levels; the substantial tax credits available to U.S. home owners for installing additional thermal insulation as a substitute for energy consumption; or the development of subsidized gasohol plants in Brazil.

[19]There may still be other, not yet well-explored, coal deposits in the country, or there may be additional hydropower sites whose costs may make them appropriate candidates for estimating the maximum economic value of additional gas supplies. However, too little is known about these potential resources and their costs to use them for evaluation purposes here.

erating capacity could be developed at that site.[20] However, since 525 MW are already programmed for development at Mae Moh in any case, only about 1500 MW of potential capacity could be looked upon as an alternative to further natural gas utilization beyond 1983. At a 70% load factor, these 1500 MW of coal-fired capacity would replace a consumption of some 245 MMCFD of natural gas.

The comparison of the real economic cost differential between additional natural gas or alternative coal utilization for electricity production has to be analyzed in terms of delivered electricity costs per kWh because of the differences in generating plant and transmission line investment costs, and the differences in O.M.&R. and fuel costs between natural gas and coal-fired powerplants. Coal-fired plants are more costly in terms of generating plant and transmission line investment and O.M.&R. costs, while they may have a substantial advantage in terms of delivered fuel costs on a btu basis.

The only other presently foreseeable alternative use of natural gas is in industrial applications as fuelstock for chemicals, fertilizers, plastic, etc. On the basis of the rather high[21] negotiated average prices per unit of gas from the Union Oil field, prospects for establishing such industries in Thailand are not high because of the relatively small domestic markets for end products (and consequent diseconomies of scale in the size of conversion plants), the inability to compete in export markets, and the likely availability at relatively low prices of such basic chemical feedstocks on the world market in the future.[22]

The Economic Supply Costs of Gas[23]

The following cost categories should be considered in the evaluation of the marginal, economic costs of natural gas: 1) The (foreign exchange) wellhead price paid to the drilling company, net of royalties: 2) the O.M.&R. costs for each field; 3) the investment costs of the pipeline network needed to bring the gas

---

[20]According to the latest power plant expansion program of EGAT at the Mae Moh site the third 75 MW unit would be completed by 1981, an additional 150 MW unit by 1984, and three 300 MW units in 1986 and 1987, respectively. However, this program also proposes total gas-fired capacities of some 3,195 MW by 1984. Such a program would essentially utilize all the gas that could be made available.

[21]In terms of natural gas prices or costs in other competing locations such as the Indonesian Archipelago, or the Persian Gulf.

[22]Given the huge surpluses of natural gas in many oil-producing regions of the world far from available natural gas markets.

[23]The analysis in this section is based on data available in early 1979. Revised cost estimates of late 1979 resulted in increased cost estimates of US$482 million for the Union Pacific section, compared to the US$395 used here, and US$220 million for the Texas Pacific connection, compared to the original US$180 million. However, the validity of the arguments presented here remain unchanged, even though all estimated costs would have to be increased in relative proportion to reflect the more recent data.

to markets, the O.M.&R. costs of the pipeline; 4) the investment costs needed to modify existing installations for use of gas instead of alternative fuels; 5) the net differentials in operating costs, including differentials in fuel-use efficiencies resulting from the use of gas instead of other fuels.

Because of the possibility to develop the overall project in three distinct steps—the pipeline to the Union Oil field, the interconnection to Texas Pacific, and, finally, the installation of additional compressors to boost pipeline capacity—the marginal costs of each of these steps must be evaluated separately.

To Thailand, the immediate economic costs of on-site field developments are zero, because they are being financed by foreign-owned corporations out of their own funds. However, these companies expect to be compensated for their present investments through the flow of future dividend and interest payments, as well as depreciation charges. These future flows of foreign exchange funds out of the country represent the true economic costs to Thailand.[24] They should be appropriately discounted to the present. In this form they are to be counted as part of the marginal economic costs of each project increment.

The projected O.M.&R. costs of operating the fields should be appropriately shadow-priced (for the use of domestic resources) and these costs included in the present value of overall marginal costs.

The estimated marginal pipeline investment costs for each step are US$395 million for the Union Oil connection, and US$180 million for the extension to the Texas Pacific Field (including initial compressor stations). Foreign exchange costs amount to about 75% of the total. Of the remaining 25%, or $145 million for domestic expenditures, about $20 million have to be netted out for taxes and transfers. The appropriate shadow prices must be applied to the various cost categories representing the remaining $125 million.

According to present proposals, the local costs of $145 million will also be financed through foreign loans. These foreign counterpart funds should not be subjected to the foreign exchange shadow price adjustment[25] because the real investment comes from domestic resources,[26] and the foreign loan is converted into freely available foreign exchange reserves at the time of the loan. After applying the appropriate conversion factors to the net financial cost of U.S. $125 million, their shadow-priced real economic costs amounts to U.S. $116.3 million.[27]

The underwater pipeline connecting Union Oil with the coastal terminal and

---

[24]Retained earnings or depreciation charges reinvested in Thailand should not be included as part of these future costs.

[25]i.e., evaluated at the numeraire's value of 1.

[26]The appropriated conversion factors are 0.69 for unskilled and 0.96 for skilled labor; 0.91 for other domestic nonlabor inputs, and 0.93 for the standard conversion factor.

[27]Evaluated at the standard conversion factor of 0.93.

overland route has been designed with a diameter[28] sufficient for a total throughput of over 700 MMSCFD (with additional booster compressors). This specification anticipates the hookup with the Texas Pacific Field and the possible expansion of output up to that capacity.

It may be argued that the appropriate cost comparison should be based on a smaller diameter, less costly line, sufficient only to transport the projected 250 MMSCFD from the Union Oil field. This argument must be rejected because the average transportation costs in such a line would be higher than in a combined one. They would amount to U.S. ¢52/MCF. The average costs of a separate, 250 MMSCFD line from the Texas Pacific field would be U.S.¢ 71/MCF. The combined line now proposed, with a total throughput of 500 MMSCFD, would result in average capital cost charges of U.S.¢48/MCF instead. Since an economically justifiable market for 500 MMSCFD is readily available, the combined line is preferable to two separate ones. The average capital costs charges for the 250 MMSCFD supply from the Union Oil field alone for the combined line will amount to ¢65/MCF. The marginal additional cost for the U.S. $200 million extension to the Texas Pacific field would result in average capital charges of U.S. ¢31/MCF.[29] In addition, the capital costs for increasing the capacity of the Union Oil field section to accomodate the Texas Pacific gas is about US$60 million or US$9.3/MCF at an average throughput rate of 250 MMSCFD from Texas Pacific.[30] However, the doubling of throughput of the line from Union Oil's field to markets would reduce the charges for the latter from ¢65/MCF to the combined average of U.S. ¢ 48/MCF for the pipeline system as a whole. If the Texas Pacific field's reserves actually amount to 3.5 TCF instead of 1.5 TCF as assumed in this analysis, the average financial pipeline capital cost charges would be reduced to U.S. ¢ 43 per MCF. Adjusted by the standard conversion factor for domestic capital expenditures, the average economic capital cost charges for the proposed combined pipeline network would be U.S. ¢45.6 per MMSCFD at a throughput of 500 MMSCFD and a total combined reserve of 3 TCF; given a total reserve of 5 TCF and an average daily throughput of 500 MMSCFD,[31] costs would fall to U.S. ¢40.7.

To arrive at the net delivered costs of the gas to customers, the costs of equipment conversion from fuel oil (or other, displaced fuel sources) to natural gas must be added to the financial and economic costs of the gas. In the case of

---

[28]34″ diameter to the shore and smaller from then onwards.

[29]Assumed recoverable reserves of each field equal to 1.5 TCF, 17 year life, 12% interest. In order to be cost equivalent to the joint line, the capital cost of the separate Union Oil line would have to be less than U.S. $310 million.

[30]It should be noted that operating costs for the Texas Pacific field will be higher than for the Union Oil field because of the higher content of carbon dioxide and the need for pressurizing the gas.

[31]Which implies an economic life of 28 years.

the South Bangkok power plant, for example, the capital conversion costs have been estimated at U.S. $20 million. Given the projected gas utilization rate of this plant of an average of 257.9 MMSCFD for a 17 year period, these additional capital cost charges would amount to about U.S. ¢3 per MSCF in financial, and U.S. ¢2.9 in economic terms.[32]

Another cost item which has to be added results from the reduction in the net output efficiency of the various powerplants and boilers when gas is used instead of fuel oil. These net reductions in output, for the electric powerplants, have been estimated to amount to 17½% for South Bangkok installations (units 1 and 2) of some 400 MW capacity.[33] The time period over which these reductions are maintained is uncertain and will depend on the quantity of gas available for power generation. On a throughput of 500 MMSCFD this loss of capacity is likely to be relevant for two years only.

While the overall analysis, so far, has been based on the assumption of constant real costs, it is likely that the relative costs of imported fuel oil and of the natural gas supplied from the Union Oil-Texas Pacific fields will diverge. Given the uncertainties surrounding future world market fuel oil demand/supply relationships, it is impossible to say whether future fuel oil prices, in real terms, will increase or decrease. If they were to rise, then the real economic value of the natural gas would also rise, while its financial costs would decline in relative terms. This is so because the long-term price escalator clauses of the Union Oil supply contract (assumed to be duplicated in the pending Texas Pacific contract) depend only to 40% on the world market price of fuel oil.[34] Everything else remaining equal,[35] this means that for each percentage point increase in fuel oil prices the gross field costs of gas increase by only 0.4%.

The opposite would be true, of course, if the world market price[36] for fuel oils falls. However, the downward risk in economic terms is much lower than the upward one, because the economic cost of the gas is significantly lower than that of the oil. This means that gas would still remain the preferred fuel, even if fuel oil prices were to drop significantly from present levels. Hence, it can be concluded that, overall, the contractual terms of the price escalator clauses of the gas supply contract provide some potentially significant additional eco-

---

[32]SCF = 0.93, one-third domestic costs.

[33]No information is available about these effects on industrial bioler installations. However, their projected consumption will be less than 10% of total gas consumption, hence these differentials are minor, and of importance only for tariff setting purposes. Such losses of capacity occur only in existing boilers whose design is exclusively based on the use of fuel oil.

[34]The specific price escalator clauses of the Union oil contract are more complex and contain specific ceiling and floor price levels, and are weighted according to domestic, foreign exchange, and fuel oil price changes.

[35]i.e., no feedback effects of fuel oil price changes on the other relative prices that form part of the price escalator formulas.

[36]The fuel oil escalator price is actually tied to ex-refinery Singapore prices.

nomic benefits to Thailand, that could be reaped without harming the gas field operators. On the contrary, the latter would make a windfall gain before taxes of some 40% if fuel oil prices increase.

The Economic Value of Natural Gas

The maximum economic value of the gas for use in the existing or committed EGAT power plants that depend on the use of either oil or natural gas, as well as in most industrial uses, is determined by the economic costs of the fuel oil that it will replace. However, for future, so far not committed powerplants the relevant value of the gas would be determined by the delivered cost of electricity from alternative coal-fired powerplants at tidewater, lignite-fired plants elsewhere in the country, or even one of the potential, large hydroplants along Thailand's borders. Judging by recent price trends for petroleum products, any one of those other alternatives is likely to deliver electricity at lower cost than fuel-oil-fired powerplants. Hence, the present and estimated future oil price sets the maximum ceiling for the value of gas only for those plants that must, over their economic life span, use fuel oil if natural gas is unavailable.

The second type of opportunity costs are given by the value of the gas in its next best alternative use. These alternative domestic uses are limited. Sufficiently large markets simply do not exist to absorb a much larger proportion of the gas presently not allocated to EGAT's powerplants.

The only other alternative, therefore, would consist of exports. However, export markets (basically LNG or ammonia-urea or methanol gas derivatives) do not appear to be too attractive because of the substantial competition from other, lower-cost gas producers elsewhere, and the gas conversion costs. An LNG system for shipment to Japan, for example, may cost, on average, $2.50 to $3.00 per MCF. With landed gas prices in Japan of between $4.50 to $5.00, the net-back to Thailand would be in the order of $1.50 to $2.50, barely sufficient to cover the costs of production and shipment to a hypothetical liquification plant.

Hence, the domestic use of natural gas as a replacement for fuel oil appears to represent its highest foreseeable value. However, this does not mean that gas prices charged to users should be set equal to the present and future costs of fuel oil. As has been argued in more detail elsewhere (see Chapters 4 and 11), the true present value of the gas is given by its discounted future replacement cost, i.e., the cost of replacing it when the available deposit has been exhausted, or when other, high value uses start exceeding available supply.

To establish the present-value equivalent of this future opportunity cost requires detailed information about future consumption needs, use rates over time, prices, costs of required infrastructure (additional pipelines, processing facilities, etc.). This information is not available. However, following the gas

Table 10.16. Illustrative Calculation of Gas Depletion Premia.[1]

ILLUSTRATIVE DATA	
1981 marginal economic production cost (supplier price) of natural gas, per MCF	$2.00
Average costs of transport and processing, per MCF	$0.45
1981 Equivalent costs of fuel oil, per MCF	$5.00
Real price increase in fuel oil prices	3% p.a.
Contractual real increase in gas prices paid to producers	$0.4 \times 3\%$ p.a.
Initial year of natural gas deficiency	1990
Consumption horizon for potential alternative gas uses (1990–1999)	10 years
Real rate of interest	8%
Present (1981) value of gas depletion premia per MCF	$1.52
1981 marginal costs of gas supply, payment to producer, transport, and processing	$2.00 $0.45
Total economic cost of gas	$3.97
Required annual price escalation in real terms	8% p.a.

[1]Fuel oil price equivalent = $29.00/bbl.

utilization projections for gas deliveries to EGAT after 1989, the user cost of the gas could be presented by the difference between the delivered future price of fuel oil and the future marginal costs of producing and delivering the gas over time, discounted to the present.

Table 10.16 presents a simplified example of the required calculations. The data used are only approximations of actual ones. The three basic assumptions are first, the best alternative for additional gas utilization would consist of new, domestic demands starting in 1990 that would have to use fuel oil instead; second, real fuel oil prices would increase by 3% per year; and third, the gas used up today would otherwise be consumed in equal amounts over a ten-year period, starting in 1990. Granting these simplifying assumptions, the gas depletion premia that must be charged in 1981 are $1.52/MCF. To this must be added the real costs of purchasing the gas, processing, and transporting, assumed equal to $2.45/MCF, resulting in a total economic opportunity cost of $3.97/MCF in 1981. This price must be increased in real terms by 8% per year each year thereafter in order to reflect the real rate of interest of 8%.[37] If large additional gas deposits were to be discovered and no new additional, and economically equal or more valuable uses be found for them, the depletion time horizon would be pushed back and the resulting depletion premia would

---

[37]In addition, of course, prices must be adjusted to reflect purely inflationary cost increases.

decline.[38] If additional gas uses were to be found without additions to known reserves, the depletion premia would have to be increased. This means that the prices and costs should be continuously reevaluated over time in the light of changing circumstances.

10.5   PETROLEUM PRODUCT PRICING

Traditionally, ex-refinery petroleum product prices have been determined by accounting costs.[39] However, prior to the OPEC price escalation, they were also subject to substantial excise taxes which ranged between 120 to 150% for gasoline, from 60 to over 80 percent for jet fuels, and to about 50 to 80 percent for kerosene and LPG, respectively. For diesel oils, tax rates were much lower—some 23 and 24 percent, and for fuel oils they were as low as 11 and 12 percent. Municipal and business taxes added another 13 to 14 percent to gasoline costs, 3 to 11 percent to jet fuel, 3 to 7 percent to kerosene and LPG, 7 percent to diesel fuels, and a modest 2 percent to fuel oils.[40]

In that period, petroleum excise and municipal taxes represented a significant amount of total government revenues. In 1970, they produced some 8.6 percent of the total, accounting for 1.2 percent of total GDP. In terms of the value of petroleum imports central government excise taxes alone amounted to almost 84 percent of import costs.

These relationships between product prices and taxes changed rather significantly in the post-1973 period. While ex-refinery prices were allowed to rise to reflect the higher market costs of crude, excise taxes vere initially held constant in absolute terms, except for a relatively modest increase in gasoline taxes from B0.80 to B1.00 per liter. Municipal taxes were increased for all products in absolute amounts, but their share relative to ex-refinery prices fell for most products. Subsequently, most excise taxes were actually reduced. These adjustments represented an attempt by the Government to reduce the overall impact of the OPEC price increases on the economy. Excise taxes for LPG and fuel oil were all but eliminated. For jet fuels, they were reduced from the former B0.33 per liter to B0.12 in 1977. For diesel fuels taxes were raised slightly from B0.12 to B0.14. Gasoline taxes were increased from B1.00 to B1.10 in September 1977, and to B1.83 in March 1978. This increased gasoline taxes to some 75 and 84 percent of (1977) ex-refinery prices for premium and standard,

---

[38]This, in fact, has happened in the meantime, with gas reserves estimated in 1982 at 16.5 trillion cu. ft. *From:* "Gas Bubbles and Oil Pools: Mainly a Buried Treasure," The *Christian Science Monitor,* March 23, 1982, p. B 2.

[39]Because the various refinery products have joint production costs, setting prices for individual products always involves a certain degree of arbitrariness. Depending on either willingness to pay principles (inelastic demand schedules) or social or economic considerations, prices for some products may be increased, thus implicitly subsidizing others.

[40]All data refer to 1971.

respectively. However, these rates were still substantially lower than the 119 and 152 percent prevailing in pre-OPEC days.

Three more price and tax changes were made in 1978 and 1979. By July 1979, most ex-refinery product prices had risen by between 50 to 70 percent over 1977 levels, and excise tax rates had been increased to about 86 percent for gasoline, 12–13 percent for jet fuels, and to around 21 percent for middle distillates. LPG, previously untaxed, also faced a modest 8 percent tax. Only fuel oil remained essentially tax free. Municipal taxes were reduced to one percent of excise taxes.

The large world market price increases in late 1979 forced the Government once again to consider substantial price increases. Average crude oil prices had increased from $13.29/bbl in 1978 to $21.89/bbl by December 1979, an increase of 65 percent, while imported fuel oil had increased by as much as 89 percent. The Government, which had had to rescind an attempted price increase in electricity tariffs of some 56 percent in November 1979, had to pay heavy subsidies to EGAT because of the high fuel oil costs. Once again it attempted to raise petroleum product prices at the retail level by some 24–51% on February 1, 1980. At the same time it increased electricity tariffs by some 38%. These increases were met with overwhelming opposition in the National Assembly and by the public, and, as a consequence, the Government had to resign.

The attempted petroleum price increases of February 1980 would have increased gasoline and other petroleum products prices to such an extent that the Petroleum Authority (PTT) could have placed some 20% of its revenues into the Oil Compensation Fund which is used to cross-subsidize the power sectors fuel and diesel oil consumption. However, the new government reduced the announced retail prices of kerosene, diesel, and bottled gas on March 19, 1980 by between 12–15% (see Table 10.17). This in turn reduced the revenues of the Oil Compensation Fund. For the balance of 1980, the government pledged to avoid further price increases.

One fundamental change was introduced in the excise tax system in 1979. Until 1978, taxes were levied in specified amounts per unit of product. This, in 1979, was changed to a percentage tax of the retail value instead. Any increase in basic retail prices, therefore, automatically increases tax revenues as well. Tax rates levied on retail prices as of July 14, 1979 were 41% on gasoline, 15% on kerosene and diesel fuels, and 5% on all other products. For fuel oil the rate remained fixed at B0.0001 per liter and for jet fuel at B0.30 per liter. The municipal tax for all products is set at 1% of the total excise tax revenues.

As can be seen from Table 10.18, in 1970, excise taxes and customs duties on petroleum products had amounted to 1.2% of GDP and 8.6 percent of total Government revenue. By 1976, owing to the Government's attempts to protect the economy from the effects of the higher world market price, these shares

Table 10.17. Maximum Retail Price of Petroleum Products in the Municipality of Bangkok.

PETROLEUM PRODUCTS	BAHT PER UNIT	1978 MAR. 10	1979				1980	
			JAN. 31	MAR. 22	JULY 13	JULY 20	FEB. 9	MAR. 19
Premium gasoline	liter	4.98	5.60	no change	7.84	no change	9.80	no change
Regular gasoline	"	4.69	5.12	"	7.45	"	9.26	"
Kerosene	"	2.68	3.06	"	5.12	4.20	6.71	5.70
High-speed diesel	"	2.64	3.03	"	4.88	no change	7.39	6.54
Low-speed diesel	"	2.50	2.93	"	4.71	"	7.12	6.27
Fuel oil								
600 second	"	1.66	1.86	1.90	3.04	"	3.78	no change
1,200 second	"	1.62	1.79	1.83	2.93	"	3.64	"
1.500 second	"	1.61	1.77	1.81	2.90	"	3.61	"
LPG								
Size 12 kg	12 kg	66	no change	no change	100	90	132.5	114.5
" 14.5 kg	14.5 kg	79	"	"	121	108.90	160	138
" 15 "	15 "	82	"	"	125	112.50	165.5	143
" 25 "	25 "	123	"	"	193	173.70	261.75	225
" 45 "	45 "	221	"	"	348	313.20	471.25	405
" 50 "	50 "	245	"	"	386	347.40	523.5	450

had fallen to 0.8 percent and 6.2 percent respectively. The gradual but systematic increase in tax rates since then has finally brought about a significant reversal of these trends. For 1980, applying the new March 19 prices and tax rates to NEA's projected total consumption for the year, excise taxes will amount to approximately 2.4 percent of estimated GDP, or double its share of GDP in 1970.

Suggested Economic Pricing Principles for Petroleum Products

Four major factors must be taken into account to estimate the real economic, as opposed to the financial, costs of petroleum products. These four factors are: first, the shadow price for foreign exchange; second, the shadow price for labor; third, the effects of externalities; and fourth, the derived effects (both costs and benefits of petroleum product uses).

The shadow price for foreign exchange has been estimated for Thailand as amounting to some 27 percent above the official exchange rate.[41] This means that, on average, petroleum product imports should be priced in Bahts at 127

---

[41]Committee for Coordination of Investigations of the Lower Mekong Basin, *Calculation of Conversion Factors for Project Appraisal,* Bangkok, Sept. 1977. This shadow price was used in the recent marginal cost pricing study for the electricity sector. However, this rate may be increasing over time because of the increasing capital intensity of new development investments.

Table 10.18. Petroleum Product Excise Tax Plus Import Taxes as a Percentage of Total Tax Revenue and Total Petroleum Imports, Selected Years 1970–1980. (Millions of Bahts and Percent)

CATEGORY	1970 ฿	1970 TAX AS A % OF	1972 ฿	1972 TAX AS A % OF	1974 ฿	1974 TAX AS A % OF	1976 ฿	1976 TAX AS A % OF	1977 ฿	1977 TAX AS A % OF	1980[2] ฿	1980[2] TAX AS A % OF
Petroleum excise taxes	1,066		1,382		2,496		2,207		3,080		15,929	
Petroleum import duties[1]	544		282		529		425		406			
Total taxes on petroleum products	1,610	100.0	1,664	100.0	3,025	100.0	2,632	100.0	3,486	100.0	15,929	100.0
GDP	136,100	1.2	164,600	1.0	269,700	1.1	332,200	0.8	370,400	0.9	652,500	2.4
Total gov't revenue	18,721	8.6	21,144	7.9	37,925	8.0	42,203	6.2	52,681	6.6	n.a.	—
Net value of petroleum imports	1,929	83.5	2,797	59.4	11,903	25.4	16,247	16.2	19,913	17.5	n.a.	—
Gov't expenditures on highways	11,500	14.0	11,700	14.2	8,400	36.0	8,800	29.9	9,000	34.7	n.a.	—

Sources: World Bank, Report No. 2059-Th, NEA.

[1]Petroleum import duties are levied at the same rates as excise tax.

[2]Based on NEA 1980 consumption estimates and tax rates of March 19, 1980. GDP estimated at 15.6% above 1979 in current prices.

percent of their actual foreign exchange costs in order to reflect the opportunity costs of foreign exchange. Also, all future interest, dividend and depreciation payments made to foreign lenders and shareholders of petroleum production and distribution facilities should be evaluated at 127 percent of their actual accounting values.[42]

The net economic employment effects of petroleum use are difficult to evaluate. They are positive because petroleum importation, refining, distribution, and consumption creates jobs for people that may otherwise remain un- or underemployed. For these jobs, shadow prices below going market rates are appropriate. On the other hand, the petroleum sector also draws on individuals with technical and marginal skills that are probably in short supply in Thailand, and who, therefore, have foregone opportunity costs that are higher than their wage rates. A premium above their wages, reflecting their net productivity in alternative employment, should be added to reflect this scarcity value. Furthermore, the use of petroleum-product-based energy sources is likely to replace less attractive alternative energy sources (wood fuels, coal, animal-drawn vehicles, etc.) all of which require manpower for their operation. While probably less efficient—at least on an accounting cost basis—these displaced energy systems also mean displaced employment, which is numerically probably greater than the new employment created by the petroleum-based technology. The overall, immediate net effect may well be a decline in overall employment. However, indirect effects, such as a greater productivity (in agriculture and transport, for example) may more than overcome these direct effects because of increases in output and sales (including exports and net value added). Short of detailed and complex studies of these consequences, little can be concluded here about the overall net employment effects of petroleum product use.

Externalities of petroleum product use can be negative and positive. On the negative side, congestion and pollution costs are probably the most significant. On the positive side there are factors as, for example, the greater mobility of the labor force, or the likely reduction in overcutting of timber resources for fuel and resulting reduction in erosion, etc.

Pollution costs, mainly through air pollution from the high-sulphur crudes utilized in Thailand, may be quite significant, at least in and around Bangkok.[43] However, because of the well-known difficulties of converting those health and human well-being related effects into economic magnitudes, no attempt will be made here to measure them or to suggest analytical methods appropriate for such measurements. This should not be interpreted to mean that these effects can be safely ignored, however. Ultimately studies should be

---

[42]To the extent that they are remitted abroad.

[43]Lower-sulphur crude oil such as that produced by Indonesia is not imported into Thailand because of its higher cost (approx. $1.00 more per bbl in recent months).

undertaken to establish whether these effects are reaching dangerous levels for health, at least in such heavily polluted areas as central Bangkok.

Congestion costs (most of which are concentrated in the greater Bangkok area) consist of three major components. The first is the additional amount of fuel consumed by vehicles held up by congestion; the second consists of the time lost by drivers and passengers and the additional costs for less than optimal utilization of the vehicle stocks (the costs of waiting plus the costs of fewer ton- or passenger-miles per vehicle). Indirect consequences related to these inefficiencies are derived effects such as reduced productivity in plants, offices, warehouses, etc., from slower and less reliable deliveries. The third category of costs relates to the additional costs of traffic control and management. A further, potential cost may be related to a higher accident rate. However, if traffic were to flow with less impediments, prevailing driving habits and higher speeds could well have the opposite result.

Without some detailed studies little can be said about the economic magnitudes of these external economic costs. However, from casual observations of traffic conditions in Bangkok it appears that congestion may well, on average, add at least 20–30 percent to driving time, fuel consumption, and reductions in vehicle efficiency in the greater Bangkok area. With, perhaps, two-thirds of Thailands's total passengers and one-half of its commercial vehicle stock operating in this area these external costs are likely to be substantial indeed. Taken together they may well exceed present retail gasoline and diesel prices by 100 percent or more. Sample studies should be undertaken to estimate the likely magnitude of these costs. In addition, projections should be made for the future, given the rapid rate of increase in population and vehicle use, particularly in and around Bangkok.[44]

10.6   THE PRICING OF LIGNITE

Lignite is a nontraded commodity. Owing to its handling characteristics, tendency for spontaneous combustion, low btu content per unit-weight, and other factors, it can be used for little else but boiler fuel in power production. One exception to this rule may be the higher quality deposit at Li which can be utilized for tobacco curing, or, in the form of briquettes, could become a useful fuel for a variety of industrial, commercial, and household uses.

The largest known lignite deposit at the present time is at Mae Moh. It is to be fully developed for power production by EGAT. The lignite can be produced

---

[44]National Energy Administration statistics indicate a 7 percent and 8.4 percent growth rate in total vehicle stocks in the municipality of Bangkok and the whole country between 1976 and 1977, respectively. Projections made by NEA to 1987 suggest annual average rates of growth of 5.8 percent for Bangkok and 6.2 percent for the country as a whole (From Dept. of Highway Statistics).

by open-pit mining methods, and production costs are low. In 1978 average mining costs per ton of all lignite mines in Thailand were only $4.20/ton. Estimated 1984 mining costs for the expanded Mae Moh mining operation are about $10.20/ton. In the past, EGAT has accounted for the cost of its lignite by using the long-run marginal extraction costs, based on established financial accounting procedures. However, because of the finite nature of the deposits it has been argued that the relevant costs to be used are the replacement costs.

At present, the replacement costs of the lignite committed to power production are not known with certainty. The most likely replacement will be additional lignite that has not yet been found. Some 50 lignite deposits have been encountered throughout the country, and at least one of them, at Mae Tip, appears to be several times larger than the Mae Moh deposits. However, nothing is known about actual reserves, likely costs of production, quality, etc.

The second most likely replacement fuel for lignite is steam coal from Australia. Its F.O.B. Australia costs have been estimated at between $30 and $38/ton.

A detailed study of the value of lignite in terms of its likely 1984 replacement costs based on Australian coal have been undertaken by consultants in the appraisal of the Mae Moh mine expansion project. The value of Mae Moh lignite, adjusted for quality differences, differences in net heat rates, power plant capital and operating costs, etc., was estimated to range between $13.20 to $15.90 per ton in 1984 prices. It was suggested that this cost should be used as the base for the transfer price of Mae Moh lignite to the powerplant.

This approach to the pricing of lignite is not appropriate. Allowance must be made for the fact that the lignite replacement will not be needed until some time in the future. In the case of EGAT's power expansion plans, new, major thermal plants (atomic or coal-fired) are scheduled to come on line in October 1990. The net opportunity cost (or depletion premia) of the lignite, committed today, and not available for future expansion, therefore, is represented by the present value of the difference in its marginal production costs and the costs of the replacement fuel, adjusted appropriately for differences in all other cost variables.

Replacement fuels, needed in 1990, would have to be produced or purchased over a period of 25 years to parallel the life-expectancy of the respective powerplant. Given a 1984 lignite mining cost of $10.20/ton, and an Australian steam coal replacement cost of $15.20/ton of lignite equivalent, the depletion premia in 1984 would amount to $1.35/ton. The transfer price to be charged for the lignite in 1984, therefore, should be $10.20/ton, the marginal mining costs, plus $1.35/ton or a total of $11.55/ton. The depletion premia must be raised by 8% per year in order to reflect its increasing value over time.[45]

---

[45]The levelized, constant annual depletion charge between 1984 and 1990 in real terms would amount to $1.75/ton.

Table 10.19. Financial Analysis of LPG Conversion
for Taxi Owners.

Cost of conversion to LPG per vehicle	B 5,000
Cost of LPG per liter (retail)	B 2.80
Cost of gasoline per liter (retail)	B 5.12
Ratio of LPG/gasoline	1.3/1.0
Cost Ratio LPG/gasoline	3.65/5.12
Net Savings per liter of gasoline consumption	B 1.48

Assume: 50,000 km/year annual driving
Taxi life expectancy: 200,000 km
Gasoline consumption: 10km/liter = 1.3 liter LPG

$$\sum_{t=1}^{4} \left[50{,}000 \ km(\tfrac{1}{10})1.3(B\ 1.48)^{t} \left(\frac{1}{1+r}\right)^{t} \right] - 5000 = 0$$

I.R.R. = 189%

If additional, lower-cost lignite were to be found in the meantime, however, the replacement cost as well as the time horizon of the depletion period would have to be adjusted accordingly.

Financial Versus Economic Costs: The Case of LPG-Fueled Taxis

LPG has become an attractive fuel for taxis in Bangkok. According to March, 1979 reports some 1000 taxis, or 8% of Bangkok's total fleet, have been converted from the use of gasoline to LPG. The financial return of this conversion is extremely attractive to the owners, as can be seen from the data in Table 10.19 which show that the internal rate of return to the owner of a converted taxi is an almost incredible 189%.

The reason for this high rate of return is, of course, the differential between the cost of the (heavily taxed) gasoline (some B 1.25 per liter in Sept. 77) and the financial costs of LPG, which is subject to only minimal taxes (B 0.18 in Sept. 77), and is sold at relatively lower regulated refinery prices.

However, the economic analysis at the national level, in contrast to the private financial one, looks quite different, even if excise and municipal tax differentials are disregarded. This can be seen from the data in Table 10.20.

In recent years, both gasoline and LPG had to be imported because local refinery capacity was insufficient to cover the demand of either fuel. Hence, on marginal cost pricing principles, the marginal import prices are the relevant measures for an economic analysis of this interfuel substitution.[46]

---

[46]Nominal, ex-refinery prices in Thailand were lower for both products. In 1977, our reference year, regular gasoline sold for B 2.1877 and premium gasoline for B2.475. This was higher than the LPG ex-refinery price of B 1.7896/liter.

Table 10.20. Economic Analysis at Level of National Economy.

Costs of imported gasoline C.I.F.	฿ 2.56/liter
Costs of imported LPG C.I.F.	฿ 2.23/liter
Ratio of LPG/gasoline consumption	1.3
Costs ratio LPG/gasoline	2.90/2.56 = 1.13
Net costs per liter of gasoline consumption	฿ 0.34

$$\sum_{t=1}^{4} \left[50,000 \text{ km}(\tfrac{1}{10})1.3(\text{฿ } 1.34)^{t} \left(\frac{1}{1+r} \right)^{t} \right] - \text{฿ } 4650 = 0^{48}$$

I.R.R. = −32%

In 1977, the latest year for which data were available at the time of the analysis, the import price of all imported gasoline amounted to ฿ 2.56/liter.[47] The average import price for LPG in the same year was ฿ 2.23/liter. Both price levels have risen since then. It is assumed here that this increase was proportional for both fuels so that the result of the subsequent analysis would remain unchanged.

Distribution costs for LPG are likely to be higher than for gasoline because of the need for handling and recharging the pressure bottles. This would reduce delivered gasoline costs relative to LPG. However, in the absence of information about actual handling and transportation costs these differences are disregarded here.

Because the import costs for 1.3 liters of LPG per equivalent liter of gasoline are higher by some ฿ 0.34 the use of LPG instead of gasoline results in a net economic loss to the economy. Expressed in terms of internal rates of return, this loss amounted to −32%. Obviously, from a national point of view, this type of fuel use conversion should be strongly discouraged. It is advantageous only to taxi owners, and only so because of the very substantial differences in excise tax levies between gasoline and LPG.

Considering that free, unencumbered government revenue from gasoline taxes has an opportunity cost greater than unity (i.e., a shadow price measured in domestic currency of greater than 1), the loss of gasoline tax revenue from conversion to LPG further increases the overall negative economic impact.

This difference between the very attractive private financial gains and the large national economic loss of this switch from a taxed to a nontaxed fuel source is a classic example of the possible magnitude of negative economic con-

---

[47]This, presumably, includes premium grade gasoline which means that the subsequent analysis overstates the import costs of (regular) gasoline. From NEA, *Oil and Thailand 1977*, p. 17.
[48]Conversion costs from gasoline to LPG of ฿ 5,000 times standard conversion factor of 0.93 = ฿ 4,650.

sequences resulting from otherwise well-meant, socially oriented energy pricing policies.[49]

Fuel Costs Versus Energy Systems Costs: Tobacco Curing[50]

The Thai tobacco industry plays a small but regionally important role in the economy of the country. Tobacco growing is concentrated in the four north-ernmost provinces where soils and climate provide exceptionally favorable con-ditions. Flue-cured Thai tobaccos, particularly the Virginia leaves, are famous for their quality. Over 60 percent of total production is regularly exported.

The tobacco-harvesting and curing season lasts through most of the winter. Two or three leaves are picked from each plant every five to seven days. The leaves then are brought to the curing stations that consist of rows upon rows of two-story high frame and brick-sided curing barns. The capacity of each of these barns is about four tons of fresh leaves, which reduce to between 300 and 500 kg when cured.

Quality differentials of fresh-picked tobacco leaves are pronounced. These differentials are apparent in the prices paid to farmers, which vary from US ¢ 3 (฿0.60) per kg, the legal minimum in 1978, to US ¢14 (฿2.70). The average price paid in 1978 amounted to US ¢8 (฿1.60).

Each barn is equipped with two simple firing boxes that connect to concrete-lined baffles and then to large diameter sheet-metal pipes, which serve as heat exchangers inside the barns. The firing boxes can accept a wide range of com-bustible materials. In recent years, by far the most important were firewood, charcoal, lignite, and low-speed diesel or fuel oil. Firewood and charcoal come from the surrounding mountain forests. However, available supplies are insuf-ficient to supply the needs of the curing industry. The lignite comes from a number of open-pit mines to the south of the tobacco-growing area. The most important source of lignite is at Li. Diesel or fuel oil is bought from local whole-salers at the government-controlled price.

The major consideration in flue-curing of tobacco is temperature control. Ideally, temperatures are not allowed to increase at rates of more than 1 or 2 degrees per hour, and maximum temperatures must be carefully controlled to obtain a high-quality product. The kind of fuel used for the curing process has a major effect on the ease or difficulty of control. For control purposes, firewood and diesel or fuel oils are far superior to lignite.

Cured leaves are hand sorted into some 26 different grades. Grades depend

---

[49]Low taxation of LPG was presumably introduced to protect households and small industrial and commercial users from the impact of the large world market price increases.
[50]For a more detailed discussion, see Gunter Schramm and Mohan Munasinghe. "Interrelation-ships in Energy Planning: The Case of the Tobacco-Curing Industry in Thailand," *Energy Systems and Policy*, Vol. 5, No. 2, 1981, pp. 117–139.

in part on the original structure and size of the green leaves but even more so on the subsequent curing process. If temperatures during the cure are too low, the leaves will stay green, if they are too high, the leaves will turn brown and crinkle. Either way, they lose value. Depending on grade, export prices (in 1978) varied between US$0.50 to US $3.00 per kg (B10 to B60); the average price amounted to US $1.50/kg.

According to industry sources, woodfuel provides the easiest temperature control, followed by diesel or fuel oil. Lignite, although it has been used extensively in recent years,[51] is very hard to control and has a strong tendency to overheat. This affects both quantity and quality. For wood or oil-cured tobacco, the fresh-to-dry-leaves weight ratio is 7–8 to 1; for lignite-cured tobacco, it is, on average, 9 to 1, a weight loss of some 15 to 20% compared to the former. Because the quality of the cured leaves is, on average, also lower, the value of the finished product from the lignite-cured shed is considerably lower; these differences may amount to between $200 and $300 per barn load, or $0.50 to $0.75 per kilo.

The use of lignite presents other problems. Deliveries from the mine are unreliable; sometimes a truck will have to wait several hours to load. The quality of the lignite supplied is poor. Only about 70% is usuable because of excessive breakage and admixtures of clay. Stored lignite has a tendency to ignite, leading to fires. In 1977/78 one curing station lost 3 barns because of lignite-caused fires. Because the lignite has a high sulfur content, baffles and heating ducts must be frequently renewed. With lignite as fuel, baffles usually last only one or two drying cycles before they have to be replaced.

Lignite firing also requires more manpower. One man can supervise the firing of two or, at the most, three lignite-fired kilns; with oil or diesel fuel, he can supervise six. Given all the problems related to the use of lignite, it is not surprising that the tobacco industry is continuously looking for alternative fuels. One option offered by the government is the use of forest stands in remote, high mountain areas. Costs of access, cutting, and transport were found to be excessive. Another alternative under consideration is the reforestation of overcut areas with fast-growing species for the production of firewood or charcoal. Such schemes are under consideration but would require a number of years for implementation and first harvest.

The only realistic fuel alternatives are either lignite or some hydrocarborn fuel because wood fuels are in increasingly short supply.[52]

---

[51]Production at Li amounted to some 100,000 tons per year, of which perhaps 80,000 tons were used by the tobacco-curing industry.

[52]In economic terms, the increasing shortage of wood, forest depletion, as well as erosion problems resulting from overcutting imply significant external diseconomies associated with wood fuel use, that is, the social cost of woodfuel use would be very high, although data are not available for quantifying this value precisely. For an analysis of the physical dimensions of the problem see FAO, *Timber Trends Study, Thailand,* Rome, 1972.

Table 10.21. Comparative Energy Prices, Northern
Region, Fall 1978.

FUEL	LOCAL PRICE, US$ PER UNIT	US$/10⁶KCAL
Wood-fuel delivered	$6.00/m^3$	14.79
Low-speed diesel	0.13/liter	14.14
Kerosene	0.14/liter	15.36
Lignite (from Li) delivered[1]	12.50/ton	3.11

Sources: National Energy Administration of Thailand and the tobacco
industry.
[1]Assumes a net heat rate of 4,000 kcal per kg of (moist) coal.

At present prices, lignite is by far the lowest-cost fuel. (See Table 10.21.)
This is also apparent from the data of Table 10.22 which indicate that the costs
of lignite, delivered, per barn per year amount to only US $450, compared to
US $1,325 for low-speed diesel fuel. However, the combined disadvantages of
using lignite compared to diesel or fuel oil far outweigh the initial price advantage. Additional labor costs increase the costs of using lignite by some $100
per barn per year.[53] However, the major disadvantages of using lignite are a
result of the poorer quality and reduced quantity of output resulting from the
use of this fuel. Overall, the value of the output of an oil-fired barn in market
prices amounts to an average of US$7,500; from a lignite-fired one, it is only
US$5,250, a difference of US$2,250, or 30 percent, of the value of the output
of the former. This differential dominates the initial cost advantages of lignite.
The residual markup between the selling price and fuel-cost related inputs plus
raw tobacco is US$3,103 for oil-fired barns but only US$1,628 for lignite-fired
ones, a net difference in favor of the former of US$1,475, or 20 percent, of the
value of the output.

It could be assumed that appropriate shadow pricing of imported fuels versus
the more labor-intensive lignite would substantially alter these results in favor
of lignite. This, however, is not the case, as can be seen from the economic data
in Table 10.22. Evaluated at the appropriate shadow prices, the net advantage
of oil curing over lignite curing is still about $1,069, or 16 percent of the value
of the output of oil-cured tobacco. This stems from the fact that about 60 percent of the tobacco—particularly the higher-quality leaves—is exported and
that the main difference resulting from the use of diesel oil is a large increase
in the total value of output. This increase is larger than the increase in the
shadow-priced economic costs of the imported diesel fuel. Hence, regardless of

[53]Additional costs are coal storage, coal waste, and higher handling costs; these costs are partially
offset by the investment costs for oil burners, tanks, and feedlines. Overall, these costs may
mutually cancel out.

Table 10.22. Fuel-Related Inputs, Outputs, and Their Market and Economic Costs and Revenues in Tobacco Curing per Barn per Season. 1978 Costs and Prices; US $1.00 = ฿ 20.00

		OIL-FIRED CURING			LIGNITE-FIRED CURING	
		VALUE OR COST AT				VALUE OR COST AT	
ITEM	QUANTITY	MARKET PRICES	SHADOW PRICED[6]	QUANTITY	MARKET PRICES	SHADOW PRICED[6]
		US $	US $		US $	US $
Uncured tobacco	37.5 tons	3,000	3,000	37.5 tons	3,000	3,000
Labor (curing only)[1]	60 man-days	72	22	144 man-days	172	52
Fuel, diesel, lignite	10,000 liters	1,325	1,257	30 tons	450	234
Subtotal		4,397	4,279		3,622	3,286
Output, cured leaves	5.0 tons[3]	7.500[2]	6,873	4.2 tons[3]	5,250[4]	4,811
Remaining margin[5]		3,103	2,594		1,628	1,525
Net gain from oil-fired curing		1,475	1,069		0	0

Source: Thai Tobacco Industry.
[1]One laborer can handle 6 oil-fired barns simultaneously but only 2–3 lignite-fired ones; assume 8-hour days at US $1.20/day, 3 shifts per day, 120-day season.
[2]This is the average selling price that, however, includes sorting, packing, transportation, and so on.
[3]Ratio uncured/cured 7.5/1 for oil-fired and 9.0/1 for lignite-fired barns.
[4]Average assumed selling price for lignite-cured tobacco $1.25/kg ($1.50 for oil-cured).
[5]The remaining margin contains a few fuel-specific costs for which no unit cost could be ascertained. Among them are the costs of burners, valves, fuel tanks, and lines for oil or diesel, the costs of the concrete-lined baffles (see text), or of the coal-storage sheds.
[6]The import costs of diesel ($1,000) are evaluated at the border price and the remaining $325 at the standard conversion factor. The shadow-price coefficient of unskilled labor in northern Thailand is equal to 0.3. The lignite-coefficient is estimated to be 0.52 on the basis of labor, transportation, and capital-goods inputs. The standard conversion factor of 0.791 is applied to the 40 percent of tobacco sold in domestic markets; the remaining 60 percent of export is evaluated at the border price equal to 1.

the magnitude of the foreign-exchange shadow coefficient,[54] it is economically always more efficient at prevailing fuel market prices to use diesel instead of lignite.[55]

This conclusion, of course, holds only within certain fuel price limits. As import prices rise for diesel fuel, the market as well as the economic advantages of diesel compared to lignite steadily shrinks. Assuming that all other costs and prices remain constant, the diesel import market price at which lignite is competitive with diesel is ¢25/liter ($39.75/bbl), or ¢28/liter, at the pump.[56] This

[54]Because foreign exchange serves as the numeraire in our evaluation, increased scarcity would result in a downward adjustment of all domestic shadow-price coefficients relative to foreign exchange.
[55]This assumes, however, that the social value of additional foreign-exchange earnings accruing to private tobacco traders is equal to the shadow price of additional imports. This may not necessarily be true.
[56]This price is 210 percent higher than the ¢13.3/liter ($21.07/bbl) used in the analysis.

translates into total fuel costs per barn per year of $2,800, a cost sufficiently high to eliminate the original advantage of the diesel. Evaluated in economic terms, the import price for diesel would have to rise to about ¢23/liter ($36.98/bbl) until equivalence with lignite is reached. Hence once diesel prices reach or exceed this level, such policy measures as taxes or direct fuel allocations should be introduced to force a switch to lignite, even though market prices would still favor the use of diesel fuel.[57]

A number of important conclusions with regard to energy use patterns in developing countries can be drawn from this specific case study. The first and most important one is that user choices among alternative energy sources will usually not only be determined by the financial costs of each alternative fuel compared on a net heat-content basis. Equally, or more important may be the auxiliary costs to the user of utilizing each one of them. These will be made up by the costs of labor, capital, and land related to receiving, unloading, storage, handling, ease of control in use, potential risks, waste material disposal, and system maintenance. Reliability of supply and transportation, uniformity of quality, and the effect on the quality of processed materials will be other important considerations. Habits and the availability and reliability of experienced operators may significantly affect use. Particularly in situations in which the potential choice lies between a known, customarily-used energy material and new alternatives of unknown characteristics, resistance to change may be formidable. Lack of knowledge and experience is perhaps one of the most formidable barriers to change in energy use patterns, particularly in developing countries (although this was not a particular issue in the case of the Thai tobacco industry). Only if all these factors are systematically and carefully taken into consideration will it be possible to bring about desirable changes.

Another important consideration from a national point of view should be that market prices to users frequently do not reflect the true economic costs of the specific energy use. Appropriate shadow pricing is necessary to determine the difference, if any, and either changes in market prices or regulations affecting alternative fuel choices should be used to bring about a more realistic choice pattern reflecting real economic costs.

EPILOGUE

Since this case study was completed in 1979 several important changes have taken place. The most significant one has been the very large increase in off-shore gas reserves from about 3 million cu ft to some 4 trillion cu ft of proven

---

[57]This assumes, of course, that all other costs and prices remain constant. If not, the appropriate switching prices have to be recalculated.

and another 16 trillion cu ft of probable reserves as of early 1981.[58] In addition, Esso has encountered gas onshore, with estimated reserves ranging between 1 to 10 trillion cu ft. So large are the gas reserves now relative to potential domestic demands that the government is seriously considering allowing Texas Pacific to commit part or all of its reserves to exports—if a market can be found. Other important events on the supply side were the discovery of fairly significant crude oil reserves by Shell which are projected to yield an annual production of some 500,000 tons by 1990. In addition, the proven reserves of lignite of Mae Moh stand now at 650 million tons, enough to sustain some 1,500 MW of thermal generating capacity.

These substantial increases in domestic energy resources have significant effects on the value of these resources in the ground, i.e., the relevant depletion premium. As was pointed out earlier (see also Chapter 4), the higher the reserves are relative to projected demand, the lower the depletion allowance becomes because exhaustion is postponed. While in Thailand the effective demand for gas may increase above previously predicted levels, given recent plans for a gas-based petrochemical and fertilizer industry as well as an LPG gas extraction plant, prospective gas reserve/production ratios are still much higher than before. If we assume, for example, that the reserve constraint becomes binding only after 20 years rather than the ten years estimated in the earlier evaluation, the 1981 depletion premium falls from $1.52/MCF to $0.99/MCF (see also Table 10.16). If the relevant time period extends to 30 years it falls to $0.87/MCF.[59] This has important implications for the assessment of the total economic costs of the gas (i.e., LRMC plus depletion allowance), and it may be economic to utilize the gas in lower-value uses for which it was considered too expensive before (for an example of such uses see Chapter 11 on Bangladesh). Another important factor reducing the economic costs of delivered gas is the extension of the economic life of the offshore pipeline. This reduces the unit transportation costs because of the greater throughput and the longer useful life.

Similar considerations apply to the valuation of the depletion premium for lignite, because the most likely replacement for lignite in the early 1990's is not imported Australian steam coal as postulated earlier, but either the extended reserves of domestic lignite or domestic natural gas. This, of course, reduces the applicable depletion allowance (from $1.35/ton in 1984 to $0.67/ton if the binding resource constraint is pushed back from 1990 to the year 2000).

The question whether these reductions in real economic supply costs for gas

---

[58]All revised data from: Petroleum Authority of Thailand (1982).
[59]This, however assumes that real prices of fuel oil would continuously rise at 3% p.a.—not a likely assumption given the potential availability of alternative fuels.

and lignite should be fully passed on to the respective energy users is, of course, a separate one that must be decided on the respective merits of lowered energy consumption and production expenses versus higher governmental income or higher energy producer profits. This question has both income distributional and efficiency dimensions which have been discussed in Part I.

A further important change that is taking place is related to the availability, use patterns and economic costs of LPG. As we have seen previously, the wide divergence between domestic market prices of LPG and alternative fuels has made the former increasingly a popular fuel in the transport sector. While prices for all petroleum products had been raised substantially in March, 1981, the retail-level spread between gasoline and LPG prices had been widened even further, from a ratio of 1.83[60] to 2.11 on a volume basis ($1.88/gallon for gasoline, $0.89/gallon for LPG). Not surprisingly, this led to a continuing increase in the demand for LPG which had to be covered by subsidized imports. In 1980, for example, total domestic LPG refinery output was 125,000 tons while imports amounted to 115,000 tons, or 48% of total consumption. As argued earlier, the subsidies required are economically inefficient.

However, in 1980 the government decided to proceed with a natural-gas based, liquified petroleum gas project which will produce large quantities of LPG and other by-products. This gas separation plant is projected to produce some 460,000 tons of LPG by 1985 and 920,000 tons by 1990. Another 60,000 tons are expected to be added from increased petroleum refinery production. These levels of LPG output are so large that Thailand will switch from an LPG importer to a net exporter, at least for a number of years after the extraction plant comes onstream. Exports alone are not particularly attractive because of depressed LPG world markets resulting from the construction of similar, large extraction plants in the Middle East and elsewhere. However, if the LPG is evaluated in terms of its import substitution potential for gasoline, diesel and kerosene fuels as well as scarce domestic charcoal resources, the economic rate of return for the gas extraction plant is estimated to amount to almost 30%.[61]

In light of the large—albeit temporary—surplus of LPG that will be created, strong marketing efforts are needed now to systematically introduce LPG into energy systems that are presently based on other fuels. This may, in fact, make the current net subsidies for imported LPG economically efficient because they increase market penetration and prevent even larger, temporary surplusses in the future that would have to be disposed off abroad at depressed prices. The theoretical rationale for such a promotional pricing strategy has been discussed in Chapter 5.

It is interesting to note that LPG is also now considered as a serious con-

---

[60]Petroleum Authority of Thailand, 1982.
[61]Petroleum Authority of Thailand, 1982

tender for tobacco curing, substituting for the much higher-priced diesel fuels and kerosene used earlier and the inferior-quality lignites.

Overall, the lessons to be learned from these rather dramatic changes in the energy supply and demand balances of Thailand are that energy planning, demand management and pricing are subject to continuing and dynamic changes. As such changes occur, policy responses must change as well. This need for flexibility and adaptability, however, must be balanced against the need for continuity and predictability of energy supplies and prices, in order to allow energy users to make rational decisions with respect to investments in long-life and costly energy-using facilities and appliances. Balancing these oftentimes conflicting goals are one of the important challenges facing energy planners and policy makers world-wide.

11

Bangladesh: Gas Pricing and Utilization in a Low Income, Oil Importing Country[1]

11.1 BACKGROUND AND OVERVIEW OF THE ENERGY SECTOR

Bangladesh belongs to the large number of developing countries that depend totally on petroleum imports to satisfy their domestic requirements. This imposes a heavy burden on their foreign exchange resources. In the case of Bangladesh, for example, the costs of petroleum imports absorbed as much as 60% of the country's total export earnings in recent years. Projections for the future are even more dismal.[2] However, it also belongs to the much more select group of countries that own substantial other energy resources instead. In its case these resources are large deposits of natural gas.[3]

However, gas is a far less convenient fuel than oil. It is generally limited to certain types of uses—mainly as a fuel for high temperature heat applications as in boilers and furnaces. It is less convenient to utilize as well as capital intensive because it requires a continuous pressurized delivery system from the gas well to the user's premises. Petroleum products, by comparison, are far more versatile. Their energy content per cu. ft. is about 900 times greater than that of gas at atmospheric pressure. They also can be moved in both large or small quantities requiring no more than simple, unpressurized containers for transport. Petroleum products are most valuable in the transport section in which natural gas is difficult to use because of its low energy density. While natural gas can be converted into a liquid fuel or adapted for use in the transport sector, this is relatively expensive and has rarely been done in the industrialized nations which have large enough conventional markets to absorb all available gas.[4]

---

[1]This case study is based in part on: Gunter Schramm, "The Economics of Natural Gas Utilization in a Gas-rich, Oil-poor Country: Bangladesh," *The Energy Journal*, Vol. 4, January 1983.
[2]T. L. Sankar and G. Schramm, Ch. 3.
[3]Other gas-rich countries with insufficient or no domestic petroleum resources are, for example, Thailand, Pakistan, Afghanistan, Morocco, Colombia, Bolivia, New Zealand, and Australia.
[4]One exception is New Zealand which presently has a large methane gas to gasoline conversion plant under construction. An active program to develop compressed natural gas as a vehicle fuel

Traditionally, therefore, natural gas, apart from its role as an important feedstock, is largely used as a boiler or furnace fuel. In this role its delivered value is equivalent to that of low-sulfur fuel oil.

Two solutions to this imbalance between domestic availability of natural gas and needed petroleum imports appear to offer themselves. The first would be to export some of the surplus gas and use the net revenue obtained for financing petroleum import requirements. The second would be to try to replace domestic petroleum consumption with natural gas as much as possible. In order to assess the feasibility of either strategy information is needed about: (a) the foregone future value of natural gas in the ground, i.e., its so-called user costs; (b) the feasibility and net value to the national economy of gas exports; (c) the potential substitutability of gas for petroleum domestically; (d) the net costs of such utilization and substitution; and (e) the appropriate level of gas prices that should be charged to domestic users in order to induce gas substitution on the one hand and to recapture as much of the potential economic rent of the gas on the other.

Bangladesh is primarily an agricultural country. It has a population of about 90 million people and an area of some 56,000 square miles, making it one of the most densely populated countries in the world. About 90% of the population lives in villages and 80% of it is dependent upon agriculture. More than 55% of the gross domestic product comes from agriculture and only about 9% from modern industry. Average per capita income is less than $100, making it one of the poorest countries in the world. Over ⅔ of the population is considered to live below the poverty line in terms of minimum caloric requirements. Poverty, malnutrition, unemployment, and illiteracy are rampant.

In spite of strong efforts to modernize and change the structure of the economy with the help of massive foreign aid, relatively little has been accomplished during the 1970s. In those years, agriculture's relative importance declined only slightly from 58 to 55% of GDP while the manufacturing sector's contribution actually declined from 10.5 to 8.7%. Between 1972/73 and 1979/90 the average annual rate of real growth in GDP was about 4.3% and the per capita growth rate about 1.4%.[5]

The foreign trade section of the country is characterized by a persistent and worsening imbalance between imports and export earnings. Overall, in 1978/79, exports amounted to only 43% of imports; for 1979/80 the projected percentage is an even lower 32%. Next to food imports, crude oil and petroleum

---

is also underway. In Northern Italy, for a number of years, compressed natural gas has been used as a fuel for some 250,000 trucks that operate close to an existing gas distribution network. For a detailed discussion of both developments see J. P. West and L. G. Brown, *Compressed Natural Gas,* New Zealand Energy Research and Development Committee, University of Auckland, Auckland, New Zealand, April 1979.

[5]*Data from:* Bangladesh Planning Commission, Dacca, various statistical series.

products imports are playing a major role in this worsening trend. In 1976/77 the net import costs of petroleum accounted for about 35% of total commodity export earnings; in 1980/81 this share had increased to some 61%.

Noncommercial fuels, such as firewood and crop residues, account for about 75% of the total consumption of primary energy. Among commercial energy sources, imported petroleum accounts for the largest share of the market, followed by indigenous natural gas, imported coal, and a small amount of hydroelectricity.

Since its discovery in the early 1960s, natural gas has supplied an increasing proportion of the demand for commercial energy. Domestic petroleum products consumption fell during the Independence War at the beginning of the 1970s, but recovered by 1973 and then grew at an average annual rate of 4.8% until 1979. More significantly, the annual rate of growth has increased sharply since 1976. Between that year and 1979/80 the average annual rate of growth was 12.3%. Hydroelectricity production remained about constant between 1970 and 1978. Coal consumption fell sharply after the War and remained about constant thereafter.[6]

A notable feature about past consumption trends is the substitution of natural gas for petroleum, particularly in power generation, where the share of natural gas increased from 56% in 1969–70 to 78% in 1978–79. Some 107,000 customers were connected to the gas distribution network in March 1980. The major users were power plants with 37% and fertilizer plants with 35%.[7]

Overall, in 1979/80, some 45,400 MMCF of natural gas were consumed. This consumption is estimated to rise to 120,000 MMCF by 1984/85. Of this total, some 35% is projected to be used by the electric power section, 38% by the fertilizer industry, 24% by other industrial, commercial, and domestic users, and 3% by cement factories and a pulp and paper mill.[8] In spite of this very rapid rate of growth, which averages about 21% per annum between 1979/80 and 1984/85, the relationship between production and available reserves will still be imbalanced by 1984/85, with a reserve-production ratio equivalent to a 100 year-plus inventory of natural gas. Presently estimated reserves amount to between 12 and 13 trillion cu. ft. of gas.[9] In addition to these proven reserves, new deposits, whose magnitude has not yet been determined, have been encountered in eight additional locations. Hence, it can be expected that the ultimate gas reserves of the country will turn out to be much larger than those established so far.

---

[6]*Data from:* T. L. Sankar and G. Schramm.
[7]*Data from:* Petrobangla, Dacca, 1981.
[8]Planning Commission, Government of the People's Republic of Bangladesh, *The Second Five-Year Plan 1980–1985*, Dacca, May 1980.
[9]*Source:* Petrobangla, 1981.

11.2   THE ECONOMIC VALUE OF NATURAL GAS

The economic value of gas is determined by five types of opportunity costs. The first consists of long-run marginal costs of supply, which include exploration, development, processing, transmission, and distribution costs.

The second represents the foregone future net value of the gas, or user costs.

The third is determined by the net value of the gas in alternative uses, such as its F.O.B. export price net of production and delivery costs plus depletion allowance.

The fourth represents the net value of the gas as a current replacement of other energy resources, net of all differences in delivery and usage costs between the alternative fuels.

The fifth, finally, is determined by the net value of the gas in uses that would not occur if alternative, higher-priced energy or feedstock materials had to be utilized. Examples are fertilizer production or gas exports whose viability depends on prices below those of alternative fuels.

The first two of these opportunity costs, the long-run marginal supply and the user costs, are additive and represent the basic economic costs of the gas. The others must be higher than the sum of the former in order to produce economic net benefits.

The following estimates of these various cost categories are based on the assumption of constant prices (unless otherwise noted) at real interest rates of 10, 12, and 15%, respectively. Fifteen percent is the rate used by the Bangladesh Planning Commission for project evaluation purposes. Twelve percent is the rate applied by major multilateral finance organizations. The 10% rate has been shown as an illustrative lower bound. These "real" rates of interest may appear high by international standards. However, they are relatively low, given available high-yield investment opportunities in various sectors of the country's economy and particularly the energy sector, where real rates of return are far in excess of the rates of interest applied here. Many of these investments cannot be realized because of an acute shortage of domestic savings and strict limits on foreign donor funds and other potential capital imports.

11.3   LONG-RUN MARGINAL SUPPLY COSTS

The long-run marginal supply costs (LRMC) of natural gas should include all exploration, development, production, transmission, and distribution expenditures. However, most of the presently known gas deposits had already been discovered in the 1960s; also, their size is such that they will support any foreseeable level of domestic demand to the year 2000 and beyond. Therefore, for the time being, it appears unnecessary to include exploration expenditures into long-run marginal cost estimates.

A special problem with LRMC estimates is that of the presently producing four fields, three are essentially single-purpose producers, supplying only one or a few industrial operations (e.g., cement, fertilizer factories). The fourth, Titas Gas Company, on the other hand, supplies a large number of domestic and commercial customers (some 107,000 in 1980). These domestic and commercial customers account for 99.6% of all connections, but only 11.1% of total sales.[10]

The cost of Titas, therefore, is heavily affected by the high cost of serving this section of the market. Hence, it would not be representative for estimates of the LRMC of natural gas supplies.

A new gas development is presently under way that consists of the development of the Bakhrabad gas field (which was discovered in 1969), a 110 mile, 24″ pipeline to Chittagong with a capacity of 350 MMCFD, and a distribution network that will supply the gas to various industrial and large-scale commercial consumers in the Chittagong region. Being an integrated project from development drilling to distribution, its costs can be taken as a representative approximation of the LRMC. Evaluated at a rate of interest of 12% and covering the projected, 37-year lifetime expenditures of the project, average costs are projected to amount to some \$0.36/MCF at 10%, \$0.43/MCF at 12% and \$0.56/MCF at 15% in constant 1980 prices. These values are used as the representative LRMC of natural gas supplies in eastern Bangladesh, exclusive of any distribution costs to small-scale domestic or commercial users.

In order to estimate the user costs of the gas allocated today, it is necessary to estimate the future costs of the alternative energy sources that must ultimately be used to replace it. Furthermore, it is necessary to project the likely depletion rate of the gas. More rapid depletion because of higher production rates today will lead to an earlier leveling off of maximum feasible production rates from a given deposit. These would then either tend to be uniformly lower for the rest of a reservoir's life, or they would tend to lead to an earlier depletion of the reservoir. Given the importance of maintaining a sufficient gas supply over the economic life expectancies of gas producing and utilization facilities, it was assumed that the gas authorities would aim at an initial 15-year inventory level of gas supplies at maximum production rates. In other words, once the reserve/production ratio has reached 15:1 no more additional gas could be allocated from a given deposit so that incremental demand would have to be satisfied from other resources. At this point in time, the depletion constraint becomes binding, and the marginal value of an additional unit of gas is represented by the potential utilization of this unit over the 15-year remaining life-expectancy of the reservoir. Figure 11.1 represents the approximate exploita-

---

[10]Titas Gas Transmission and Distribution Company, Ltd., *A Technical Brochure,* Dacca, 1981.

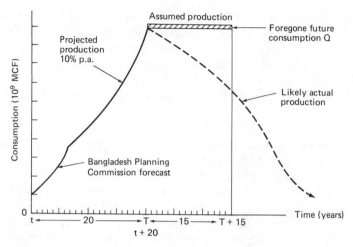

Figure 11.1.   Projected production profile.

tion path, with the units approximating estimated gas reserves and demands in Bangladesh. The cost differential between the gas and its replacement fuel can be measured by the equivalent energy content of either fuel. The latter must be adjusted for differences in net use efficiencies (i.e., differences in net heat rates.)[11] Adjustments are also necessary for differences in delivery costs. These may be low between gas and fuel oil (delivered), but high between gas and coal. Most important, adjustments must be made to allow for differences in energy-systems costs. A gas-fired power plant, for example, will generally have 20–30% lower capital costs than an equivalent sized coal-fired one.[12] Also, coal fired plants have higher operating, maintenance, and repair expenditures than gas-fired ones.

Table 11.1 shows the 1980 user-costs or depletion premia that should be added to the LRMC of a MCF of gas in order to arrive at the total marginal economic cost per MCF of gas used in that year. These costs have been shown for several potential replacement fuels at various real price levels and for real rates of interest (or discount) of 10, 12, and 15%. Fuel oil prices are shown at $194/ton, which reflects 1980 delivered costs, and $350/ton, which assumes an average annual increase in real world oil prices of 3% for the 20-year period until the depletion constraint becomes binding. Coal prices are shown using

---

[11]For example, due to technical factors, the net efficiency of a dual-fired boiler is somewhat higher with coal than with natural gas, given btu equivalence of the two fuels.
[12]In locations with strict environmental pollution standards, such as the USA, Europe, or Japan, capital costs between gas and coal-fired boiler plants may vary by as much as 100 percent.

Table 11.1. 1980 User Costs per MCF of Natural
Gas for Various Replacement Fuels. (US¢/MCF)

REPLACEMENT FUEL/RATE OF INTEREST	10%	12%	15%
Fuel oil at $194/ton	34.4	21.5	10.9
Fuel oil at $350/ton	64.8	40.5	20.5
Coal at $58/ton	17.5	10.9	5.5
Coal at $71/ton	21.3	13.3	6.7
Natural gas at $1.50/ton	8.1	5.0	2.5

Source: Appendix 11.1.

1980 F.O.B. prices for Australian steam coal of $36/ton plus $16 freight plus
$6 delivery charges, for a total of $58/ton. Alternatively, it was assumed that
these prices would increase at 1% per year in real terms, resulting in a price of
$71/ton after 20 years. The natural gas price of $1.50/MCF is based on current costs of delivered offshore gas in Thailand.

It is common practice both in Bangladesh and elsewhere to assume that the
only future replacement fuel for natural gas used today would be imported fuel
oil or other petroleum products. This, however, is unrealistic. The present gas
reserves in Bangladesh had all been found by 1969. No systematic additional
exploratory drilling has taken place since. Offshore gas deposits, although pos-
itively identified, have not been evaluated because of a lack of markets. Seismic
and geological surveys undertaken recently indicate that the likelihood of finding additional gas reserves are excellent. Already, several new fields have been
discovered. Hence the most likely replacement for gas used from the presently
known deposits is not imported petroleum, but gas from as yet not delineated
additional deposits.

Alternatively, the most likely replacement fuel, at least for power plant and
large industrial furnace use, is not high-priced petroleum, but lower-priced
imported coal. Coal is available in practically unlimited quantities from suppliers such as India, Australia, China, or others.

Hence likely future gas finds, or coal in boiler applications, are the most
likely replacement fuels for natural gas used today.

What these estimates indicate is that the user costs of the presently
underutilized natural gas reserves of Bangladesh are quite low, being in the
range of US¢3/MCF to US¢65/MCF.[13] The most likely range is between 5
and 13 cents, given a 12% rate of interest and additional gas finds and/or coal
as the likely replacement fuels. The total economic costs of natural gas, con-

[13]However, it should be remembered that over time, this value will rise at the given real rate of
interest, provided no additional gas reserves are identified.

Table 11.2. Long-Run Marginal Supply Plus (1980)
User Costs at Different Rates of Discount and with
Alternative Replacement Fuels. US¢/MCF

REPLACEMENT FUEL	10%	12%	15%
Fuel Oil at $194/ton	70.4	64.5	66.9
Fuel Oil at $350/ton	100.8	83.5	76.5
Coal at $58/ton	53.5	53.9	61.5
Coal at $71/ton	57.4	56.3	62.7
Natural Gas at $1.50/MCF	44.1	52.6	58.5

Source: Table 11.3 and text.

sisting of the sum of the long-run marginal and the user costs, therefore, range between US$0.44/MCF and US$1.01/MCF in 1980, with the most likely range between US$0.48/MCF and US$0.56/MCF. (See Table 11.2.)

11.4   THE VALUE OF GAS IN ALTERNATIVE USES

As can be seen from the data in the Appendix, domestic gas consumption is projected to increase rapidly during the next few years, at rates of over 21% until 1984/85. The major reason for this rapid growth is that determined efforts are underway to utilize gas as much as possible as a replacement for petroleum and as a feedstock for new fertilizer plants that will displace presently imported supplies. New electric generating plants in the eastern part of the country are all gas-based, and a high-tension transmission line across the Jamuna River, to be completed in 1982, will make it possible to substitute gas-based electric energy for the, at present, exclusively oil-based generation in the western part of the country. A large rural electrification program, designed to supply about 25% of all villages by 1990 (as against 2% of present), will help to reduce the demand for kerosene for lighting and the demand for diesel fuel for irrigation pumps. Nevertheless, by about 1985 most of these conventional uses of gas (fertilizers, power generation, brick burning, and industrial boiler uses) will have been converted to gas, so that further expansion of gas demand will slow down to a rate more in line with overall rates of economic growth (expected to be about 5% p.a. during the 1980s). Hence it is projected here that after 1984/85 the rates of growth will slow down to about 10% per year. This will leave large reserves of gas underutilized or uncommitted until at least the turn of the century. Given the precarious financial and economic conditions of the country, a more rapid utilization of this potentially valuable resource must have the highest priority.

Four potential markets offer themselves. The first consists of conventional gas uses in the western part of the country. This would require a pipeline crossing of the Jamuna River, whose cost alone is estimated at US$120 to $150 million,[14] plus another twelve major river crossings. While detailed cost studies are lacking, rough estimates of total pipeline costs range around US$600 million. With a market size unlikely to exceed 200,000 tons of oil equivalent, this would mean cost per MCF of gas alone would be about $8.40 per MCF, which would render the gas noncompetitive with oil. In fact, given the electric interconnector across the Jamuna, the feasible, gas-supplied uses are likely to be considerably less than the 200,000 tons of oil equivalent assumed here.[15]

The second potential market are export sales by pipe to India. Such exports would have the additional advantage of providing gas supplies to the western part of Bangladesh as well. Hopes were high some years ago that an agreement would be reached, but by now have all but disappeared, mainly for political reasons.[16]

A third set of alternatives would consist of gas or gas-derived exports in the form of liquified natural gas (LNG), methanol, urea, or similar products. Such exports would require substantial investments in capital-intensive facilities. Furthermore, these exports would have to compete in world markets against similar products coming from countries with huge gas surpluses that presently are unutilizable, or flared, but are now being developed into such transportable commodities. Huge, gas-derived, export-oriented plants are either already on stream or under construction in Indonesia, Malaysia, Australia, New Zealand, Iran, Saudi Arabia, Bahrain, Algeria, Mexico, Trinidad and Tobago, and Canada. Others may follow soon. As a result, competitive pressures are likely to lead to deteriorating world market prices that could well fall to levels closely approaching the zero opportunity cost of much of that gas to its producers. In any case, recent attempts by the Bangladeshi Government to attract such gas-based, export oriented plants have been unsuccessful so far, with the exception of plans for a partially export-oriented urea plant that are under active consideration. The quantities of gas involved for this plant, some 17,000 MMCF/year, are relatively small, however. No firm price has been set for the gas needed by this plant. However, it is unlikely to be higher than $1.20 to $1.50/MCF, which would result in a net income to the Government of between US$0.75 to US$1.00, given the range of gas costs estimated above.

The fourth, and potentially highest-value alternative is the use of gas as a

---

[14]The costs for an inherently much less costly 220kV electric transmission line crossing presently under construction is US $85 million. The high costs are a result of the width of the river (7–15 miles) and the several hundred feet of unstable river bottom sediment.

[15]Most of the western half of the country is agricultural, with few industries that would be potential gas users.

[16]However, economic considerations may be equally important for India. Since the gas would have to be imported (i.e., require foreign exchange) and replace mainly low-cost Indian coal and relatively low-priced fuel oil rather than high-cost middle distillates and gasoline.

replacement fuel of petroleum products in nonconventional uses and sectors, such as transportation. On a btu-equivalent basis, the value of gas as a replacement for petroleum products ranges from $5 to $8 per MCF, depending on the specific petroleum product involved. With a real cost of gas of less than $1/ MCF, the potential for absorbing high cost-penalties for using gas instead of oil is substantial. Hence, replacing petroleum by natural gas, even if it would appear to be costly by conventional (i.e., industrialized nations) standards, could be economically highly efficient in Bangladesh (or similar gas-rich, oil-poor countries).

Petroleum product consumption, while very low by international standards,[17] imposes a major burden on the country's foreign exchange earnings. In 1979/ 80, some 1.44 million barrels were consumed (see Table 11.3). Forecasts for 1984/85 are for some 1.9 million barrels,[18] which would absorb over 90% of the country's foreign exchange earnings, increasing to some 110% by 1990.[19]

It is estimated that 85% of kerosene is used for lighting. Gas-generated electricity, through the urban and rural electrification programs, appears to be the only practical substitute. This also applies to irrigation pumps, although for the smaller ones LPG-powered engines may be a viable substitute. Industry will generally switch to lower-priced gas whenever it can be made available by pipeline or could switch to gas-derived LPG. Conversion of the power sector to gas-fueled plants is well underway. This leaves the transport section as the major one in which natural gas, or gas-derived products, could bring about major substitutions of petroleum products.

Several technically well-known options offer themselves. The first is the use of liquid petroleum gases (LPG). Such gases (butane and propane) represent around 1% of the gas coming from producing wells. Given projected gas production rates, close to 100,000 tons of LPG could be extracted in the late 1980s. Similar extraction plants are already under construction in Egypt and Thailand. While specific cost studies for Bangladesh have yet to be made, unit extraction costs in Thailand, exclusive of gas costs, are estimated at about $2.20/MCF-equivalent, and in Egypt at about $1.25/MCF-equivalent.[20] The Egyptian cost data are likely to be more representative for Bangladesh because the Thai plant faces significant additional costs for separating-out substantial amounts of carbon dioxide. Gas in Bangladesh is free of this contaminant. In

---

[17]Its 1979 average per capita consumption of 41 kg of coal equivalent made it the second lowest in Asia (after Nepal); by comparison, consumption in Thailand was 376 kg, in Japan 4,260, and in the USA 12,350 kg. From World Bank, *World Development Report 1981,* Washington, D.C., Table 7.

[18]Bangladesh Petroleum Corporation.

[19]T. L. Sankar and G. Schramm.

[20]At a discount rate of 12% and a plant life expectancy of 20 years.

Table 11.3. Bangladesh Actual
Petroleum Product Consumption by
Sector, 1979/80. (in thousand bbls)

Gasoline		62
Kerosene		384
Domestic	383	
Power	1	
Jet Fuel		44
Jute Batching Oil		33
Diesel		446
Power	104	
Agriculture	92	
Road vehicles	117	
Motor transport	78	
Bunker (foreign)	16	
Railway	31	
Industry	8	
Fuel Oil		470
Power	96	
Railway	59	
Industry	269	
Bunker (foreign)	45	
Bitumen	—	
Others	2	
		1,441

Source: Bangladesh Petroleum Company.

either case, average economic costs would be less than $3/MCF of gas equivalent (this includes gas production and depletion costs), or less than half the costs of imported petroleum products. LPG is an excellent fuel for spark-plug engines and conversion costs from gasoline are low—a few hundred dollars at most.[21] LPG can also be used as a partial fuel for diesel engines, but this requires a dual fuel system and is technically complex. This would probably make it unsuitable for use in diesel engines in Bangladesh, where technical maintenance is a pervasive problem. LPG, however, may be even more valuable as a household, commercial and industrial fuel in parts of the country without access to natural gas.

A second alternative is the use of compressed natural gas (CNG). The advantages of CNG are that it is clean-burning, nonpolluting, and extends engine life. Disadvantages are the weight penalty of the high-pressure (2400

---

[21]See also Chapter 10.

Table 11.4. Representative Capital and Operating Costs of a CNG-Powered
Vehicle Fleet Operation.

	$
Operating Costs:	
Compressor Electricity Costs,[1] 11 kW × $0.07/kWh = $0.77/h	
$\dfrac{\$0.77/h}{2\ MCF/h} = \$$	0.34
Maintenance Costs: $\dfrac{\$2000/yr}{2\ MCF/h \times 6000\ h/yr} =$	0.17
Total Operating Costs	$ 0.56

Costs for conversion and refueling
Equipment per vehicle $1,000
Pro-rated costs of refilling
Facilities (for 500 vehicle fleet)
   per vehicle $2,000
Total pro-rated costs per vehicle $3,000
Monthly equipment costs (12% interest, repayment period 48 months for vehicles, 96 months
   refill station) $59/month   pro-rated equipment cost per MCF $ 1.97
Average monthly consumption = 30 MCF (259 gal. of gasoline equivalent):

Total systems costs per MCF	$2.53
Economic opportunity costs of gas	$0.60
Opportunity costs of gasoline (8.7 US gals. equivalent)	
($350/ton F.O.B. Chittagong)[2]	$ 7.18
Net economic gain from using CNG	$ 4.05/MCF

Data source: Gas Development Corp., Chicago, 1981, mimeo.
[1]Electricity costs were adjusted to account for higher Bangladesh costs but lower kWh requirements due to
assumed take-off from high-pressure (960 psi) gas lines.
[2]Bangladesh is a net exporter of naptha (i.e., gasoline) stocks at present, because of insufficient domestic
demand for gasoline.

psi) cylinders and the limited range of vehicles which is about one-third to one-half of that of similar gasoline-powered ones.[22] This requires closer spacing of refilling stations. Also, depending on the technology employed, refilling may require up to 10 minutes.

Table 11.4 summarizes cost data adapted from an existing U.S., 500 vehicle CNG system. The apparent savings in fuel costs, after allowance is made for the higher CNG fuel supply system costs, are quite significant, amounting to some $4 per MCF of gas used, reducing fuel costs from comparable, gasoline-driven vehicles by some 56%.

However, allowances also have to be made for the reduced carrying capacity

[22]This is a consequence of the low energy density of gas, even under such high pressure. In U.S. Systems, the two-cylinder storage capacity of a vehicle contains 650 cf. of gas, equivalent to 5.6 gallons of gasoline. In New Zealand various test vehicles carried fuel tanks with capacities ranging between 325 and 748 cf. *From:* J. P. West and L. G. Brown, p. 22.

Table 11.5. Net Gains and Losses in Total Operating Costs From the Use of CNG(2-ton vehicle). (in percent)

	%
Total operating costs of a gasoline-powered vehicle	100.0
Fuel costs as a share of total vehicle operating costs	40.0
Net savings from CNG (0.56 × 0.40)	−22.4
Net loss in carrying capacity, 200 lb/average utilization factor of carrying capacity, 60% (0.6 × 200/4000)	+3.0
Net loss in operating hours, ½ h per 10 hour day	+5.0
Net reduction in total operating costs from using CNG	−14.4

of the CNG-powered vehicles, and the added time needed for refueling. These, for a 2 ton capacity vehicle, may amount to about 8% of total operating costs. The net gains from using CNG, then, would be reduced from some 22% of total costs to a still quite attractive 14% (see Table 11.5). These gains would be higher, if appropriate shadow prices for the foreign exchange and labor components of the two alternatives were included in the analysis. It appears, therefore, that the use of CNG in road transportation, at least for shorter-range and urban oriented vehicles, could be an attractive option to replace petroleum by natural gas.

Even more interesting would be the use of CNG for rail transport. This sector consumes about 90,000 tons of petroleum products per year. In rail transport, the weight penalty of the high-pressure cylinders would matter little. Only a few refilling stations would be needed since trains could always haul sufficient fuel to return to one of them. Both diesel and steam-driven locomotives could be converted to CNG.[23]

Another potential alternative of using gas would be the conversion of gas into methanol, a liquid fuel that is an excellent replacement for gasoline, but can also be adapted to diesel engines.

From the foregoing it can be concluded that for gas-rich, petroleum-poor countries such as Bangladesh, the introduction and systematic promotion of gas substitution for petroleum in unconventional uses could bring about long-term relief from the debilitating pressures of high petroleum import bills and could result in lowered costs of energy overall. Import substitution, more so than gas or gas-product exports, appears to be the economically most efficient use of surplus gas (although the latter are not ruled out by the former).

---

[23]The technology of using CNG in diesel engines is well established. It depends on the use of 10–15% of diesel fuel as a starter-fuel for the combustion process.

Table 11.6.   Natural Gas Tariffs. Titas Gas Company[1]
US¢ per MCF*

	EXISTING Feb. 6, 1979	NEW AS OF JUNE 1, 1980
1. Wellhead price	30.4	40.0
2. End-users price:		
a. Fertilizers, power, cement,		
pulp and paper	42.4	48.9
b. Other industry:	108.5	146.1
c. Commercial	115.3	147.4
d. Domestic (metered)	108.5	105.3
Burner (single) per mo.	135.6	131.6
Burner (double) per mo.	244.1	236.8
3. Exercise duty component	20.3	31.6/73.7

*Converted at the Fall 1979 rate of exchange US$1.00 = Taka 14.75 for
1979 and the September 1981 rate of US$1.00 = Taka 19.00 for 1981.
Source: Bangladesh Planning Commission.

11.5   NATURAL GAS PRICING

Natural gas is by far the lowest priced commercial fuel in Bangladesh, even
though average price levels to various consumer classes have been raised several times in recent years. Past and June 1981 tariffs for various customer
groups have been shown in Table 11.6, while the costs of gas relative to competing petroleum products have been calculated in Table 11.7. As can be seen,
the delivered cost of gas to consumers ranges between 11 and 28% of the wholesale price of substitute petroleum products.[24]

Most of the more than 107,000 domestic gas consumers are billed at a fixed
rate per month based on the number of burners on their premises, because their
connections are not metered. This billing system, although perhaps necessary
initially because of the cost of metering relative to consumption, will probably
become more and more unsatisfactory because it will be difficult, if not impossible, to control the actual number of gas-using appliances and burners in use.[25]

Questions have been raised whether to above price levels for natural gas are
sufficient, given the long-run costs of supply, or appropriate, given the high cost
of alternative fuels. These issues are addressed below.

---

[24]For some industrial applications such as cement or brick manufacturing, the appropriate substitute fuel would be lower-priced imported coal, rather than fuel oil.

[25]Built-in flow restrictions in pipes can easily be removed or bypassed. Furthermore, there is no
incentive to control gas use because the marginal cost of the gas to the user is zero. Reported
average consumption data for domestic connections are indicative of the potential waste: These,
in 1980, amounted to 60CF per day per connection, or about four times the estimated average
cooking gas requirements. Domestic gas consumption data from Titas Gas Company, Dacca.

Table 11.7. Comparison of Natural Gas, Fuel Oil, and Kerosene Prices.[1]

SECTOR	NATURAL GAS REPLACEMENT FUEL	NATURAL GAS COSTS PER MCF US$	REPLACEMENT FUEL COSTS PER MCF EQUIVALENT US$	NATURAL GAS PRICE AS A % OF REPLACEMENT FUEL PRICE
Fertilizer, power, cement	Heavy Fuel Oil	0.59	5.26	9.3
Other industry	Heavy Fuel Oil	1.46	5.26	27.8
Commercial	Kerosene	1.47	7.74	19.0
Domestic metered	Kerosene	1.05	7.74	13.6
Domestic unmetered[2]	Kerosene	0.91	7.74	11.8

[1]Based on September 1981 prices.
Exchange rate: US$1.00 = Taka 19.00
[2]Based on weighted average price per MCF.

11.6   A COMPARISON OF GAS COSTS AND PRICES

All gas supply systems in Bangladesh are owned and operated by government companies. Several of them are single-purpose operations that serve only a few large-scale users such as power plants, cement or pulp and paper mills, etc. The only system with a substantial utility load is the Titas Gas Company which serves the greater Dacca area. Titas supplies approximately 70% of all gas used in the country.

In the Titas system three power and one fertilizer plant account for some 73% of total gas sales (see Table 11.8) some 450 industrial plants utilize another 16%, the commercial sector with about 2,500 connections, accounts for 3% of sales, and the domestic one with more than 112,000 connections for about 8%. As can be seen from a comparison of the average economic costs of

Table 11.8. TITAS GAS COMPANY; Number of Customers by Group, Sales in MMCF, and Revenue in TK $\times$ 10[6] and Percent, 1979/80

GROUP	NO.	%	CONSUMPTION	%	SALES	%
Power	3	0.003	11,018	35.1	70.274	25.9
Fertilizer	1	0.001	11,975	38.1	75.860	28.0
Industry	451	0.4	4,923	15.7	73.913	27.2
Commercial	2,508	2.2	1,021	3.3	17.082	6.3
Domestic	112,285	97.4	2,472	7.9	34.140	12.6
Total	115,248	100.0	31,409	100.0	271.269	100.0

Source: Titas Gas Company.

Table 11.9. Assumptions Underlying the Calculation of the Joint Costs of Supplying Domestic and Commercial Customers of the Titas System.

---

1. 1979/80 Titas total operating costs = TK72.97 million[1]
2. Operating costs related to commercial/domestic customer = 77.2% of TK72.97 million = TK56.33 million (see text)
3. Total gas sales to these customers = 3,493 MMCF.
4. Unit operating costs of these customers = TK16.13.
5. Number of such customers = 102,116 (Average full-year equivalent)
6. Unit connecting cost for commercial domestic users = TK24.06

---

[1]*Source:* Titas Gas Company.

supply (see Table 11.2) with the tariffs of Table 11.6, the power and fertilizer sectors presently pay a price that is roughly within the ranges of the long-run marginal costs of supply including users' costs (i.e., economic costs of $0.56/ MCF compared to a tariff of 0.49/MCF). Industrial users pay almost three times as much.

The situation is different for domestic and commercial users, however. Taken together, they account for only about 11% of total gas consumption, but 49.6% of all connections. Therefore, their costs of service (distribution lines, services, maintenance and billing) are far higher than the respective shares of the larger users. This can be seen from the following analysis which breaks down the operating costs according to customer groups and compares them with the respective revenue flows.

In 1979/80, the average operating costs of the Titas System amounted to US¢15.7/MCF,[26] while those estimated for the Bakhrabad System are US¢ 3.6/MCF.[27] Because the only essential difference between these two system is that Titas serves a large number of domestic and commercial customers, while the Bakhrabad System will sell initially only to large industrial accounts, the difference of U.S.¢12.1/MCF, or 77% of Titas' total operating costs of U.S.$3.84 million can be assumed to be caused by its commercial and domestic services. While it would be desirable to separate these two users groups this unfortunately is not possible because the gas distribution lines to these two groups are inextricably intermingled. This analysis, therefore, considers these two groups jointly. The results have been summarized in Tables 11.9 and 11.10.

As can be seen from the data, the two groups caused estimated net accounting losses of $2.12 per MCF sold, or a total of some US$7.4 million in 1979/

---

[26]Titas Gas Company, 1979/80 Annual Report.
[27]*Source:* Petrobangla, project files.

Table 11.10. Average Revenues and
Costs of Service to Domestic and
Commercial Customers 1979/80. (per
MCF)

COST CATEGORY	JOINT COMMERCIAL AND DOMESTIC CONSUMERS US¢*
Cost of gas	31.5
Cost of meters	7.1
Connecting costs	163.1
Operating costs	109.4
Total costs	311.1
Average revenue	99.4
Loss	(211.7)

*Exchange rate during 1979/80, TK14.75 =
US$1.00.

80. This loss compares to a total 1979/80 accounting profit of the company of
$2.01 million dollars.

These estimated losses from sales to domestic and commercial customers are
underestimated, however. This is so because accounting capital cost charges
for the Titas System are based on heavily subsidized rates available under
terms of international loans or on nominal 5% interest charges for government
loans. In 1979/80, for example, total capital charges (interest plus debt repay-
ments) paid by Titas amounted to only 5.1% of all outstanding loans. A more
appropriate charge would have been 5% for amortization plus 12% for interest
(not counting inflation). This would have increased the nominal debt charges
of US$1.9 million in that year to US$5.8 million. Unfortunately, data are not
available that would make it possible to allocate these actual, additional capital
costs to the various gas consuming categories. It is clear, however, that the
inclusion of these economic opportunities costs of capital would increase the
net loss of serving the domestic/commercial sectors even further.

Few arguments can be made in favor of maintaining such a heavy subsidization of the domestic/commercial sectors.[28] The number of gas users favored
by these low prices represents less than 0.7% of the total number of households

---

[28]It should be noted, however, that the major subsidy apparently goes to domestic users. Commercial ones not only pay higher unit costs, they also have to pay significant meter rental charges.
Also, their relative service cost per MCF of consumption is much lower because of their much
higher usage rates.

in Bangladesh. All others here to pay the much higher market prices for wood-fuel or kerosene (at least if they live in urban areas without their own fuel supply). Furthermore, a majority of gas users belong to the very small middle and higher income classes. Hence, no equity argument can be raised in favor of such subsidies.

If, instead, prices to these users were to be raised by a factor of 3.5, that is from approximately $1/MCF to $3.50/MCF, the actual costs of service to them would be recovered by Titas Gas and a small profit made. The company, as a result, would show a significant accounting profit overall, and come much closer to cover its real economic costs; domestic/commercial gas users on the other hand would still receive gas at costs that would only amount to 65 to 75% of the costs of alternative energy supplies. Hence, they would still derive significant net benefits.

The question whether gas prices should be raised substantially to the power and fertilizer sectors is less clear. As has been seen, present charges roughly cover the economic costs of supply. The government-owned electric utility company and the fertilizer factories are presently heavily subsidized because their controlled prices are far below their operating costs.[29] Increased gas prices would increase the losses of these companies. It could be argued, however, that the required governmental subsidies to these organizations should be increased accordingly, if they are not allowed to raise their prices. While it may appear to be little more than account juggling if gas prices were to be raised in order to increase governmental revenues while on the other hand higher subsidies would have to be paid by the same government to the power and fertilizers sectors, there could be two advantages to such a policy. First, the subsidies would have to be made directly to the sectors receiving them. This would make them more obvious and subject to more scrutiny. Second, higher net revenue flows to the various gas supply organizations would enable them to undertake some urgently needed tasks that are presently neglected. Among them are improved maintenance and a revision of salary scales for technical personnel. The latter are far too low at present to retain technically skilled people who have attractive alternative job opportunities in middle eastern countries. As a result, the gas supply companies have serious operating problems which in turn reduce supply reliability, causing substantial losses to gas users as a consequence.[30]

---

[29]For example, the net losses of the Bangladesh Power Development Board amounted to 42% of gross revenues in 1979/80. *From:* Bangladesh Power Development Board, 1979/80 Annual Report.

[30]Gas supply difficulties, for example, led to the shutdown of a major fertilizer factory in 1979/80 for several months.

ANNEX 11.1   METHODOLOGY AND DATA BASE FOR ESTIMATING THE DEPLETION COST ALLOWANCE[1]

Methodology

The depletion allowance can be calculated by the following set of equations:

$$\sum_{t=1}^{T} c\ (1 + i)^t + 15c_T = R - K \tag{1}[2]$$

where:

c_t = gas consumption in year t

t = time in years, $t = 0, 1, 2 \ldots$,

T = year in which the reserve production ratio reaches 15/1 (the assumed maximum rate of annual production)

$i_{1,2}$ = rates of projected annual increase in gas consumption

R = total known gas reserves in year $t = 0$

K = additional commitments of gas over and above projected rates of growth in consumption

Given c_t, i, R and K (if applicable) the equation can be solved for T.

Once T has been found, today's required depletion allowance can be found by the following expression:

$$V = (P + M - G)\left(\frac{1}{1 + r}\right)^T \sum_{t=1}^{15} \frac{1}{15}\left(\frac{1}{1 + r}\right)^t \tag{2}$$

where:

V = the present value equivalent of the foregone future net value of an additional MCF of gas sold today

P = the future price of the appropriate replacement fuel per MCF equivalent, adjusted for net changes in transport costs compared to gas

M = the additional capital and handling costs of alternative fuels (e.g., coal), pro-rated per MCF of gas

G = the long-run marginal supply costs of gas

r = the real rate of interest

Data:

The following data were utilized for estimating the appropriate depletion allowances:

$R = 10.1 \times 10^{12}$ standard cubic feet (scf)

$c_t\ (t = 0) = 45,400 \times 10^3$ MCF (base year consumption)

$i_1\ (t = 0, 1, 2, \ldots 6) = 21.5\%$ (annual rate of growth in consumption)

$i_2\ (t = 7, 8, \ldots 19) = 10.0\%$ (annual rate of growth in consumption)

---

[1] An earlier version of this model was presented in T. L. Sankar and G. Schramm, Annex 6.

[2] If i cannot be represented by a constant rate of annual increase, discrete values have to be substituted instead.

r = 10%, 12%, 15%

K = 0

P = (a) Fuel oil at $194/ton

    (b) Fuel oil at $350/ton ($194 plus 3% p.a. for 20 years)

    (c) Australian steam coal at $36/ton F.O.B. plus $16/ton freight plus $6/ton delivery costs = $58/ton

    (d) Australian steam coal as above, but increasing at 1% p.a. for 20 years, equal to $71/ton delivered

    (3) New natural gas at $1.50/MCF

M = US$0.55 per MCF equivalent (for coal only)

G = US$0.43 per MCF equivalent

T = 20 years.

12

United States: Energy Management Issues in a High Income, Energy Surplus Region-Alaska

12.1 INTRODUCTION

In contrast to most other regions of the world the state of Alaska is a veritable storehouse of energy resources. Its proven and probable oil reserves amount to over ten billion barrels or more than 25% of total U.S. reserves. Its current production of about 1.6 million barrels/day is greater than Canada's total output. Natural gas reserves in the main Prudhoe Bay deposit alone are estimated at 29 trillion cubic feet or some 15% of total U.S. reserves, while the reserves of the producing fields around Cook Inlet south of Anchorage contain an additional 5 trillion cubic feet.

The known coal reserves of the state are immense, even though they have never been delineated. Some claim that they may contain as much as 5 trillion tons, more than those of all other 49 states combined.[1] Alaska's undeveloped hydro potential is estimated to be at least 19 million kW, or about 50% of the total installed hydro capacity in the lower 48 states.

The overall wealth of energy resources in Alaska is such that only a very small fraction of it can ever be utilized by the state's indigenous economy. To utilize its potential, therefore, energy must be exported, either directly in the form of raw energy (i.e., crude oil, gas, coal, electricity) or indirectly in the form of energy-derived products, such as hydrocarbon-based chemicals or energy intensive products such as aluminum metal or magnesium.

The major energy management issues for Alaska, therefore, are related to such direct and indirect energy exports.

The following discussion analyzes two of them: The potential for the development of large, low-cost hydropower resources, and the marketing of Alaskan natural gas.

---

[1] Alaska Pacific Bancorporation, *Alaska Business Trends, 1982 Economic Forecast,* Anchorage, 1982, p. 21.

Alaska's Economy

Alaska is huge. Its 586,000 square miles of territory could be called a subcontinent rather than a state. Alaska is remote. It is thousands of miles away from more populated areas both in North America and Asia. It is empty. When it was granted statehood in 1959 its population consisted of less than 200,000 people, a quarter of which were Indians and Eskimos. Even today, in spite of its massive oil boom, fewer than 500,000 people live in the state and many of them prefer to leave temporarily during the long and cold winter months.

Alaska is far from being a storehouse of mineral wealth. Gold and copper production, once the mainstay of the economy, have all but disappeared. Today fewer than 200 family-size placer mines are in operation in the state. There is no large mining operation of any metal. The only mineral mined, apart from sand and gravel, is coal for use in two local power plants and, perhaps in 1982, for modest exports to Korea starting at a level of 200,000 tons per year. Alaska's forest products industry is largely limited to the southeastern panhandle and consists of a number of timber and pulp operations whose major market is Japan. The overall size of the industry is modest, employing some 3,000 workers in the early 1980s. Fishing is a major industry, with salmon, king crab, halibut, and herring the major products. However, the employment potential of the industry is limited.

In terms of employment, out of a total labor force of less than 250,000 in the early 1980s, government accounted for almost 40% of all jobs, followed by services and trade with some 34%. Mining, including oil, gas, and coal production, accounted for only 4% of total employment. In 1965, prior to the discovery and development of the huge Prudhoe oil deposits, over 80% of total personal income in Alaska was directly or indirectly derived from government expenditures and only 20% from the commodity sectors.[2]

Most of Alaska's population lives in a few cities, with Anchorage, the state's commercial capital, accounting for almost one-half of total state population.

Prior to the North-Slope oil developments, the state's economy was in a rather precarious condition and largely dependent on federal spending, much of it related to military facilities that were established during and after the Second World War.

Oil in significant quantities was first discovered in 1957 in the Cook Inlet area and at the Kenai Peninsula south of it. By the middle of the 1970s the various fields had been developed. Production had reached a plateau of between 70 to 80 million barrels a year, or about 6% of that of Texas, the largest pro-

---

[2]Gunter Schramm, *The Role of Low-Cost Power in Economic Development,* Arno Press, N.Y., 1979, p. 13.

ducer. Substantial gas deposits had also been discovered in the same region adjacent to tidewater and the ice-free ports of southcentral Alaska. Gas production remained relatively limited because of the small regional market available. By the end of the 1960s, a LNG plant utilizing some 135 million cubic feet/day started shipping liquified natural gas to Japan. A gas-based urea and ammonia plant was also constructed selling its products overseas.

However, the impact of the oil and gas sectors was relatively modest until the discovery, development, and marketing of the Prudhoe oil deposits. Oil from this region at the north coast of Alaska is transported through the Transalaska Pipeline System to the ice-free port of Valdez in southcentral Alaska, where it is shipped to the lower forty-eight states. The impact of this discovery and development on the state's economic fortunes was dramatic. In 1960, the state's total annual budget had been less than $200 million, growing to $660 million by 1976. For fiscal 1982, however, it is an almost incredible 6.2 billion dollars.[3] Eighty-nine percent of the budgetary funds will be supplied by the state's petroleum sector.

Alaska is not about to run out of oil soon. Expectations are high that the presently proven reserves will at least be doubled as more exploration activity is carried out. Most of this oil will have to be exported. This is true also for the immense reserves of coal, natural gas, and hydropower that potentially could provide additional sources of income in the future, provided the formidable problems of distance and high transport costs can be overcome. The interrelated problems of lack of local markets and high transport costs to export markets (which include the lower 48 States) lie at the heart of the problems of Alaska's energy resource development and marketing discussed in the following two case studies.

12.2 LOW-COST ELECTRIC POWER AS A REGIONAL DEVELOPMENT CATALYST[4]

Introduction

This case study addresses two questions: The first is to what extent low-cost, but location-bound energy resources can be used to attract industrial development. The second looks at the potential financial and economic effects of direct or indirect government subsidies to bring about such development in such locations.

---

[3]Alaska Pacific Bancorporation, p. 42.
[4]The material in this case study is based on Gunter Schramm, *The Role of Low-Cost Power in Economic Development.*

Background

In the early 1960s hopes were high that a massive hydropower development in Alaska could be used to establish a sound economic base for the state's development. The overall economic situation at that time was precarious. The military construction and employment boom of the Cold War years of the 1950s had run its course. Ballistic missiles had replaced manned bombers as the main threat to the continent and the expensive early warning and fighter bases in Alaska were no longer needed. The miniboom of earthquake reconstruction work in Anchorage had passed as well and unemployment in the state was running at high levels.

Planners in the Federal Army Corps of Engineers and Bureau of Reclamation, together with enthusiastic supporters in the U.S. Congress as well as at the state level, seized upon development of some of the huge inventory of low-cost hydropower sites as a means to bring about longer-term, sustainable development and economic growth.

The Economic Role of Electricity

Modern industry would be unthinkable without electricity. Yet, while its use is all-pervasive in most activities, its share of overall production costs is still surprisingly small. This is quite apparent from the data of Table 12.1, which show that for most industries the costs of electricity do not amount to more than a few percent of their total value added and, on average, for only 1.2% of their value of output. This means that locational decisions of most industrial activities are unlikely to be affected by electricity costs. Other factors, such as the costs and availability of labor, other inputs, and access to markets are usually far more important.

This leaves a numerically small, but in terms of total value of output important, group of industries for which electricity as an input plays a much larger role than for others. Most of these "power-intensive" industries utilize electricity for the purpose of smelting, refining, or electrolytical processes.

Table 12.2 lists the most important of these products together with their electricity requirements per ton of output and the value of output per ton. Figure 12.1 converts these data into graphical form and shows the sensitivity of these production processes to the costs of electric energy. As can be seen, at average United States industrial power rates, the costs of electricity for a number of these processes would become prohibitive. In the past, such industries had located in areas of exceptionally low power cost (e.g., the Pacific Northwest and the Tennessee Valley region). However, most low-cost sources of electricity had already been developed by the time the Alaska power development strategies were evolving. These strategies were based on the assumption that

Table 12.1. Average Costs of Electricity and as a Percent of Value Added and of Sales, Major U.S. Industries, 1979 Data.

SIC CODE	INDUSTRY GROUP & INDUSTRY	AVERAGE COSTS OF ELECTRICITY PURCHASED, US CENTS/KWH	IMPUTED COSTS OF ELECTRICITY[1] AS A PERCENT OF	
			VALUE ADDED	VALUE OF SHIPMENTS
20	Food and kindred products	3.33	2.0	0.6
21	Tobacco products	3.17	0.8	0.4
22	Textile mill products	2.92	4.3	1.7
23	Apparel, other textile products	4.26	1.2	0.6
24	Lumber and wood products	2.90	2.4	1.0
25	Furniture and fixtures	3.81	1.4	0.7
26	Paper and allied products	2.62	7.1	3.0
27	Printing and publishing	4.05	1.0	0.6
28	Chemicals, allied products	2.45	5.5	2.6
29	Petroleum and coal products	2.70	3.5	0.7
30	Rubber, misc. plastic products	3.41	3.4	1.7
31	Leather, leather products	4.17	1.3	0.6
32	Stone, clay, glass products	3.05	4.2	2.2
33	Primary metal industries	2.21	8.1	3.0
34	Fabricated metal products	3.66	1.7	0.9
35	Machinery, except electric	3.57	1.2	0.7
36	Electric, electronic equipment	3.31	1.4	0.8
37	Transportation equipment	3.32	1.3	0.5
38	Instruments, related products	3.69	0.9	0.6
39	Misc. manufacturing industries	4.12	1.3	0.6
	All industries, total	2.79	2.8	1.2

[1]Includes the imputed costs of self-generated electricity evaluated at the average costs of purchased electricity. Self-generation is particularly important in the paper and allied products group (37.0% of total consumption) and in petroleum and coal products (14.6%).
Sources: Calculated from U.S. Department of Commerce, Bureau of the Census, 1979 *Annual Survey of Manufacturers,* (a) *Industry Statistics:* 1981, Table 2 (b); *Fuels and Electric Energy Consumed,* Table 3.

such industries would have to migrate to the far corners of the earth in their search for low-cost power.

Alaska's Power Resources

The total identified prime power capacity of potential hydro sites in Alaska has been estimated at 18.6 million kilowatts, or 161 billion kilowatt hours annually (see Table 12.3). As one source has pointed out: " . . . As long as Alaska's hydropower resource remains unharnessed, once every nine years the electric energy equivalent of the Prudhoe Bay oil flows out to sea."[5]

[5]Alaska Pacific Bancorporation, p. 41.

Table 12.2. Electric Energy Requirements Per Ton of Output and Value of Output, for Various Power-Intensive Products.

PRODUCT	ELECTRIC ENERGY REQUIRED PER TON OF OUTPUT KWH	VALUE OF OUTPUT PER TON $
Titanium sponge	30,000	7,960
Ferro nickel 55%	24,000	5,900
Magnesium electrolytic process	19,000	2,180
Aluminum	15,000	1,325
Elemental phosphorus	13,000	1,089
Magnesium, silicothermic process	12,000	2,180
Ferrosilicon 75%	10,000	925
Ferrosilicon 50%	6,000	842
Electrolytic zinc	3,600	750
Ferro-manganese	3,000	465

[1]U.S. average prices, December 1979.
Sources: American Metal Market, *Metal Statistics 1981,* Fairchild Publ., New York, 1981; U.S. Bureau of Mines, *Minerals Yearbook, 1980,* Washington DC, 1981, Gunter Schramm, "Effects of Low-Cost Hydropower on Industrial Location," *The Canadian Journal of Economics,* Vol. II, No. 2 May 1969.

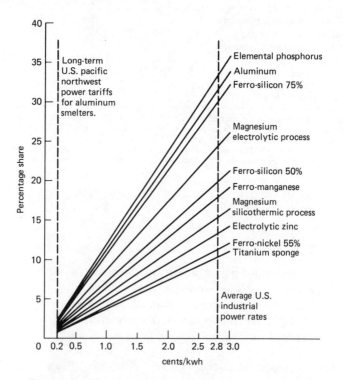

Figure 12.1.   Electric energy costs as a percent of output value.

Table 12.3. Summary of Estimated Prime Power Capacity of Identified Potential Hydroelectric Developments over 7,500 kW.

REGION AND SITE		PRIME POWER CAPACITY IN kW
Southeast		2,872,000
Yukon-Taiya	2,400,000	
30 Other	472,000	
Southcentral		5,111,000
Wood Canyon	2,500,000	
Upper Susitna (4 reservoir plan)	801,000	
Copper River No. 2	513,000	
Chakachamna	188,000	
Tokichitna	129,000	
38 Other	980,000	
Interior		6,318,000
Rampart	3,450,000	
Woodchopper	1,620,000	
Kanuti	385,000	
Hughes	130,000	
Campbell	160,000	
Tanana River No. 2	137,000	
15 Other	436,000	
Southwest		3,850,000
Kaltag	1,500,000	
Crooked Creek	1,170,000	
Holy Cross	700,000	
Holokuk River	140,000	
11 Other	340,000	
Northwest		487,000
Noatak (2 reservoir plan)	278,000	
Ikpikpuk	103,000	
6 Other	106,000	
Gross total all sites less decrease in Yukon River		18,638,000
power potential represented by Yukon-Taiya Project		1,225,000
Net total all sites		17,413,000

Source: U.S. Dept. of The Interior, *Rampart Project: Alaska Market for Power and Effect of Project on Natural Resources: Field Report,* Juneau, Alaska, January 1965, Table 52.

Given this huge inventory of potential electric energy the question has often been asked how these unutilized resources could be developed in the most efficient way for the benefit of Alaska and the United States as a whole. There are three uses to which they could be put:

First, to supply incremental needs of the existing and future Alaskan economy in such a way that energy costs are kept to a minimum level, contributing thereby to the minimization of the costs of living and the costs of doing business in the state.

Second, to develop these energy resources in excess of normal Alaska needs with the aim to attract, through very low prices, industries that otherwise would not settle in Alaska. This would require that energy costs for these industries would be low enough to overcome the disadvantages and higher costs of an Alaskan location.

Third, to develop these energy resources for the purpose of exporting part or all of them to areas outside of Alaska.

The first option led to the development of a few, modest-sized hydroplants near load centers, to serve the modest power needs of existing activities.

Until the early 1960's, however, no active development had taken place to utilize the huge, really low-cost hydro-power resources of the state for either exports or for the purpose of attracting power-intensive industries to the state. The location of the most important of these sites has been shown in Figure 12.2.

The potential site on which attention was focused at that time was the Rampart Project downstream of the Yukon Flats. With a capacity in excess of 5 million kW, Rampart would have been the largest hydro-power site in the United States. Its full capacity output would have been almost 40 times larger

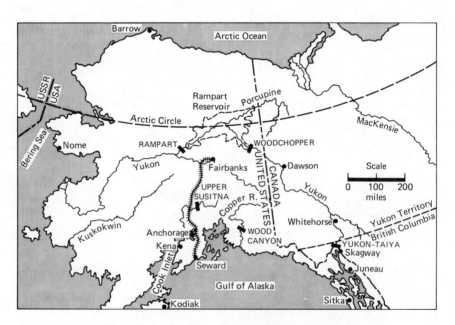

Figure 12.2.   Alaska's major potential hydro projects.

than the state's total power demand in 1962. Given this discrepancy in potential output and preexisting demand this project could only have been developed if either firm industrial commitments of unusual magnitude within Alaska itself could be obtained, or if a transmission intertie to large enough outside markets would have been possible. The latter option had to be ruled out because of the length of transmission lines to the next available larger market, which was located in the Pacific Northwest and was approximately 2,500 miles away from Rampart. While technically possible, transmission costs simply were too high to make this a viable option. Furthermore, the Yukon-Taiya Site, which straddles the Alaska-Canadian border, was far better located for the purposes of electric energy exports to the lower forty-eight states.[6] The remaining alternative was to attract additional industry to the state. Assuming this to be feasible, a number of studies had come to the conclusion

"... that requirements for power in 1990 (which was the first year in which the full output of the project was expected to be available) in the area to be served by the Rampart project can range from its full potential power outlet to considerably larger amounts."[7]

Projected Power Demand

The basic demand projections were prepared by the Bureau of Reclamation and published in the Rampart Field Report.[8]

Five load projections were prepared by the Agency. These were based on different assumptions regarding the degree and type of likely future resource and industrial developments. Here we will concentrate on one only which was considered to be the most likely one. The industrial demand components of this forecast have been shown in Table 12.4. To simplify the discussion only the projected 1995 loads are shown. It can be seen that slightly more than 13% of the projected demand was to come from industries based on Alaskan raw materials, whereas the other 87% would have had to come from processing operations based on imported materials. What is even more important, however, is that over 90% of these import-based industries were projected to consist of one particular type, namely aluminum smelting. This meant that any economic evaluation of the project could concentrate on the likelihood of attracting alu-

---

[6]See also G. Schramm, "The Economics of an Upper Yukon Basin Power Development Scheme", *The Annals of Regional Science*, Vol. II, No. 1, December 1968, pp. 214–228.

[7]U.S. Senate, *The Market for Rampart Power, March 1962*, Committee Print, 87th Congress, Seventh Session, June 30, 1962, p. 7.

[8]U.S. Department of the Interior, *Rampart Project, Alaska Market for Power and Effect on Natural Resources, Field Report*, Vols. I–III, Juneau, Alaska, January 1965.

Table 12.4. Rampart Field Report Projections of Industrial Power Demand,[1] Demand Forecast for 1995, Case II.

PRODUCT	ANNUAL PRODUCTION CAPACITY	MILLIONS/ kwh	PER CENT	PER CENT BASED ON	
				ALASKAN RAW MATERIALS	IMPORTED RAW MATERIALS
Electric furnace steel, tons[2]	100,000	50	0.43		0.43
Magnesium (from seawater), tons	30,000	500	4.37	4.37	
Chlorine and caustic soda, tons	40,000	115	1.01		1.01
Calcium carbide, tons	10,000	31	0.27		0.27
Nitrates, tons	75,000	135[3]	1.18	1.18[3]	
Aluminum, tons	600,000	9,000	78.73		78.73
Cement barrels	2,000,000	60	0.53	0.53	
Ferro-alloys, tons	100,000	650	5.69		5.69
Silicon carbide, tons	8,000	68	0.		0.59
Petroleum and gas liquefaction, million barrels	11,300,000	82[3]	0.72	0.72[3]	
Wood pulp, tons	792,000	692	6.05	6.05	
Timber, million board feet	655	48	0.42	0.42	
Total		11,431	100.00[4]	13.27	86.73[4]

[1] *Rampart Field Report,* Table 67, p. 574, Table 85, p. 682, Table 92, p. 727. Where the *Field Report* assumed increasing power demands over time the projections for the year 2000 have been shown.
[2] Tons mean short tons throughout.
[3] Industries likely to use own natural gas resources for power generation.
[4] Does not total exactly due to rounding.

minum smelting to Alaska. Without this consumption, the project would not have been viable because large blocks of energy would have remained unutilized for many years.

The Economics of Aluminum Smelting in Alaska

In order to investigate the economic feasibility of aluminum production in Alaska it was necessary to evaluate the relative cost structure of the industry there vis-a-vis other possible locations. If production costs including the costs of transportation of raw materials to the plant plus shipping costs of finished products to markets would have been equivalent or lower than at other possible locations a case for aluminum processing in Alaska could have been made. If such cost equivalence or advantage was not achievable, government policies, or even subsidies, could have been used to possibly attract the industry to the state. Both possibilities are evaluated here.

The Rampart Field Report projected annual aluminum production capacities in Alaska of 600,000, 800,000, and 1,200,000 tons, respectively.[9] This compared with a 1966 capacity of 5.7 million tons in the west, and a projected demand of some 11.4 million tons by 1980. Viewed in the light of this projected global demand, an aluminum smelter capacity of 0.6 to 1.2 million tons in Alaska appeared rather large, but not impossible. However, while aluminum is generally traded on a worldwide basis (for example, there exists a considerable cross-traffic of ingots between Europe and the North American continent) transportation economics, in part, dictate the most economical sources of supply. Other factors influencing the location of new smelter capacity are import tariffs and import restrictions[10] as well as the ownership patterns of major fabricators of which many are owned by primary aluminum producers.

Almost all aluminum is produced from bauxite, a mineral easily mined and found in many, mostly tropical, locations. A wide variety of other minerals could be used as well (most of the minerals forming the earth's crust contain aluminum in some chemical combination), however, costs of processing would be considerably higher. Since world bauxite reserves are plentiful[11] there has been little incentive so far to use other base minerals. The most suitable supply sources for an Alaskan smelter would have been Australia (which accounts for about 45% of total known world resources), the Dominican Republic, Haiti, Jamaica, Guyana, Surinam, Indonesia, Malaysia, or Sarawak. Basic transportation costs from any of these countries to Alaska would not have varied greatly, although harbor conditions and loading facilities, availability of return cargo, as well as individual mineral rights by specific companies could have led to considerable differences in costs and availability from individual sources.[12]

While bauxite forms the basic raw material, the production of aluminum actually takes place in two distinct stages. The first is the refining of the bauxite into alumina. This step is necessary in order to prevent impurities from being

---

[9]The projections were based on the assumption that aluminum would take up any free power capacity available. Lower projections for other uses, therefore, allowed greater aluminum capacity.

[10]Significant tariff or quota restrictions against aluminum imports existed in Japan, the European Economic Community, Australia, and India.

[11]Total world reserves (excluding marginal and submarginal reserves) are estimated at 5,760 million tons of bauxite.

[12]For a detailed analysis of the effects of transportation costs on the economics of aluminum production see Sterling Brubaker, *Trends in The World Aluminum Industry,* Johns Hopkins Press, Baltimore, 1967.

carried through to the metal. An important factor in the location of alumina plants is that about 2.1 to 2.5 tons of bauxite are needed to produce one ton of alumina (which is aluminum oxide in an almost pure state). Both bauxite and alumina are processed and shipped in powder form. Significant transportation cost savings, therefore, are possible if the alumina refining takes place close to the source of the bauxite. This would mean that tonnages to be transported to a distant smelter are less than half than if the bauxite itself is moved. Experience shows that the costs of moving alumina over equal runs are actually approximately twenty-five per cent higher. If the raw materials, bauxite or alumina, for an Alaska smelter would have originated in Australia, about 6,500 miles distant from Alaska, bauxite transportation costs would probably have amounted to $7.00 per metric ton, given good loading and unloading facilities and a large Alaskan smelter. Alumina transportation costs then would have amounted to approximately $7.90 per short ton (all in 1965 prices). With identical costs of alumina production in either Australia or Alaska, these transport cost relationships would have favored alumina production in Australia.

In order to analyze the feasibility of establishing aluminum smelters in Alaska the likely cost structure of these smelters had to be compared with the costs at the most likely alternative sites.

This, first of all, required an analysis of prospective markets for a U.S. West Coast smelter, because freight costs of metal have a significant effect on overall costs. An Alaskan smelter, therefore, would have had to look primarily for markets within reasonable ocean-haul distance. This meant markets around the rim of the Pacific Ocean, of which the most important were the U.S. West Coast and Japan.

Japan's additional requirements until 1980 were estimated at 590,000 short tons. Demands from the U.S. West Coast had been projected for the same year of ranging between 1,280,000 and 1,920,000 tons;[13] given an estimated 1967 Pacific Northwest smelter capacity of about 1 million tons, this indicated a net demand of 280,000 to 920,000 tons.[14] Other markets around the Pacific Basin were assumed to grow as well, although overall tonnages were projected to be considerably smaller. This indicated an overall net demand in the primary marketing area for an Alaskan smelter in the neighborhood of between ¾ and 1½

---

[13]Samuel Moment, Ivan Block & Assoc., *The Aluminum Industry of the Pacific Northwest,* Preliminary Report to the Bonneville Power Administration, Portland, Oregon, 1966, p. III-A-82.

[14]Not all of this demand would have been satisfied from West Coast smelters. In the mid 1960s, about ⅓ of California's ingot demand was supplied from eastern or midwestern sources. On the other hand, about sixty per cent of the output of Pacific Northwestern plants was shipped to markets outside of the western states (California, Oregon, Washington, Idaho, Utah, Montana, Arizona, and Nevada). Samuel Moment, Ivan Block & Assoc., p. III-A 76.

millions tons of ingot by 1980, net of existing capacity. From a demand point of view, therefore, a sizeable Alaskan smelter appeared to be feasible. However, it must be recognized that any single smelter owned by one or two firms would have captured only a fraction of this market, given the firmly established marketing interests of the other existing aluminum producers. This would have been the case even if the Alaskan plant would have had a much more favorable cost structure than any other competing plant. For this reason alone, the best that could have been hoped for would have been a plant in the 100,000 to 200,000 ton capacity range, a plant large enough to capture most of the available economies of scale, but not large enough to absorb the output of any of the large Alaskan hydro-power projects.

Looked at from the vantage point of the mid-1960s, what would have been the likely locations, other than Alaska, for new smelter capacity serving these markets? There was, first of all, the Alcan smelter at Kitimat, British Columbia which probably had the most favorable cost structure of all aluminum producers in the area. To reach its ultimate design capacity of 550,000 tons per year Kitimat could have added another 330,000 tons of capacity; this expansion was almost certain to take place.[15] Prospects for further expansion of primary aluminum smelting in British Columbia were more doubtful despite the wealth of undeveloped hydro resources. Most of the more economical power sites were farther inland, which would have meant additional transmission costs to a tidewater smelter.

The most logical area for further expansion would have been the Pacific Northwest, where six producers had an estimated combined production capacity of one million tons. Historically, the major incentive for the establishment of these plants had been the exceptionally low power rate of about two mills per kilowatt hour (see Table 12.5). However, there was some question whether this favorable rate structure could have been maintained or made available for new producers or further plant expansions.[16] This depended on the future marketing policy by the Bonneville Power Authority which faced higher incremental power generating costs from the remaining marginal hydro sites and future, higher-cost atomic and thermal power plants.

There were many other areas along the Pacific rim that could have been

---

[15]This expansion finally got underway in 1980.

[16]Low-cost, long-term block sales by the Bonneville Power Administration to industry were severely rationed throughout most of the post-war period. During 1954–60, no further plans were made to increase sales to industry over and beyond previous commitments (Moment, p. II-A 11). However, industrial supplies were increased by about 1 million kilowatts in the 1971/72 period. This new power availability led to the establishment of a new 151,000 ton smelter by Intalco (owned jointly by American Metal Climax, Howe Sound Company, and Pechiney of France) at Bellingham, Washington. Energy costs to this plant reportedly were in the 2 mill range. Other existing producers had indicated plans to increase capacity by about 200,000 tons by 1972.

Table 12.5. Price of Power Paid by Aluminum
Industry in the Pacific Northwest.[1]
1963

	PLANT CAPACITY SHORT TONS	MILLS/kWh
Alcoa, Vancouver Plant	98,000	2.07
" , Wenatchee Plant	109,000	2.10
Anaconda[2]	68,000	1.66
Harvey[2]	82,000	1.66
Kaiser, Spokane Plant	176,000	2.02
Tacoma Plant	41,000	n.a.
Reynolds, Longview Plant	61,000	2.00
" , Troutdale Plant	92,000	2.48

[1]Samuel Moment, p. II-F 18 and Table II-A 3.
[2]Direct revenues from "at site" rate charged to plant. Wheeling and other charges had to be added to this rate.

considered as potential sites for future capacity. Japan, with her fast growing market for metal, was a leading candidate. However, her major disadvantages were power costs (average costs of thermal power in the early 1960s were 12.5 mill/kWh, with 8 mills generally charged to aluminum smelters).[17] Other areas were Australia, which had the advantage of large bauxite deposits but lacked low-cost hydro power. However, Australia owned large, low-grade coal deposits. Several smelters had already been built, but all of them were relatively small, high-cost producers, designed mainly to supply the home market (which was protected by import quotas). The fact that the large international companies active in Australia resisted considerable government pressure to build large export smelters seemed to indicate that the economics of Australian metal production for world markets was not encouraging. Another candidate was New Zealand which has large, undeveloped hydro sites on the South Island.[18] Many of the less developed nations around the Pacific have large hydro resources which could be utilized for aluminum production. However, in most of them risk factors resulting from political instability were high. This has always been an important deterrent for the privately-owned, international aluminum companies. For the time being, therefore, construction of aluminum plants in these areas appeared unlikely.

The areas most likely to see new smelting capacity established, therefore, appeared to be Japan and the Pacific Northwest. Unfortunately, data on Japanese production costs were not detailed enough to allow comparisons. Drawbacks for Japan were higher power costs, presumably higher land costs, and

---

[17]U.S. Senate Committee on Public Works, *The Market for Rampart Power,* p. 124.
[18]Construction of a smelter started in the late 1970s.

Table 12.6. Estimated Costs Excluding Costs of Energy of Aluminum Ingots Delivered to Southern California and Japanese Markets from Plants in the Pacific Northwest and Alaska in 1965 Prices.

	PACIFIC NORTHWEST	ALASKA
	DOLLARS PER SHORT TON	
CAPITAL COST	650	845
Capital charges per ton at a 15% gross rate of return, 85% capacity factor	122.17	158.82
Alumina costs 1.9 tons, F.O.B. alumina plant	95,00	95.00
Transportation costs alumina to smelter, 1.9 tons	15.00	15.00
Carbon electrodes and pot lining, 0.55 tons	38.00	38.00
Electrolyte and aluminum fluoride	15.00	15.00
Water, 30,000 gallons	3.00	2.00
Production labor, 14 hrs.	49.98	67.47
Wages and salaries	10.00	13.00
Social security 3.3%	1.98	2.67
Operating and maintenance supplies	16.00	18.00
Local taxes and insurance	6.00	6.00
Transportation, ingot to markets:		
½ ton to Southern California	6.50	8.00
½ ton to Japan	12.50	12.50
Total costs (excluding costs for power) at 15% gross rate of return	395.63	457.35

Source: Gunter Schramm, *The Role of Low-Cost Power in Economic Development,* Table 41.

the considerably higher costs of capital. Advantages were the rapidly growing domestic demand, lower labor costs, and the existing tariff protection of the home market.[19] However, it was assumed that a new smelter, located either in Alaska or the Pacific Northwest, would have shipped half of its output to Japan.

Table 12.6 shows the estimated costs of the plants in Alaska and the Pacific Northwest excluding the power cost component. While a number of input costs were estimated to be the same at either location, capital investment costs and labor were significantly higher in Alaska, reflecting the much higher Alaskan cost structure. The profit expectations of the industry were expressed in terms of a gross rate of return of 15% on equity plus long-term debt. This resulted in a 23% pre-tax rate of return on equity, given the prevailing sixty-forty equity/

[19]In the past import tariffs had ranged as high as 15%. In 1967, they were 13%. Under the new Kennedy Round agreements of the General Agreement on Tariffs and Trade (GATT) they were reduced to 9%. *The Financial Post,* Toronto, July 8, 1967, p. 8.

Table 12.7. Estimated Net Cost Differences Between Aluminum Plants in the
Pacific Northwest and Alaska and Required Mill Rates per Kilowatt-Hour to
Equalize Overall Costs Between Them.[1]

	PACIFIC NORTHWEST		ALASKA	
	COST DIFFERENTIAL EXCLUDING POWER COSTS, DOLLARS/ TON	REQUIRED MILL RATE DIFFERENTIAL, MILLS/kWh	COST DIFFERENTIAL EXCLUDING POWER COSTS, DOLLARS/ TON	REQUIRED MILL RATE DIFFERENTIAL, MILLS/kWh
At 15% gross rate of return	0	0	61.72	−4.11

[1]Based on data in Table 12.6. Assumed power requirements of 15,000 kWh per short ton.

debt ratio of the industry. The actual average operating results of the industry had been about 10% in previous years. However, it was assumed that the expected rate would have to be higher; therefore, the data based on the fifteen % rate were assumed to be representative for industry evaluations of new plant locations.

Production cost differences between the Pacific Northwest and Alaska were marked, with the latter showing a considerably higher cost structure. Table 12.7 shows the net differences in costs, as well as the needed difference in energy costs to make an Alaskan location cost-equivalent to the Pacific Northwest.

Alaska's cost differential amounted to $62 per ton of metal. To equalize costs, this would have meant that the Pacific Northwest's power rates would have had to be 4.11 mills higher than those obtainable in Alaska. If we were to assume that the lowest rate obtainable in Alaska would have been about 1.5 mills[20] this would have meant that energy costs to additional aluminum producers in the Pacific Northwest would have had to rise to between 5 and 5.6 mills, or by 250% compared to existing rates. This was not a likely prospect as seen from the vista of the mid-1960s (apart from the fact that power rates in all other competing areas would have had to rise in a similar fashion). On the basis of these data it was concluded, therefore, that aluminum production in Alaska was uneconomical.

The Potential Effects of Government Loan Guarantees

While economic considerations spoke strongly against the establishment of an aluminum plant in Alaska at that time, this did not necessarily mean that a

[20]Since none of the Alaskan power projects was to produce power at such a low rate, a 1.5 mill rate would have meant that energy would have been sold at less than average cost. This would have been rational, however, if it would have meant that overall average costs to all Alaska electricity users would have been lowered by economies of scale compared to the costs of a smaller power plant designed to serve a market without an aluminum smelter.

producer could not have been attracted to the state. The foregoing analysis was based on the existing financing practice and capitalization pattern of the U.S. industry. Since this pattern had prevailed over a number of years it could be assumed that it reflected the industry's needs and operating experiences in the face of long-term overall market conditions and risk factors. However, under the banner of regional development objectives many governments attempt to attract new plants by offering special financing arrangements, free land, lease-back of buildings, or tax concessions. Quite often such assistance takes the form of government credit guarantees or direct government loans.

Let us assume, therefore, that either the state of Alaska or the U.S. federal government would have been sufficiently interested in attracting an aluminum plant to Alaska to make such assistance available. For illustrative purposes let us suppose that forty percent of total long-term debt (as in the case of nonass-isted producers), and another forty percent, or alternatively forty-five percent, would have been provided by direct government loans or government credit guarantees. This would have reduced equity requirements to twenty or fifteen percent, respectively. The assumed borrowing rate (including the government supported portion) was six percent, and the minimum profit expectation of the producer twenty-five percent pre-tax on equity.

The effects of these changes in the equity/debt ratio are shown in Table 12.8. Costs to a nonassisted Pacific Northwest producer—excluding costs of energy, but including an expected pre-tax rate of return on equity 25%—would still have been $395.63 per ton. Costs to the Alaskan producer, given the much smaller share of equity capital, however, would have been lowered to $413.29 for a twenty percent and $405.54 for a 15% equity share, respectively. The net

Table 12.8. Estimated Differences in Costs and Mill Rates for Alaskan and Pacific Northwest Aluminum Plants under Assumption of Varying Debt to Equity Ratios for the Alaskan Producer.[1]

EQUITY/LONG-TERM DEBT RATIO OF ALASKAN PRODUCER	COST EXCL. ENERGY FOR PACIFIC N.W.[2] (DOLLARS/TON)	COST EXCL. ENERGY FOR ALASKA (DOLLARS/TON)	NET COST DIFFERENCE (DOLLARS/TON)	REQUIRED DIFFERENCE IN MILL RATE TO MAKE ALASKAN PRODUCER COMPETITIVE (MILLS/kWh)
20/80	395.63	413.29	17.66	1.18
15/85	395.63	405.54	9.91	0.66

[1]Data from Table 12.8. Calculations based on 6% borrowing rate, postulated minimum profit goal by industry: 25% pre-tax return on equity.
[2]From Table 12.6. Equity/debt assumed unchanged at 60/40.

cost differentials between the Pacific Northwest and Alaska would have declined to \$17.66 or \$9.91 per ton, respectively, meaning that power costs in Alaska would have had to be no more than 1.18 mills or 0.66 mills, respectively, lower in order to establish cost equivalence between both areas. But a range of one mill or less in power costs between Alaska and the Pacific Northwest was quite within feasible range.

On the basis of this illustrative example, therefore, it can be seen that an aluminum producer could have been enticed to settle in Alaska, given the suggested government support. Such an offer would have been particularly attractive to firms that wanted to avoid substantial increases in equity capital requirements, or firms that wanted to maintain, or obtain a share of the overall aluminum market, but lacked the financial means to enter the market by financing a large-scale smelter on their own.

The question must be asked, however, whether such support would have been in the best interest of the government entity providing it. A case could probably have been made for it if the power consumption of the smelter would have been a prerequisite for the construction of a low-cost source of electric energy.[21] This would not have been the case in Alaska.

In southcentral and interior Alaska, gas-fired thermal power plants, even if built on a relatively small scale, would have brought generating costs down to an attractive level. Addition of an aluminum plant would have resulted in additional savings. But these would not be any greater than 0.5 to 1 mill per kilowatt-hour. Since generating costs account only for a fraction of total energy costs to the ultimate consumer, these savings would not have been worthwhile, considering the very large loans or credit guarantees involved. The only other major benefit for Alaska would have been increased employment, amounting to, perhaps, 2,000 permanent jobs for a 200,000 ton smelter.

Would these combined benefits have been worth the price? The price would have been a credit liability to the supporting government of some 68 million dollars,[22] a rather sizeable sum for a state whose total budget in 1966 amounted to only 152.2 million dollars.[23] Given the fact that even the employment benefits would have gone largely to skilled workmen who would have had

---

[21]This, for example, was the case in Ghana where the Volta River hydro project would not have been feasible without the construction of the Kaiser-Reynolds aluminum smelter. The latter obtained large credit guarantees from the U.S. Government. This was, in fact, a form of economic development aid from the United States to Ghana. Ghana, however, was probably justified in using its (limited) U.S. aid in this fashion since the overall result was not only the construction of an aluminum smelter with its limited employment opportunities but also the coming into being of a low-cost source of energy not otherwise obtainable.

[22]Assuming a 200,000 ton smelter, capital costs of \$845/ton and a 40% credit or credit guarantee.

[23]State of Alaska, Division of Finance, *Annual Financial Report 1966*, Juneau, Alaska.

to be brought in from the contiguous forty-eight states, rather than to under-employed Alaskans, this would have been questionable indeed. If direct credits, or credit guarantees had been used as means for the speeding-up of economic development it would probably have been wiser, and much more beneficial in terms of employment, if these credits would have been reserved for less capital, more labor-intensive activities that would have relied on the employment of needy Alaskans. Commercial fishing and cannery operations which could have tapped the large, underdeveloped fishing banks along Alaska's coast were prime examples.

In 1967, the U.S. Department of The Interior decided that the Rampart Project was not economically justified, in effect tabling the project indefinitely.[24] Many reasons, such as heavy damages to fish and wildlife and the availability of competitively priced power from alternative sources such as natural gas were responsible for this decision. But one of the most important ones was that no electroprocessing producer had come forward and expressed serious interest in an Alaskan location.[25] At the same time, however, a number of large, new smelters utilizing higher-cost electricity from coal-fired power plants were built close to markets (as in Great Britain and West Germany, for example).

Conclusions

The main reason why electroprocess industries could not be attracted to Alaska on the basis of low-cost power availability alone was one of economics. Total production plus delivery costs of outputs to markets were higher there than those from alternative locations in spite of existing differences in the costs of electricity.

This situation has basically not changed since the mid-1960s. While the costs of coal-fired power plants—the closest competitor to low-cost hydro power sites—have increased greatly since then,[26] so have the costs of construction of large-scale hydro power developments, particularly in remote locations.[27] The

---

[24]U.S. Department of The Interior, *Alaska Natural Resources and The Rampart Project*, Washington, D.C., 1967.
[25]The same was, incidentally, the case for the Yukon-Taiya Project, which had been investigated by a number of aluminum producers such as ALCOA, Venture-Frobisher, and others. See G. Schramm, "The Economics of an Upper Yukon Basin Power Development Scheme."
[26]From $100 to $120 per kW of capacity in 1965 to $400–$500 in 1980, provided the plant does not require elaborate pollution control measures such as scrubbers for sulphur removal, in which case costs may be as high as $1000/kW.
[27]Average construction costs of large water resources projects in the Western United States have increased from 100% in 1967 to 310% in 1981, while the Department of Labor's Cost of Living Index increased from 100% to 260% in the same time span. *From:* U.S. Dept. of The Interior, Bureau of Reclamation, *Construction Cost Trends*, Denver, CO, July 1981.

latter suffer in particular from the much higher rates of interest that prevail today compared to the mid-1960s.[28]

Certain lessons can be learned from this case study that are applicable to many hydro-rich countries. Most of the undeveloped, low-cost hydro resources are located in remote locations of developing countries. For them it would be difficult, if not impossible, to attract export-oriented smelters, unless the owners of these plants receive elaborate capital risk protection and, possibly, financial support or other concessions. Such concessions are costly to the host country, however, because they affect its overall credit rating and the availability of funds that usually have high opportunity costs in alternative uses. The basic economic aim of a developing country should be to increase income and employment of its population. Capital-intensive, electroprocessing plants that employ little and generally mostly high-skilled labor are little suited to promote these goals. Attracting such plants to such countries and locations, therefore, generally will make sense only if:

a) There is a large enough domestic market available to absorb most of the output (as in India, for example);
b) or, the output can be produced domestically at competitive world-market costs (applying appropriate shadow prices to the costs of capital and other inputs);
c) or, the construction and power consumption of the smelter is a precondition to reduce overall power costs for the economy as a whole to such an extent that the high investment costs, properly shadow-priced, can be justified on these grounds.

12.3   MARKETING ALASKA NORTHSLOPE GAS

Introduction

The 29 trillion cu ft of natural gas contained in the Prudhoe Bay deposits on Alaska's Northslope represent the United States' largest single gas reservoir, accounting for about 15% of total U.S. reserves at the end of 1979.[29] The remote location of the deposit combined with the difficulties of bringing this

---

[28]The Federally mandated discount rate applied in the evaluation of Rampart was a lowly 3%. High discount rates affect hydro plants proportionately more than thermal plants because for the former capital costs amount to between 80–90% of total costs, whereas for the latter this share is only around 35–40%.

[29]United States Department of Energy, *Principal Findings of the 1978 U.S. Crude Oil, Natural Gas, and Natural Gas Liquids Report,* DOE/EIA-0216(79)EX, Washington, April 1981.

gas to markets in the continental U.S. via a pipeline through Canada have prevented its utilization so far, in spite of the U.S. Government's efforts to make the gas more marketable by relaxing several regulatory requirements.

The pipeline construction cost upon completion in 1986 is estimated at 43 billion dollars. This would make it by far the world's largest private investment project ever undertaken. Apart from its monumental costs, the pipeline raises and, by its very size, magnifies some troublesome issues that are common to the economic evaluation of regulated energy industries.

This case study does not attempt to pass judgment upon the respective merits or demerits of the pipeline as such, but discusses a number of issues that are common in the economic evaluation of regulated energy carriers and energy supplies. These specific issues are:

a) The distorting effects of government regulations on pipeline owners' profit optimization decisions;
b) The problems of average versus marginal cost pricing (i.e., the "folding-in" issue);
c) The question of optimum timing for gas pipeline construction.

The Alaska Gas Pipeline

Of the various means to bring the Northslope gas to markets, only one so far has been found to appear practical: This is to build a pipeline from the Prudhoe Bay deposit through Alaska, the Yukon Territory, northern British Columbia to Alberta, where the line would be split into an eastern and western leg to serve Midwestern and Pacific markets in the lower 48 states respectively. Figure 12.3 shows the proposed routing of the system.

The sections from Alberta to Western and Midwestern markets are under construction now and will initially be used to carry surplus Canadian gas from Alberta to U.S. markets. However, it is the understanding of both the United States and Canadian Governments that this prebuilding is a temporary expedient only. For its economic viability the pipeline system depends on the supply and throughput of North Slope Alaskan gas. Ultimately, the line is expected to carry also Canadian gas from already known, sizeable gas fields around the lower MacKenzie River and, perhaps, the Beaufort Sea to Canadian and U.S. markets south. This would require the construction of a MacKenzie Valley spur-line that would join the ANGTS line near Whitehorse in the Yukon Territory.

So far, it has not been possible to secure financing for the Alaskan leg of the line. This is by far the most expensive section with recent cost estimates of

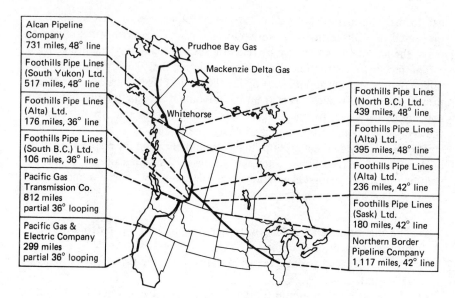

Figure 12.3.   Map of ANGTS route.

twenty-seven billion dollars.[30] In 1977, when Congress approved construction of the overall pipeline in principle, total costs for all segments together were estimated at ten billion dollars, and operation was expected to start by 1983. The reason for the huge cost increases is the need, not fully recognized earlier, of installing the forty-eight inch line over frozen tundra through most of the Alaskan section. This will also require the continuous refrigeration of the gas to prevent thawing out of the foundations and buckling of the line.

According to the 1977 President's decision that was approved by Congress,[31] the pipeline is to be fully financed from private sources. The order also stipulated that no charges could be levied against gas consumers until the line was actually operational. However, because of the difficulties that have been encountered by the pipeline sponsors in raising the necessary funds, Congress and the Reagan Administration have recently made further concessions by agreeing that:

---

[30]United States Department of Energy, *Principal Findings of the 1978 U.S. Crude Oil, Natural Gas, and Natural Gas Liquids Report*. This includes $6 billion for the required gas conditioning plant at Prudhoe Bay and $21 billion for the pipeline itself.

[31]Executive Office of The President, *Decision and Reports to Congress on The Alaska Natural Gas Transportation System*, Washington, D.C., September 1977.

a) Long-term sales contracts for the gas to be transported can include an all-events clause which would require payment by consumers even if no gas ever were to be delivered by the line;

b) Construction costs of the line can be prebilled to ultimate gas consumers as they accrue, i.e., they can be "folded-in" into the rate base of the gas distribution companies as soon as a partial component of the line is completed.

The 1977 order determined that the ceiling price for Prudhoe gas at the wellhead would be $1.45/MCF plus an annual inflation adjustment.[32] It should be noted, however, that this is a maximum price. It could well happen that actual wellhead prices would have to be lower to make the gas saleable in lower forty-eight markets. This, for example, happened to crude oil from Prudhoe Bay for a number of years.[33]

The fundamental problem is that gas delivered through the line would cost substantially more than gas from other sources. It would also be more costly than competing energy resources such as #6 fuel oil for industrial and utility use, and probably cost about as much as #2 distillate oil for residential and commercial users. This can be seen from the data in Table 12.9 which compare the levelized costs of Alaska gas per MCF with those of #2 and #6 fuel oil at wholesale price levels. Hence, the marketing risk for Alaskan gas is formidable, which explains the pressures by potential financing sources to shift as much of the marketing risk forward to gas consumers who would have little choice (at least in the short-run) but to pay the high cost of folded-in Alaskan gas.

The National Economic Costs of Alaska North Slope Gas

The national economic costs of Alaska's North Slope gas consist of four major components: (1) field development and production; (2) gas conditioning (i.e., reduction of the carbon dioxide content to 3% or less); (3) pipeline transport

---

[32]This adjustment is estimated to raise the wellhead price to $1.88 by 1986. *From:* Douglas B. Fried and William F. Hederman, Jr., "The Benefits of an Alaskan Natural Gas Pipeline," *The Energy Journal,* Vol. 2, No. 1, Table 1.

[33]Prudhoe crude oil prices were controlled under the now expired oil price regulations and set at a maximum of $11/bbl at the wellhead (in 1978). However, the crude had to compete in U.S. markets where imported oil delivered to Gulf Coast ports was priced at $14/bbl. Transportation costs for Alaskan oil were $7/bbl or more depending on the point of destination. As a result wellhead prices ranged from as little as $1 to not more than $7, depending on transportation costs to ultimate markets. *From:* Arlon R. Tussing and Connie C. Barlow, *An Introduction to the Gas Industry With Special Reference to the Proposed Alaska Highway Gas Pipeline,* Institute of Social and Economic Research, University of Alaska, Anchorage, Alaska, October 25, 1978, pp. 1–54.

Table 12.9. The Economic Costs of Alaska North Slope Gas. (in 1986 present values, billion US$)

COST CATEGORY	CAPITAL COSTS[1]	ANNUAL AVERAGE CAPITAL COSTS[2]	OPERATING COSTS[4]	CANADA TAXES[5]	TOTAL ANNUAL COSTS
Field development and gas conditioning	6.0	0.480	0.029	—	0.509
Alaska pipeline	21.0	1.680	0.101	—	1.789
Canada pipeline	13.0	1.040	0.062	0.210	1.312
Lower 48 pipelines	3.0	0.240	0.014	—	0.254
Totals	43.0	3.440	0.206	0.210	3.856
Average costs per 10^6 btu[3]					$4.40

[1]Source: Alaska Pacific Bancorporation, p. 15.
[2]25 year life, 7% real average rate of interest.
[3]Annual average throughput of 876 billion cu. ft./year.
[4]6% of fixed costs.
[5]2% of fixed capital in Canada.

to gas distributing companies in the lower forty-eight states, and, (4) taxes payable to Canadian authorities. Not included are the costs of finding the gas and developing the fields because these are sunk costs which are irrelevant for the economic evaluation. The latter should take account only of the long-run marginal costs of the gas, i.e., the future economic costs of supplying the gas to users. These should include user costs that reflect the depletion value of the gas. However, because of the size of the deposits and their long life expectancy as well as the high-cost characteristics of the gas itself and the uncertainty of future prices of replacement fuels (which may mean that the potential user costs are insignificant or even negative) no attempt has been made here to assess their magnitude.[34]

Cost estimates were based on the recent published cost projections in 1986 dollars for the project which assumed start-up of construction in 1982 and completion by 1986. Because no detailed information was available about applied interest rates during construction, projected inflation allowances and other factors prices in this analysis here are expressed in 1986 dollars, the year of completion.

Table 12.9 summarizes the results. Total costs of the project upon comple-

[34]See also Chapter 4.

tion in 1986 are estimated at US$43.0 billion. Applying a 7% real rate of inter-
est and a life expectancy of 25 years, the average annual economic costs of this
investment in constant 1986 prices amount to US$3.856 billion, resulting in
average capital costs of $4.40 per 10^6 btu, based on an annual average through-
put of 876 billion cubic feet.[35] Actual pipeline capital costs per unit of gas could
be somewhat lower if additional deposits were to be connected to the system.
This would extend either its economically useful life or its annual throughput
or both. Increasing throughput would increase operating costs, however,
because of the required higher pipeline pressures.

The Economic Value of Alaska Gas in The Lower Forty-Eight States

The economic value of Alaska gas delivered to the lower forty-eight states can
be measured by the long-run marginal costs of the fuels which the gas would
replace. If gas from other sources were in plentiful supply in the marketing
regions of the Alaska gas, this value would be determined by the marginal costs
of these competing gas resources. However, the lower forty-eight state gas
resources are limited. They amounted to only 165 trillion cubic feet at the end
of 1979, or about 8.5 times then current production rates. Moreover, in recent
years new finding rates have been lower than production with the result that
net, proven reserves have continued to decline. It remains to be seen whether
the sharply higher drilling activities of recent years will be sufficient to reverse
this trend.

The economic costs of imports of gas from Mexico, Canada, or in the form
of LNG from overseas could provide another measure of the value of Alaska
gas. Following established export policies of these potential suppliers, however,
the costs of such gas imports would be closely linked to—or could possibly
exceed—world fuel oil prices.[36] In any case, such imports are relatively small
at present and are projected to remain so in the foreseeable future.[37]

The ultimate replacement fuel for Alaska gas, therefore, would be imported
fuel oil. Some observers have claimed that high-priced #2 distillate fuel is the
appropriate substitute which provides the true measure of the value of Alaskan

---

[35]This is the announced design capacity of the pipeline and is equivalent to 21.9 trillion cubic
feet over 25 years, representing an estimated 75% recovery rate of the 29.2 trillion cubic foot
deposit.
[36]Both the existing Canadian and Mexican export contract prices are directly tied to eastern
seaboard fuel oil prices.
[37]Total 1979 imports amounted to 1.2 trillion cubic feet or about 6% of total U.S. consumption.
From: Department of Energy, EIA,DOE/EIA-0131(79), *Natural Gas Production and Consumption 1979*, January 1981, Washington, D.C., Table 1.

Table 12.10. Quantity of Gas Delivered to Consumers by Type of Consumer.
1980 Total USA 10^9 cu ft and percent

	RESIDENTIAL	COMMERCIAL	INDUSTRIAL	ELECTRIC UTILITIES	OTHERS	TOTAL
10^9 cu ft	4.752	2.441	7.172	3,682	170	18,216
%	26.1	13.4	39.4	20.2	0.9	100.0

Source: U.S. Dept. of Energy, EIA, Natural Gas Annual 1980, DOE/EIA-0131(80), February 1982, Table 11.

gas.[38] This claim must be rejected, however, because #2 distillate fuel is the appropriate replacement fuel only for gas users in the residential and commercial markets but not for the considerably larger industrial and utility markets. For the latter the appropriate replacement fuel would be the heavier and lower-priced #6 fuel oil which is used as boiler and furnace fuel in most industrial and utility applications. As can be seen from Table 12.10, in 1980 the latter user categories accounted for some 60% of total U.S. gas consumption as against 40% for domestic and commercial user groups.

If, for example, gas prices were set at levels equivalent to #2 distillate fuel oil, the industrial and utility gas market would quickly all but disappear. This, in turn, would more than double gas reserves relative to remaining gas consumption, with the consequence that the implicit economic value of existing gas reserves in the ground would fall by about one-third.[39]

It is clear, therefore, that Alaskan gas, at the margin, would largely replace (or supplement) industrial uses in which it would replace #6 fuel oil rather than #2 distillate oil. As can be seen from Table 12.11 below, the average U.S. wholesale price of the former in September 1981 was $4.44/per 10^6 btu, or only about 62% of the price of #2 distillate fuel.

However, the appropriate economic comparison must be made on the basis of expected 1986 to 2010 prices for #6 fuel oil. To estimate such prices is an impossible task, given the volatile nature of oil prices in recent years, the substantial difference between long-run marginal production costs and world market prices since 1973, and the uncertainties of market conditions, finding rates, and other factors that may affect world energy and particularly world oil prices. Because of these difficulties, the analysis here presents only a low and a high price projection. Both serve illustrative purposes only and should not be interpreted as actual forecasts. Table 12.11 indicates these two 1986 price ranges for fuel oils and natural gas; the lower estimate of fuel oil prices is based

---

[38]See, for example, Douglas B. Fried and William F. Hederman, Jr., p. 25.
[39]Assuming a decline in annual production to 40% of previous levels, an initial reserve/production ratio of 10 and a real rate of interest of 7%.

Table 12.11. September 1981 and Projected 1986 Alternative Fuel Costs,[1] $/ 10^6 btu.

FUEL	SEPT. 1981	1986 CONSTANT NOMINAL PRICES 1981–86	2% REAL ANNUAL AVG. INCREASE[2]
#6 Fuel oil, spot, U.S. avg.	4.44	4.44	6.52
#6 Fuel oil, spot, Pacific	3.98	3.98	5.85
#6 Fuel oil, spot, Midwest	5.09	5.09	7.48
#2 Fuel oil, U.S. avg.	7.14	7.14	10.49
#2 Fuel oil, Pacific	7.28	7.28	11.81
#2 Fuel oil, Midwest	7.28	7.28	11.81
Natural gas, firm, U.S. avg.	2.56	—	—
Natural gas, firm, Pacific	4.48	—	—
Natural gas, firm, Midwest	3.67	—	—
Natural gas, interruptible, U.S. avg.[1]	3.67	4.00	5.87
Natural gas, interruptible, Pacific[3]	4.16	4.40	5.27
Natural gas, interruptible, Midwest[3]	1.63	4.00	6.73
Coal, U.S. average	1.55	1.55	2.28

[1]Sept. 1981 prices represent the delivered average prices to U.S. electric power utilities. *Source:* U.S. Dept. of Energy, EIA, *Costs and Quality of Fuels for Electric Utility Plants,* DOE/EIA-0075(81109) Washington, D.C., Sept. 1981, 25 and 27.
[2]Assumed annual average rate of inflation 6%.
[3]1986 gas prices are assumed to reflect unregulated market price levels, with averages remaining about 10% below #6 fuel oil prices.

on the assumption that world oil prices would remain constant in nominal terms (i.e., decline at the rate of US$ inflation). Such a development would be a continuation of the declining price trends that set in in early 1981 and are continuing at the time of writing. The other reflects an assumed real price increase of 2% per year plus an underlying inflation rate of 6%, or an annual average increase in nominal terms of 8%. The projected 1986 gas prices reflect the assumption that prices for gas would be deregulated by that time and would range around a level of approximately 10% below projected #6 fuel oil prices. A somewhat lower price than #6 fuel oil appears appropriate given the lesser flexibility and longer-term nature of most gas supply contracts.

If the low price projection were to materialize, the economic value of delivered Alaskan gas would be just barely equal to the average costs of #6 fuel oil; hence the national economic net benefits of Alaskan gas would be zero, at least initially. It could become positive, however, if no major gas finds in excess of market demands (at #6 fuel oil prices) were made in the lower forty-eight, Canada, or Mexico, or if oil prices were to trend upward again towards replacement cost levels from oil shales, coal gasification or liquefaction, tar sands, or heavy oils.[40]

On the other hand, if oil prices were to resume their past increases in real terms, from a national economic point of view the development of the Alaskan gas resources for early utilization would be attractive because replacement fuel costs would exceed the long-run marginal costs of Alaskan gas. For example, if the relative prices of other fuels were to rise at a real rate of 2% per annum over the 25 year life expectancy of the project the national economic benefit/cost ratio of utilizing Alaskan gas would be equal to 1.77 (evaluated at a real interest rate of 7%).

From the foregoing it must be concluded that considerable uncertainty surrounds the net economic value of early Alaskan gas development. Net benefits, given present prices of competing fuels, may be equal to zero or even slightly negative. However, they could be significantly positive, if real world oil prices were to rise again and if no substantial additional gas deposits were to be found in more accessible locations on the continent. In either case, however, this analysis suggests that little will be lost in terms of net national economic benefits if the decision to proceed with the project is delayed further.[41] Because the creation of such net national income benefits depends basically on relative price increases of competing oil products, a strong case can be made for postponement of construction until such an increasing price trend has once again been firmly established. Given significantly higher alternative fuel cost the real value of Alaskan gas would be enhanced accordingly, making it more valuable in economic terms.

The Market Costs of Alaskan Gas

The real market costs of Alaskan gas consist of the sum of the long-run marginal supply costs of the gas plus the payments to the owners of the gas deposits and the severance taxes payable to Alaska, all evaluated at the appropriate real market rate of interest. The latter consists of the weighted sum of the regulated rate of return on equity plus the market rate of interest on borrowed funds, adjusted for inflation.[42]

Table 12.12 summarizes these costs in terms of constant 1986 prices. As can

---

[40]However, such upward price movements would not necessarily affect heavy oil prices if coal were to take over an increasing share of the industrial and utility fuel market and if sufficient light-oil petroleum fractions were available from conventional sources, so that costly hydrocracking of fuel oils would remain unattractive, resulting in a depressed market for heavy fuel oils.

[41]A different conclusion derives, of course, with respect to the specific economic benefits to the deposit owners and the state of Alaska, the two primary beneficiaries of Alaskan gas production.

[42]Here estimated to be 6% per annum.

Table 12.12. Costs to Distributing Companies of Alaska North
Slope Gas in Constant 1986 Prices. ($/10^6 btu)

COST CATEGORY	LEVELIZED COSTS	1986	1995	2010
1986 Wellhead price[1]	1.92	1.88	1.92	1.98
Severance tax[1]	0.19	0.19	0.19	0.20
Levelized capital cost charge[2]	4.67	6.51	4.69	0.00
Canadian taxes[3]	0.21	0.21	0.21	0.21
Operating expenses[3]	0.21	0.21	0.21	0.21
Total average costs	7.20	9.00	7.22	2.60

[1]*From:* Douglas B. Fried and William F. Hederman, Jr., "The Benefits of an Alaskan Natural Gas Pipeline," *The Energy Journal,* Vol. 2, No. 1, January 1981, Table 1.
[2]Based on $27 billion capital costs for the Alaska portion financed with 25% equity at the congressionally sanctioned rate of return of 20.4% (*Federal Energy Regulatory Commission Order No. 33,* "Order Setting Value for Incentive Rate of Return, Establishing Inflation Adjustment and Change in Scope Procedures, and Determining Applicable Tariff Provisions," Docket No. RM78-12, June 8, 1979) and $16 billion remaining capital costs with 25% equity at a real rate of interest of 6%. Annual average throughput, 876 bcf; 25 year life.
[3]From Table 12.9.

be seen, the levelized costs over the 25-year life expectancy of the pipeline are projected to amount to $7.20 per million btu. However, if the customary capital-cost repayment schedules for pipeline operations (which are based on straight-line depreciation schedules) are used, the initial charges per million btu in 1986 would amount to $9.00; by 1995 they would fall to $7.22 and by 2010 to $2.60.[43]

A special problem arises from the fact that present credit markets are in a highly unsettled state. Market rates of interest for long-term, high-grade corporate bonds range between 14–18% (in early 1982), while the average annual rate of inflation has fallen to less than 8%. This difference results in a historically high real rate of interest. Given large and growing federal budget deficits, it is not certain at this time whether these high costs of capital will continue to persist or will fall substantially, particularly if inflation rates keep declining. Even at the assumed real rate of 6% the capital costs of the Alaskan Pipeline Project will amount to 65% of the total costs of Alaska gas (see Table 12.9). However, if financing were to be arranged under present credit market conditions, nominal, long-term rates of 14–18% would have to be accepted. Such high rates would matter little, if inflation were to continue at rates equal to the

[43]These estimated costs are close to those most recently announced by the pipeline promoter, North-West Alaskan Pipeline Company. Its chairman testified in a hearing of the House Energy Committee on October 21, 1981 that gas from the system would cost $9.25 per thousand cubic feet the first year, with declining prices thereafter. Over the 20-year life of the project he claimed that the average price would be $4.85 in 1980 dollars. *From: Energy Users Report,* October 22, 1981, p. 2529.

Table 12.13. Total Average Costs of Alaskan Gas in
Constant 1986 Prices with Borrowing Costs of 14%
p.a. and a Long-term Average Rate of Inflation of 3%.

	LEVELIZED COSTS	1986	1995	2010
Total average Costs per 10^6 Btu	$7.88	$10.66	$8.23	$2.60

Basic data from Table 12. Assumed rate of inflation, 3% p.a., and costs
of borrowed funds, 14%.

difference in real and nominal rates of interest. Under such conditions, the
nominal value of the gas would rise at the rate of inflation and the capital costs
charged would simply reflect this fact. However, if inflation rates were to continue to decline and level-out at a much lower rate, then the present borrowing
costs would make the gas extremely expensive in real terms. This can be seen
from the illustrative data in Table 12.13, which illustrate the consequence of a
long-term average rate of inflation of 3% after 1986 and a long-term borrowing
rate of 14%. This combination would result in a real rate of interest on borrowed funds of 10.7%. It can be assumed that the congressionally sanctioned
rates of return on equity would remain unchanged. Under these conditions,
with all other costs remaining unchanged, the levelized, real costs of delivered
Alaskan gas would amount to $7.88 while initial costs in 1986 would be as
high as $10.66 per million btu.

   However, this is, hopefully, only a temporary risk of financing such a capital
intensive project at this time. If in the not-too-distant future credit market conditions and rates of inflation return to a more sustainable and reasonable relationship to each other this added risk for undertaking the Alaska Pipeline
would disappear. Nevertheless, until such conditions prevail it would probably
be unwise to finance the pipeline. Regardless of the conditions that apply to its
financing, it is clear from a comparison of the projected market costs of Alaskan gas and those of competing alternative fuels (see Table 12.11) that the
former will be substantially more expensive than either gas from alternative
sources or #6 fuel oil. Given free market conditions and the absence of regulation, Alaskan gas could not be marketed in the lower forty-eight states for
quite some time to come—at least until the costs of competing fuels have
reached the projected cost levels of delivered Alaskan gas. This would probably
not be the case until the mid-1990s or beyond.

   Alaskan gas could be marketed only if the buyers were to be willing to aver-
age-out the high cost Alaskan gas with low-cost gas from other sources. However, this would require that the resulting average price would remain competitive with the price of alternative energy sources. Such a pricing strategy

would be uneconomic and inefficient under free market conditions. However, it could be optimal in terms of profit maximation objectives for regulated utility companies. These peculiar effects of public utility regulations on company behavior is the subject of the discussion in the following sections.

Optimum Utility Marketing and Pricing Strategies Without Regulation

If the gas utility industry were unregulated, optimum marketing and pricing strategies would induce it to expand sales to the point at which marginal revenues, or price, would be equal to long-run marginal costs. This is illustrated in Figure 12.4. The available gas supply for a given gas distributing company is assumed to consist of four components: 10 billion MCF annually of "old gas" available under long-term contract at $1/MCF, 20 billion of "old gas" under long-term contract at $2/MCF, 20 billion MCF of "new gas" at $4/MCF, and an unlimited amount of Alaskan gas, synthetic gas, or imported LNG at $9/MCF. This marginal supply curve is represented by line ABCDEFGH.

Market demand is assumed to comprise three classes of customers:

(1) Existing, or "old," residential and commercial users whose appliances are all based on the use of gas; their willingness to pay is assumed to be

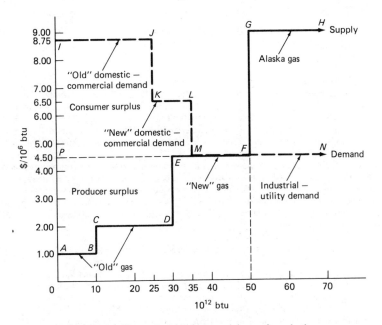

Figure 12.4.   Unregulated industry pricing and marketing.

as much as 20% higher than the assumed market price of $7.30/million btu of #2 distillate fuel oil, or $8.75/MCF. The total market demand of this consumer group is assumed to be 25 billion MCF annually.[44]

(2) New potential domestic and commercial users, for a total of 10 billion MCF, would be willing to pay as much as $6.50/MCF as an alternative to #2 fuel oil (or electricity).

(3) Industrial users or utilities. They are assumed to be willing to take up any amount of gas provided the price is not higher than the price of #6 fuel oil, assumed to be $4.50/MCF.[45]

Total market demand, then, is represented by line IJKLMN.

If the utility were restrained by regulations or other reasons to charge a uniform price for its gas to all customers[46] it would set the price equal to $4.50/MCF and limit throughput to 50 billion MCF/year. It would not purchase any of the high-priced Alaskan gas or other supplemental gas. This strategy would maximize the utility's net profit[47] which would amount to $95 million/year. Consumers, in turn, would be better off than if they had to use alternative fuels because the cost of gas at $4.50/MCF would be lower than the costs of alternatives. This would be true for the domestic and commercial users. The total "consumer surplus" accruing to these users groups would amount to $126.25 million/year. Industrial customers, which would have to pay $4.50/MCF, or exactly the price of competing #6 fuel oil, would be indifferent between using gas or oil, unless nonprice considerations such as lower pollution control costs would entice them to prefer gas.[48]

---

[44]For simplicity of exposition, the discussion is conducted in terms of megacubic feet of gas (MCF). The diagram, however, is shown in million btu. This implicitly assumes that the heat content of all available gas is equal to one million btu per MCF. This is not quite true, however. Alaska North Slope gas, for example, is expected to contain some 1.07 million btu/MCF. *From:* Arlong R. Tussing and Connie C. Barlow, *Financing of the Alaska Gas Highway Pipeline. What Is To Be Done?*, Final Report to the Alaska State Legislature, Juneau, Alaska, April 1979, p. A-18.

[45]To simplify the exposition, the discussion does not consider such added complexities as differential distribution costs to different classes of customers, the problems caused by uneven demand profiles (i.e., daily and seasonal peak loads), and the problems of long-run versus short-run supply and demand considerations, market shifts, growth, etc. While these factors would modify the optimal pricing and marketing strategies of a given utility company they would not invalidate the argument presented here.

[46]Presumably net of differential distribution costs.

[47]Net profits here are defined as the difference between the marginal costs of the gas and the price charged times the quantity of gas sold.

[48]In reality, experience in Ontario, along the Pacific Coast, and in Texas has shown that many companies prefer to use low-sulphur fuel oil when average market prices are roughly equal because of the greater flexibility offered by oil and the potential to select between competing suppliers. For example, Canadian export gas sales to the United States, which are priced on an oil-equivalent basis, have dropped to less than 50% of their authorized levels in 1981. *From:* "Recovery of Natural Gas in U.S. hurt by International Oil Surplus," *Financial Post,* April 10, 1982, p. S7.

If the utility were totally uncontrolled and could charge discriminatory prices, it would presumably raise its prices for domestic and commercial users to levels close to the costs of alternative fuels—setting them just sufficiently lower to retain its existing and attract new customers. In terms of our numerical example, this would mean a price to domestic users of about $6.50 MCF. Such a price would convert a significant amount of consumer surplus into producer surplus or company profits. However, it would not change the optimum scale of sales which would remain at 50 billion MCF/year. Whatever the distribution between consumer and producer surplus, the sum of the two would remain to be $221.25 million/per year.

In neither case would Alaska gas be purchased by the utility because its costs would exceed the supply costs of alternative fuels and, hence, the willingness of gas customers to pay its marginal costs. If the utility were to increase its sales to industrial users at the maximum ceiling price of $4.50/MCF it would incur a net loss of $9/MCF–$4.50/MCF for each additional MCF of gas sold. As a consequence total net profits would decline.

The Effects of Regulation

There are no unregulated gas utility companies of any significant size in North America or other parts of the world. The reason for this is simple. Gas utilities, just as electric utilities or water supply organizations, are natural monopolies. Because of the high costs of the fixed distribution network required, a single company can always supply the commodity in question at much lower unit costs than two or more competing organizations could. However, this natural monopoly position also gives a franchised distributor substantial market power. Therefore, such companies are usually regulated by public supervisory bodies.

The central principle of underlying regulation is that rates of profit should be held to a "fair and reasonable" rate of return on investment.[49] This, in practice, means that companies are not allowed to earn (or retain) any producer surplus. If such potential producer surplus is generated, they have to reduce their prices, generally following some form of average cost pricing rule. This has two consequences. The first is that the firms have no immediate, overriding incentive to seek out the lower-cost suppliers, as long as prices of gas are low enough to assure them a growing market vis-a-vis alternative fuels. Given the low production costs of natural gas in North America this was usually the case in the past. Even today, average gas prices in the USA are substantially lower than prices of alternative fuels.[50]

---

[49]Arlong R. Tussing and Connie C. Barlow, p. II-60.
[50]In October, 1981, for example, U.S. average energy prices paid by electric utility companies were $5.12/$10^6$btu for heavy fuel oil, $7.43/$10^6$btu for light fuel oil and only $2.99/$10^6$btu for natural gas. *From:* U.S. Department of Energy, *Electric Power Monthly,* November 1981, p. 99.

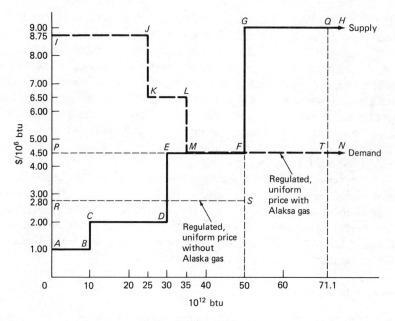

Figure 12.5.   Regulated industry pricing and marketing.

Second, to maximize profits, it is generally more important for such regulated utility companies to increase volume and sales provided this leads to an increase in the rate base (i.e., the amount of invested capital).

These effects of regulation have been illustrated in Figure 12.5 which uses the same assumptions regarding market demands and supplies as those utilized in Figure 12.4. If the utility were to refrain from purchasing Alaskan or other, high-cost supplemental gas whose marginal costs exceeds the marginal willingness to pay, it would have to limit its sales to 50 billion MCF/per year as in the unregulated case. However, if the nonregulated price of $4.50/MCF were to be charged, this would result in substantial excess profits (i.e., the $95 million of producer surplus). This would be deemed unacceptable by regulatory authorities, hence, they would force the utility to reduce its selling prices until these excess profits were eliminated. In our example, the resulting average price amounts to $2.80/MCF. However, at $2.80, substantial industrial demands for additional gas would remain unsatisfied, given the existing market supply differences between the marginal costs of alternative fuel supplies and the regulated average price of gas.[51]

---

[51]This difference was one of the major contributing factors to the recurring U.S. gas shortages during 1970s. In that period many regions, such as the Washington-Baltimore metropolitan

Because of the regulatory-induced incentive to increase sales rather than maximize net economic rent in the form of producer surplus, it becomes attractive to a gas company to purchase higher-cost gas at the margin and to "fold-in" these high-priced supplies into the average tariff structure, up to a level at which the latter approaches the marginal willingness to pay of additional customers.[52] This, in our example, results in annual sales of 71.1 billion MCF and total purchases of 20.1 MCF of gas at $9/MCF. The average annual price to all users will be $4.50/MCF. If the respective utility regulatory commission permits discriminatory pricing,[53] prices, in our example, to domestic and commercial users would be increased further to $6.50. The resulting increase in revenue from this reduction in consumer surplus could be used to subsidize additional sales of high-priced gas to low-value industrial users. In our example, such discriminatory pricing would permit the marketing of an additional 15.6 billion MCF of high-priced gas.

The economic consequence of such cost-averaging, folding-in pricing practices is that net economic returns to the economy as a whole are reduced because of the reduction of consumer surplus and the elimination of producer surplus. Net national income declines because gas is utilized whose costs at the margin far exceeds its value in use as measured by the costs of alternative, competing fuels.

In the case of Alaskan gas it is quite possible, however, that the folding-in potential will be substantially reduced in the near future. This could be the result of the proposed, complete deregulation of all wellhead prices. This would dry-up the supply of new low-cost, price-controlled gas from alternative sources. It would also substantially increase the costs of a considerable quantity of gas under those existing contracts that contain indefinite pricing provisions. These allow gas-field owners to raise prices to all customers in concert with the highest negotiated sales price from a given field.[54]

The Gold-Plating Problem

Price regulation and the dependence of utility company profits on the level of undepreciated capital investments cause another serious problem. Under these

---

areas, for example, suffered from complete prohibitions of new gas hookups, and elaborate schemes were devised to reduce or eliminate industrial demands when peak demands started to exceed available system supplies.

[52]In the Natural Gas Policy Act of 1978, Congress explicitly granted potential purchasers of Alaskan gas to "roll-in" these high-priced supplies with lower-cost, price-controlled sources.

[53]Euphemistically called "value of service" pricing in the utility regulatory jargon.

[54]Under existing legislation established in the 1978 Natural Gas Policy Act, gas from wells deeper than 15,000 feet is already decontrolled. Gradual deregulation of additional categories will result in the deregulation of almost 60% of all domestic gas supplies by 1985. Total deregulation of all gas categories is under active consideration by the Reagan Administration. *From: Energy User Report,* "Current Report," January 28, 1982, p. 93.

rules, utilities can increase their income only by increasing their rate base. They do not earn any additional profits on throughput, or total sales. This creates strong incentives to increase the amount of capital investments, or to "gold-plate" them. As Tussing and Barlow have stated:

> A regulated gas company would prefer to purchase higher cost gas that requires rate base expansion (new construction) than to purchase lower cost gas that can be transported through existing facilities.[55]

This is a particularly serious problem with respect to the Alaska Gas Pipeline, given the very high investment costs required, the relatively low costs of the gas at the wellhead, and the special, above-average rates of return on equity granted by Congress. This combination provides a powerful incentive to utilities to forego the purchase of competing gas from nearby sources which would require little additional investment but have high unregulated wellhead prices.[56] While such a choice could be optimal to profit-maximizing companies, it would not be economically efficient from a national point of view because of the resulting loss of producer surplus accruing to competing domestic suppliers.

The case would be less clear if the competing gas is imported. If delivered, imported gas prices were equivalent to delivered Alaskan gas prices, the latter would be preferable. However, if imported gas were lower priced and cost equivalent to imported #6 fuel oil, while Alaskan gas were to be substantially more costly, then national economic income maximization would call for imports rather than Alaska gas. Such gas is available now. As one Canadian source points out: "Even if the Alaska Highway line isn't built the pre-build (i.e., the pre-built sections connecting Alberta field with the U.S.) will transport surplus gas to the U.S. markets, helping to reduce the ever increasing gas bubble."[57]

Overall, it appears that no real economic justification exists to build the Alaska Gas Pipeline in the near future. If it is built anyway, its commercial viability will be entirely the consequence of utility price regulations that induce nonoptimal pricing behavior by utility companies. The resulting net losses to the economy as a whole might well be greater than any income-distribution equity gains resulting from such regulation. Regulatory pricing structures that would force utilities to charge long-run marginal costs instead would be more efficient. They could also be designed in such a way that they would be equitable as well.

---

[55]Arlong Tussing and Connie Barlow, *Marketing and Financing Supplemental Gas,* p. II-25.
[56]This assumes that these utilities would become part-owners of the Alaska pipeline system.
[57]Dunnery Best, "Rising Costs Threatening Completion of Alaska Gas Line," *The Financial Post,* April 10, 1981, p. S3.

13

Costa Rica and Brazil: The Effects of Poor Supply Quality

13.1 INTRODUCTION

In this chapter, we explore the effects of poor quality of supply on energy consumption. There are two such principal effects of particular relevance. First, as mentioned in Chapter 7, a low quality of supply for a particular form of energy will tend to reduce the demand for that energy source, especially in the longer term, as consumers turn to more reliable or convenient alternatives. Second, if the supply quality is below the expected level, then energy consumers are likely to incur unforeseen short-run costs before they are able to adapt their energy using equipment and consumption patterns to match the lower quality source.

The electricity subsector is used to illustrate quality of supply effects, because available data and models are better than in other energy subsectors. This section continues with a brief summary of some useful background concepts, especially the links between demand, optimal investment planning, and pricing policy.

In the next two sections, an explicit econometric model of electricity demand that includes quality of supply as an explanatory variable is developed and estimated for Costa Rica, to demonstrate the long-term negative impact of poor supply quality or reliability on electricity consumption. It is also shown that such quality induced shifts in the demand curve have significant policy implications for demand forecasting and measurement of the benefits of electricity consumption, especially in countries where the supply reliability is poor.[1] In the final section of this chapter, the economic costs imposed on electricity consumers due to unforeseen shortages (or outages) are described, and some quantitative estimates are summarized using data from Brazil.

In the electricity subsector, both the investment and pricing decisions are

---

[1]For a more detailed analysis of these results, see: Romesh Dias-Bandaranaike and Mohan Munasinghe, "The Demand for Electricity Services and the Quality of Supply," *The Energy Journal,* Vol. 4, January 1983,

closely interrelated (see also Chapters 4 and 8). Thus, when a power utility wishes to expand output, a demand forecast is first made for electricity consumption in the service area, assuming (sometimes implicitly) a certain future evolution of prices, incomes, and other variables that will affect electricity demand. Next, the cheapest or least-cost program of investments that will meet the future demand is determined. Both the demand forecast and investment program must be long-range (typically 10 to 20 years), because of the large and lumpy (or indivisible) investments involved as well as the significant lead times required for completing capacity additions. Finally, the price of the output is established based on the marginal cost of supply as well as financial and other constraints. For consistency, this price must agree with the expected price used in the demand forecast. In other words, the data used in the sequence of steps: assumed future price, demand forecast, least cost investment program, marginal cost and output price, are interrelated and must be mutually self-consistent.

Detailed theoretical models have shown that the net economic benefits of electricity consumption will be maximized only if optimality conditions relating to both investment (or capacity) and price are simultaneously satisfied.[2] The optimal price is the marginal cost price.[3] The optimal quality of supply is reached by improving the supply system (e.g., adding to capacity, reducing risks of failure, etc.) until the marginal cost of these improvements exactly offsets the marginal benefits realized by consumers because of the enhanced quality of supply.[4] This quality of supply also determines the optimal investment level, as well as other supply related expenditures that affect the output quality or reliability.[5]

When the supply quality changes, consumption benefits are affected in two ways. In the short-run, if the quality drops below the expected level, planned consumption is affected and consumers experience inconveniences and costs (e.g., no lighting for households or loss of output for industries, due to a power outage).[6] In the longer-run, if consumers anticipate a change in the supply

---

[2]See, for example, M. Crew and P. Kleindorfer, *Public Utility Economics,* St. Martin's Press, New York, 1979, Chapter 7.

[3]For details, see M. Munasinghe, "Principles of Modern Electricity Pricing," *Proceedings of the IEEE,* Vol. 69, March 1981, pp. 332–48.

[4]M. Munasinghe, "Optimal Electricity Supply Reliability, Pricing, and System Planning," *Energy Economics,* August 1981.

[5]The terms reliability and quality of supply are used synonymously and they encompass a variety of different aspects of electricity such as the frequency and duration of power failures (or outages), the deviation of delivered voltage from the nominal standard, deviation of mains frequency from the nominal level, and so on.

[6]See for example: M. Munasinghe and M. Gellerson, "Economic Criteria for Optimizing Power System Reliability Levels," *The Bell Journal of Economics,* Vol. 10, Spring 1979, pp. 353–365; M. Munasinghe, "Costs Incurred by Residential Electricity Consumers Due to Power Failures," *Journal of Consumer Research,* Vol. 6, March 1980, pp. 361–369; and B. von Robenau and W.

quality they will adjust their planned demand and behavior patterns. While the first aspect has been investigated in recent papers, the second has not. Next, therefore, we develop and estimate models to investigate the long-run changes in residential, commercial, and industrial electricity consumption due to variations in the quality of supply. In particular, it is shown that when the output quality is improved, the planned level of consumption will increase at any given price. This induced demand effect will shift the electricity demand curve outwards and increase the consumer surplus thus providing significant additional consumption benefits. The empirical results indicate that such induced demand shifts are of significant magnitude and will, therefore, be of critical importance in both demand forecasting and investment planning.

The demand functions derived and estimated in this study have the empirically convenient double-log form. In contrast with many earlier studies they are based on explicit utility maximizing theory. Furthermore, they are unique in that they explicitly contain the price of substitute energy as an independent variable. In addition, the methodology used in this study to derive electricity demand functions can be used to derive similar double-log functions (which explicitly include the price of substitutes) for other goods and services besides electricity.[7] This ability to extend the methodology is especially useful because many empirical demand studies use the double-log form and include the price of substitutes, without deriving this from a sound theoretical model.

13.2 SUPPLY QUALITY AND ELECTRICITY DEMAND

The dependence of demand on quality of supply will vary by type of electricity consumer. Therefore, for convenience, we divide consumers into three broad categories: residential, commercial, and industrial. Since behavioral characteristics within each group are similar, we will assume that the conditions for aggregation of individual demand functions within each category are satisfied.[8]

Empirical estimates of the demand functions (derived below) were made using data from Costa Rica. Before describing the results of these estimations, we will first discuss how the various attributes of the quality of electricity supply were incorporated into a simple 'dummy' variable.

The quality of electricity supply encompasses both voltage variations and outages. First consider voltage variations. Although the term variation has

L. Gorr, "Short Term Curtailment of Natural Gas," Ohio State University, Columbus, Ohio, August 1979 (mimeo).
[7]For example, these models may be applied to a much wider range of outputs and services such as gas, water, gasoline, mass transportation, etc., that are delivered by large centralized institutions or natural monopolies.
[8]See, for example, Henri Theil, *Linear Aggregation of Economic Relations*, North Holland Publishing Company, New York, 1955; H. A. J. Green, *Aggregation in Economic Analysis, An Introductory Survey*, Princeton University Press, New Jersey, 1964.

been used up to now the relevant voltage variation for a typical rural consumer is a voltage drop. Consumers close to substations usually have good supply voltage (that is, get service at the nominal voltage) while those further away usually have supply voltages below nominal voltage. Therefore, on average, longer lines (or feeders) usually have poorer voltage. Outages occur for a variety of different reasons, mainly due to natural occurrences like a tree branch falling across a distribution line or a truck colliding with a utility pole. The frequency of such outages for a particular line is, therefore, likely to depend on the length of the line. The durations of outages are also likely to depend on the length of a particular feeder, as the time taken to locate a fault will be greater if a feeder is longer. Therefore, in general there will be significant correlation between rural electric systems which have poor voltage on average and those which have high outage rates.

The detailed methodology described below can be used to estimate the relationship between the quality of electricity supply and the demand for electricity. This method does not require a precise quantification of the quality of electricity supply. It is only necessary to be able to distinguish between fairly broad "levels" of quality. Since it has just been argued that there is a strong correlation between voltage variation levels and outage rates we will define three broad quality of supply categories, medium, low, and very low by merely specifying voltage variation (in practice, voltage drop) levels.

Before describing these voltage drop levels in detail, consider the difference between the voltage drop level for a single consumer and that for a whole network. In general, as described earlier, all consumers being served by a particular rural electric system do not suffer the same voltage drop. Since we are only interested in a broad categorization of the quality of electricity supply, we define the quality of supply for the whole network (during any given year) based on the "average" voltage drop for the whole network as specified by a voltage drop index V, where $V = \sum_{i=1}^{n} V_i L_i / \sum_{i=1}^{n} L_i$, and L_i is the total load experiencing a voltage drop V_i percent from the nominal level (and there are n such levels V_i).

The three quality of supply levels were defined according to the voltage variation parameter V:

medium supply quality —voltage variation less than or equal to $\pm 5\%$ from nominal voltage

low supply quality   —voltage variation between $\pm 5\%$ and $\pm 10\%$ from nominal voltage

very low supply quality—voltage variation greater than or equal to ± 10 from nominal voltage

The methodology described below was used to estimate separate demand functions for the residential and commercial consumer categories for rural networks in Costa Rica. Due to the unavailability of adequate data it was not possible to make a similar estimate for industrial consumers, but some inferences may be drawn as discussed later.

Residential Electricity Demand

For residential electricity consumers in Costa Rica, where the quality of electricity supply is divided into the three categories, medium (M), low (L) and very low (VL), the demand for electricity is given by the equation:[9]

$$\ln X_{er} = C + a_L D_L + a_M D_M + t_1 \ln p_s \\ + \delta_0 \ln p_x + \delta_1 (\ln p_x)^2 + t_3 \ln Y + u \quad (13.1)$$

where
$D_L = 0$, $D_M = 0$ for very low supply quality
$D_L = 1$, $D_M = 0$ for low supply quality
$D_L = 0$, $D_M = 1$ for medium supply quality
X_{er} = domestic electricity consumption per capita per year
P_x = price of electricity
P_s = price of substitute energy (kerosene)
Y = income per capita per year
and u is a random disturbance term.

The homogeneity condition discussed in the Annex also imposes the linear restriction

$$t_1 + \delta_0 + \delta_1 \ln \mu_{p_x} + t_3 = 0 \quad (13.2)$$

where μ_{p_x} is the mean price of electricity (in the sample).

The data sample consisted of 86 observations from 9 different rural electric networks in Costa Rica, each having approximately 10 years of observations. The networks were selected after discussion with officials of the Instituto Costarricense de Electricidad (ICE), the central electricity authority in Costa Rica. They were picked such that three of them had very low supply quality, three had medium supply quality, and three had low supply quality. The determinations of the supply quality for each network for each year were made by the power engineers involved in the actual operation of the networks considered, based on the criteria described previously.

---

[9]For details, see the Annex to this chapter (equation A.13.10).

The ex post residential price charged for electricity and the per capita residential electricity demand were calculated using data from the annual publication of electricity statistics which is published by the Technical Department of the Government of Costa Rica.[10] For the sample considered the mean price of elasticity μ_{px}, was ¢0.256.[11]

Data on per capita incomes for persons in each network were not readily available. It was however possible to derive the per capita incomes by county.[12] Costa Rica which is a relatively small country (about 20,000 square miles) is divided into approximately eighty counties. The networks considered in the study typically lay in one or two counties. Therefore, if a particular network provided service in a county (or counties), the per capita income of this county (or the weighted average per capita income in the case of several counties) was taken as a proxy for the per capita income of persons being served by this network. There could be small errors resulting from this procedure as the boundaries of networks and counties do not typically coincide. These errors would not, however, lead to a systematic bias in the results.

Since pooled cross-section and time series data are used to estimate equation (13.1), a systematic time trend in the data could lead to biased results. In order to test for this possibility and to eliminate any bias, an additional set of independent dummy variables, $D_{68}, D_{63}, D_{70}, \ldots, D_{77}$, were introduced in the estimation where

$$D_{68} = D_{69} = D_{70} = D_{71} = \cdots = D_{77} = 0 \text{ in } 1968$$
$$D_{69} = 1, D_{68} = D_{70} = D_{71} = \cdots = D_{77} = 0 \text{ in } 1969, \text{ and so on.}$$

Table 13.1 shows the coefficients and their corresponding t-values obtained when the demand function (13.1) together with a set of "time dummies" was estimated with the linear restriction (13.2).

The validity of the linear restriction (13.2) was tested by reestimating the demand function (13.1) (with time dummies) without the linear restriction. The resulting coefficients and their corresponding t-values are also shown in Table 13.1. The R^2 values of the restricted and unrestricted estimates (see Table 13.1) were then used to calculate the statistic

---

[10]Several authors have shown that the use of ex post average price introduces a bias into the estimated coefficients. However, Betancourt (1981) shows that the bias is of uncertain sign and could even be zero.

[11]¢ = colones (US$1 = ¢8.5 in 1979).

[12]This was done by summing commercial, industrial, and agricultural value added by county and dividing by population. These data in turn were derived from detailed periodic national surveys conducted by the Costa Rican government.

Table 13.1. Coefficients for Residential Demand Function
(Restricted and Unrestricted).

SYMBOL	RESTRICTED[1]		UNRESTRICTED[2]	
	COEFFICIENT VALUE	t-VALUE	COEFFICIENT VALUE	t-VALUE
Coefficient of D_{69}	−0.115	−1.23	−0.097	−1.02
Coefficient of D_{70}	−0.152	−1.63	−0.107	−1.00
Coefficient of D_{71}	−0.213	−2.29	−0.376	−1.76
Coefficient of D_{72}	−0.201	−2.15	−0.358	−1.72
Coefficient of D_{73}	−0.333	−3.40	−1.534	−1.21
Coefficient of D_{74}	−0.463	−4.32	−1.534	−1.21
Coefficient of D_{75}	−0.402	−4.14	−0.885	−1.53
Coefficient of D_{76}	−0.446	−4.29	−1.084	−1.42
Coefficient of D_{77}	−0.490	−4.75	−1.073	−1.53
C	1.597	—	4.306	—
a_L	0.200	3.28	0.203	3.33
a_M	0.358	5.87	0.353	5.79
t_1	0.507	3.23	3.524	0.99
δ_0	−1.217	−9.89	−1.222	−10.17
δ_1	−0.137	−4.03	−0.139	−4.06
t_3	0.523[3]		0.514	5.29

[1] $R^2 = 0.9596$.
[2] $R^2 = 0.9600$.
[3] Calculated using the linear restriction $t_1 + \delta_0 + \delta_1 \ln (0.256) + t_3 = 0$.

$$\frac{(R^2 \text{ unrestricted} - R^2 \text{ restricted})/\text{No. of restrictions}}{(1 - R^2 \text{ unrestricted})/(\text{No. of observations} - \text{No. of independent variables})}$$
$$= \frac{(0.9600 - 0.9596)/1}{(1 - 0.966)/(86 - 6)} = 0.80.$$

It can be shown that this statistic has an F-distribution, with 1 and 80 degrees of freedom, which can be used to test the null hypothesis that the linear restriction is valid.[13] Since F = 0.8, this would mean that the null hypothesis would not be rejected even at the 99% level. This is reassuring as the restriction makes good theoretical sense.

Several observations can be made with respect to the coefficients estimated in Table 13.1. First, the estimated values of the coefficients a_L, a_m, δ_0, δ_1, and t_3 do not change by much in the restricted and unrestricted equations. In the restricted equation, unlike in the unrestricted one, the coefficients of many of the time dummies are significant at the 95% level. Finally, the coefficient representing the substitute price elasticity of demand (t_1) is significant in the

---

[13]See, G. S. Maddala, *Econometrics* McGraw-Hill, New York, 1977, pp. 194–201.

restricted equation while it is insignificant in the unrestricted one. The last observation is further reason to accept the restricted equation as the more correct one since the theory presented in the Annex indicates that the price of substitute energy plays a part in the consumer's demand decision.

Commercial Electricity Consumers

For commercial electricity consumers in Costa Rica, where the quality of electricity supply is divided into the three categories, medium (M), low (L), and very low (VL), an equation similar to (13.1) may be postulated. (See Annex.)

$$\ln X_{ec} = C + b_L D_L + b_M D_M + r_1 \ln p_s$$
$$+ \gamma_0 \ln p_x + \gamma_1 (\ln p_x)^2 + r_3 \ln V_c + u \quad (13.3)$$

where
X_{ec} = commercial electricity demand per customer
V_c = value added per commercial establishment
$D_L = 0, D_M = 0$ for very low supply quality
$D_L = 1, D_M = 0$ for low supply quality
$D_L = 0, D_M = 1$ for medium supply quality

Note that there is now no linear restriction as there is no homogeneity condition to be satisfied as in the case of residential consumers. The sample consisting of 86 observations that was used for residential consumers was also used for commercial consumers. As for residential consumers all data pertaining to electricity prices and consumption were derived from information available in the annual publication of electricity statistics. The quality of supply was the same as that used for residential consumers.

The value added per commercial establishment within each rural electric network, especially for past years, is a datum which is difficult if not impossible to obtain. Therefore, it was necessary to use a proxy variable in its place.

$$\frac{\text{Total commercial value added}}{\text{Total number of commercial establishments}}$$
$$= \frac{\frac{\text{Total commercial value added}}{\text{Total population}}}{\frac{\text{Total population}}{\text{Total number of commercial establishments}}}$$

If the ratio (total number of commercial est./total population) or the number of commercial establishments per capita is assumed to be approximately constant across networks and time, it is possible to use (total commercial value added/total population) or commercial value added per capita as a proxy variable for V_c. Again, as for residential consumers, the commercial value added per capita for the county or counties within which the network provides service was used as the value for the network. Finally, as for residential consumers, a set of time dummies, D_{68}, D_{69}, ... D_{73}, was introduced to accommodate any systematic time trend in the data.

Table 13.2 shows the coefficients and t-values obtained when the demand function (13.3) together with a set of time dummies was estimated.

Industrial Electricity Demand Estimation

As already mentioned, it was not possible to estimate an industrial demand function similar to the residential and commercial demand functions, for the following reasons. The number of industries in a typical rural network, unlike the number of commercial establishments or the number of residential consumers, was very small. Some rural networks did not have any industrial consumers. As a result of this the variation in the available industrial use data was extremely large. Another problem was that unlike residential and commercial consumers who were scattered fairly uniformly across a county, industrial consumers (whose number was very small) could be concentrated in particular areas. As a result, it was not possible to use county values as proxies for the corresponding network values.

Table 13.2. Coefficients for Commercial Demand Function.

SYMBOL	COEFFICIENT[1] VALUE	t-VALUE
Coefficient of D_{69}	0.057	0.24
Coefficient of D_{70}	−0.175	−0.65
Coefficient of D_{71}	−0.987	−1.85
Coefficient of D_{72}	−0.997	−1.92
Coefficient of D_{73}	−3.234	−1.53
Coefficient of D_{74}	−4.652	−1.47
Coefficient of D_{75}	−2.275	−1.58
Coefficient of D_{76}	−2.287	−1.20
Coefficient of D_{77}	−2.687	−1.54
C	14.100	—
b_L	−0.181	−0.92
b_M	0.542	3.11
r_1	11.750	1.32
γ_0	−1.552	−5.17
γ_1	−0.302	−3.60
r_3	0.450	2.46

[1]$R^2 = 0.7650$

13.3   POLICY IMPLICATIONS OF SUPPLY QUALITY EFFECTS

The results of the econometric estimations described in section 13.2 may be used in two principal ways:

  (i) for demand forecasting; and
  (ii) for estimating the increase in consumption benefits when the quality of supply improves.

Demand Forecasting

Unlike in traditional demand functions, the effects of alternate qualities of electricity supply have been included in the demand functions estimated in this study. This enables the analyst to make not one, but three separate demand forecasts corresponding to systems that provide electricity at very low, low, and medium quality. These forecasts can be made as follows.

Suppose the existing system in the region for which a demand forecast is to be made provides electricity of medium quality. Select this quality as that corresponding to the base case. For this case, $D_L = 0$ and $D_M = 1$ in equation (13.1) for residential consumers. The coefficients estimated in Table 13.1 (in the restricted case), equation (13.1), and projected future values of real incomes and energy prices can then be used to forecast electricity demand in the traditional manner. This forecast would be the one corresponding to a system providing electricity of medium quality. If necessary, as is often done in traditional demand forecasting, another demand forecasting technique, such as a time trend analysis, can be used to derive an alternate demand forecast for the same medium quality system. Finally, discussions with persons familiar with the system under consideration and the two separate forecasts can all be incorporated to yield a single most reasonable forecast for a medium quality system.

The estimated coefficients for D_L and D_M can then be used to make demand forecasts corresponding to systems providing electricity of low and very low qualities, respectively, assuming that other exogenous variables such as income and the price of electricity do not change. From (13.1) we can show that

$$X_{er}^L = e^{-(a_M - a_L)} X_{er}^M$$

and

$$X_{er}^{VL} = e^{-a_M} X_{er}^M$$

where X_{er}^M, X_{er}^L, and X_{er}^{VL} represent the per capita residential demands (in any given year) corresponding to top medium, low, and very low supply quality systems, respectively.

If we assume that changing the quality of supply does not significantly affect the number of consumers (N_r) in any given year,[14] we can write

$$E_{res}^L = N_r X_{er}^L = e^{-(a_M - a_L)} N_r X_{er}^M = e^{-(a_M - a_L)} E_{res}^M$$

and

$$E_{res}^{VL} = N_r X_{er}^{VL} = e^{-a_M} N_r X_{er}^M = e^{-a_M} E_{res}^M$$

where E_{res}^M, E_{res}^L, and E_{res}^{VL} represent the total residential demands (in any given year) corresponding to medium, low, and very low qualities of supply, respectively.

Using equation 13.3, it is possible to derive similar expressions for commercial consumers.

That is,

$$E_{com}^L = e^{-(b_M - b_L)} E_{com}^M$$

and

$$E_{com}^{VL} = e^{-b_M} E_{com}^M$$

where E_{com}^M, E_{com}^L, and E_{com}^{VL}, respectively, represent the total commercial demand for systems of medium, low, and very low supply qualities.

Using these equations, and the coefficients estimated for the residential and commercial consumer demand functions, we have[15]

---

[14]N_r depends on several factors. The most important among these are the price of electricity, connection charges (charges for initially connecting on to the electricity supply, which may be quite high for rural consumers), the total population in the region covered by the network, the per capita income of the population in the region, and the electricity authority's (or government's) policies for promoting the use of electricity. It is argued that in the context of rural electrification in the LDC's all of these variables are exogenous to our model and outside the control of the planner. It is conceivable that the quality of electricity supply could also affect N_r, a higher quality leading to more consumers. This effect is likely to be quite small compared to the combined effect of all of the other exogenous variables and it will thus be difficult to measure. We will therefore neglect this effect, recognizing that by doing so we will be understating the difference in total demand resulting from an improvement in the quality of electricity supply.

[15]Since b_L (in Table 13.2) was not significantly different from zero it was set equal to zero in the calculations presented below.

$$E_{res}^L = e^{-(.358-.200)}E_{res}^M = 0.85E_{res}^M \tag{13.4}$$
$$E_{res}^{VL} = e^{-.358}E_{res}^M = 0.70E_{res}^M \tag{13.5}$$
$$E_{com}^L = e^{-.542}E_{com}^M = 0.58E_{com}^M \tag{13.6}$$
$$E_{com}^{VL} = e^{-.542}E_{com}^M = 0.58E_{com}^M \tag{13.7}$$

The above conversion factors and the projected demand corresponding to medium quality supply can be used to calculate residential and commercial energy demands corresponding to systems of low and very low qualities.

Equation (13.4) shows that total residential demand when the quality of supply is low is 85 percent of the demand when the quality of supply is medium. Similarly, equation (13.5) shows that when the quality of supply is very low residential consumption is only 70 percent of the energy consumption when the quality is medium. In contrast, for commercial consumers, equations (13.6) and (13.7) indicate that total demand for both low and very low quality systems is 58 percent of demand when the quality is medium.

Estimation of Differential Benefits When the Quality of Supply Changes

As mentioned at the start of this section, the demand functions estimated in section 13.2 can also be used to calculate changes in consumption benefits that occur as a result of demand changes. To do this, consider Figure 13.1, which depicts residential electricity demand curves X_{er}^M and X_{er}^L corresponding to systems of medium and low supply qualities, respectively. Suppose $\overline{p}$ is the price of electricity and the corresponding demands for medium and low quality systems are $\overline{X}_{er}^M$ and $\overline{X}_{er}^L$, respectively.

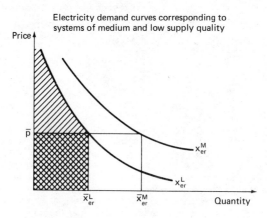

Figure 13.1.   Electricity demand curves corresponding to systems of medium and low supply quality.

Total consumption benefits when the quality of supply is low will then be represented by the "shaded" area in Figure 13.1. Of this amount, the "cross-hatched" area will represent the consumers electricity bill or the total revenue to the electricity producers. The remaining portion of total benefits is the consumers' surplus.

As shown in the Annex, total benefits are given by:[16]

$$TB_M = (e^{a_M - a_L}) \cdot TB_L$$
$$TB_M = e^{a_M} \cdot TB_{VL}$$
$$TB_M^C = (e^{a_M - a_L}) \cdot TB_L^C$$

and
$$TB_M^C = e^{a_M} \cdot TB_{VL}^C$$

where TB_M, TM_L, and TB_{VL} represent total benefits to residential consumers when the quality of supply is medium, low, and very low, and TB_M^C, TB_L^C and TB_{VL}^C represent total benefits to commercial consumers when the supply quality is medium, low, and very low, respectively.

Using equation (13.8) to (13.11) and the residential and commercial coefficients estimated in section 13.2, we have

$$TB_M = (e^{.358 - .200}) TB_L = 1.176 TB_L$$
$$TB_M = e^{.358} TB_{VL} = 1.429 TB_{VL}$$
$$TB_M^C = e^{.542} TB_L^C = 1.719 TB_L^C$$
$$TB_M^C = e^{.542} TB_{VL}^C = 1.719 TB_{VL}^C$$

Summary

The rapidly rising costs of energy supply have caused more attention to be paid to the economics of electric power in recent years. This study has focused on an aspect that has not been previously explored explicitly in the literature; that is, the relationship between the quality of supply or output reliability and the demand for electricity services.

Unique double-log demand functions were explicitly derived for residential and commercial (and industrial) consumers, starting from the microfoundations of consumer utility maximizing and producer cost minimizing behavior, respectively. In particular, electricity demand was expressed as a function of the supply quality, own price, price of substitutes, and some measure of economic activity such as household income or commercial value added. The basic

---

[16]Note that $e^{g(RM)}/e^{g(RL)} = e^{a_M}/e^{a_L} = e^{a_M - a_L}$

(excluding the introduction of the quality of supply variable) formulation of the models permits much wider application to other outputs or services such as natural gas, gasoline, water, and mass transportation, supplied by large centralized institutions or natural monopolies.

These theoretical models were estimated using pooled cross-section and time series data for several rural electric systems in Costa Rica. Good fits were obtained with values of corrected R squared of 0.96 and 0.76 for residential and commercial consumers, respectively. The coefficients of the key variables were significant at least at the 95% level, and had the signs expected from theoretical considerations. For the residential data, a linear restriction that imposed the homogeneity of degree zero condition on the price and income variables was accepted. The coefficients in this regression also made better economic sense than those in the corresponding unrestricted estimation.

In addition to the uniqueness of the theoretical formulation and wider potential applicability to other sectors, from a practical viewpoint the results of the study are important for forecasting and investment planning in two principal ways. First, they indicate that the quality of supply could have a significant impact on planned electricity consumption, and thus must be taken into account in demand forecasting. For example, ceteris paribus, residential consumers who faced low and very low qualities of electricity supply in Costa Rica were found to consume only 85% and 70%, respectively, of the electricity they would have consumed if the supply quality was medium. For commercial consumers, electricity demand was reduced to 58% when the supply quality was lowered from the medium to the low level. However, very little further cutback in demand was observed with further deterioration of reliability.

The second important practical consideration is that shifts in the planned demand curve due to changes in reliability imply that the benefits of electricity consumption (represented by the area under the demand curve) must also change. For example, in the Costa Rica case, the total benefits of electricity consumption of households at the medium reliability level were 1.176 and 1.429 times the total benefits at low and very low qualities of supply, respectively. Similarly, the total benefits at medium reliability for commercial electricity users was 1.719 the corresponding value at the low and very low reliability levels.

These results are particularly important in investment planning where the determination of optimal power system reliability levels require trading off the increased costs of strengthening the system (to improve the quality of supply), against the increased benefits of consumption to users. In brief, when system planners design systems that deliver electricity at different supply qualities, they should take into account both the changes in the demand forecast that this implies, as well as the corresponding changes in the benefits derived from consuming the output.

13.4   SUPPLY QUALITY AND OUTAGE COSTS

Reliability is an important aspect of the quality of electricity supply. The concept of reliability in engineering may be simply defined in terms of the probability that a component, or a system, will perform its intended functions satisfactorily, over some period of time, subject to actual operating conditions. The purpose of an electric power system is to supply power to consumers at some required rate, at the time and place of their choosing, while maintaining acceptable voltage and frequency levels, i.e., lying within specified limits.

Thus, an ideal electric power system that unfailingly supplies power to consumers, whenever required, is by definition a perfectly reliable one, and conversely, a system that is never able to deliver electricity to users could be termed totally unreliable. All real world power systems lie between these two extremes, and furthermore, as the system reliability level is improved, there is an inherent trade-off between the increased costs of supply, and the reduction in the inconvenience and costs imposed on consumers due to power shortages. Therefore, it is important to develop criteria and methods of assessing and ranking systems according to reliability level.

As in the case of any other good or service, electric power shortages occur when the demand exceeds the supply, due to the failure of planners to correctly predict all the uncertainties in supply and demand. The stochastic nature of demand manifests itself through unforeseen increases in the load level, e.g., the sudden buildup in the air conditioning load due to unusually warm weather. Similarly, the randomness of supply is characterized by the unexpected failure or outage of the various components which make up a power system, the unavailability of water for hydroelectric generation, and so on.

From the consumers' viewpoint, power shortages manifest themselves in a number of ways, including complete interruptions of supply (or blackouts), frequency and voltage-reductions (or brownouts), and instability effects such as erratic frequency fluctuations and power surges. While all these phenomena are likely to inconvenience and impose costs on electricity users, the consequences of supply interruptions are the most severe, and probably also the easiest to define. For example, a blackout will disrupt productive activity, and prevent the enjoyment of leisure, whereas a voltage reduction may have lesser effects such as the inconvenience caused by the dimming of lights and so on. Mains frequency variations affect a power system's own automatic control equipment, as well as other synchronous devices such as electric clocks, while power surges could also cause some damage to equipment, but these consequences are difficult to isolate. Therefore, although the term outage cost (which is used throughout) is meant to encompass the expenses incurred by consumers due to all types of shortages, in practice, the principle emphasis will be on the impact of supply interruptions. Moreover, the effects of random or unexpected interruptions will be stressed, since they are likely to impose much greater hardships on users than known or planned power cuts.

Within the economics literature there appear to be two broad approaches to measuring outage costs. One method is to estimate these costs on the basis of observed (or estimated) willingness-to-pay for planned electricity consumption.[17] The other approach, which is the one discussed below, estimates such costs in terms of the effect of outages on the production of various goods and services.[18]

As described earlier, attempts to estimate outage costs according to willingness-to-pay are actually part of a more general literature on optimal pricing for public utilities under conditions of uncertainty, i.e., stochastic demand and supply. In this approach, the variable to be maximized is net welfare which is generally set equal to the expected area under the demand curve corresponding to planned electricity consumption, minus the sum of the expected costs of supplying electricity, plus the expected costs (if any) of rationing available electricity among consumers, if an outage occurs. Thus outage costs are measured by the expected reduction in net welfare, or the amount those who are deprived of electricity would be willing-to-pay for it, minus the costs saved by not supplying it.

Studies that estimate outage costs on the basis of willingness-to-pay assume that electricity provides direct satisfaction to its consumers. Outage costs are therefore estimated in terms of lost consumers' surplus. Other studies that estimate outage costs in terms of the effect of outages on various types of production do not make this assumption. Instead, they treat electricity as an intermediate input used to produce various goods and services that provide consumers with satisfaction, rather than as a final good, which itself provides satisfaction to consumers. Outage costs are therefore measured primarily by the costs to society of outputs not produced because of outages.

The studies that estimate outage costs on the basis of willingness-to-pay have several important shortcomings: (i) it seems clear that observed willingness-to-pay for *planned* electricity consumption at a given reliability level (as analyzed

---

[17]See G. Brown, Jr. and M. B. Johnson, "Public Utility Pricing and Output Under Risk," *American Economic Review*, March 1969, pp. 119–128; R. Sherman and M. Visscher, "Second Best Pricing with Stochastic Demand" *American Economic Review*, March 1978, pp. 41–53; and M. A. Crew and P. R. Kleindorfer, "Reliability and Public Utility Pricing," *American Economic Review*, March 1978, pp. 31–40.

[18]See M. L. Telson, "The Economics of Alternative Levels of Reliability for Electric Generating Systems," *The Bell Journal of Economics*, Vol. 6, Fall 1975, pp. 679–694; R. Turvey and D. Anderson, *Electricity Economics*, Johns Hopkins University Press, Baltimore, 1977, Chap. 14; M. Munasinghe, "The Costs Incurred by Residential Electricity Consumers Due to Power Failure," *Journal of Consumer Research*, March 1980; and M. Munasinghe and M. Gellerson, "Economic Criteria for Optimizing Power System Reliability Levels," *The Bell Journal of Economics*, Spring 1979, pp. 353–65.

earlier in this chapter) is not a satisfactory indicator of what one would be willing-to-pay to avoid an *unplanned* outage. Such an unplanned outage is apt to disrupt activities which are complementary with electricity consumption, and therefore, actual short-run outage costs may be greatly in excess of observed (long-run) willingness-to-pay; and (ii) these studies measure losses of consumers' surplus on the basis of the assumption that load shedding takes place according to willingness-to-pay, i.e., those with the lowest willingness-to-pay are the first to have their electricity cut off. Thus lost consumers' surplus is the triangle-shaped area defined by the downward sloping demand curve, the horizontal price line, and the vertical line representing the quantity of electricity supplied in spite of the outage. However, in many instances it is not reasonable to assume that load shedding takes place according to willingness-to-pay. For example, if the outage is caused by a distribution system failure (rather than a generation system failure), the ability to carry out load shedding according to some predetermined order may be severely limited. If such is the case, the resulting loss in consumers' surplus is a trapezoid-shaped area defined by the short-run demand curve, the price line, and the intramarginal units of electricity not supplied. The implication of both the above arguments is that actual outage costs will exceed outage costs estimated on the basis of the marginal consumers' surplus lost. Finally, with this approach, the empirical estimation of outage costs requires exact knowledge of various consumers' short-run demand functions for electricity. Given the problems associated with empirically estimating such demand functions (see Chapter 7), it is likely that any estimates of outage costs obtained in this fashion will be subject to considerable error. Given these criticisms, outage costs would be more appropriately measured in terms of the effects of outages on various kinds of productive activity.

Production is a process in which capital and labor are combined with other inputs such as raw materials and intermediate products to produce a time stream of outputs. Under conditions approximating those of perfect competition, the net social benefit of a marginal unit of output in a given time period equals the value of the ouput minus the value of inputs. For intramarginal units of output, producers' and consumers' surplus must be included when measuring the net benefits of production. When market distortions are present in the economy, appropriate shadow prices have to be used to value inputs and outputs. The net social benefit resulting from a time stream of marginal outputs equals the present value of the resulting stream of net social benefits. Thus the opportunity cost of supplying electricity with less-than-perfect reliability can be measured in terms of the resulting reduction in the present value of this stream of net social benefits.

When an outage disrupts production, the net benefits derived from such activities are reduced, i.e., direct outage costs are incurred, since the costs of inputs are increased, and/or the value of outputs is reduced. Specifically, an outage can cause raw materials, intermediate products, or final outputs to spoil, and it can also result in productive factors being made idle. The spoilage effect leads to an opportunity cost equal to the value of the final product not being made available as a result of the outage, minus the value of additional inputs not used because the final product was not produced. However, if the value of the output is not easily determined, as in the case of household and public sector outputs which are not directly sold on the market, then it is necessary to use the cost of producing the spoiled product or output as a minimum estimate of the resulting cost of the outage.

Case Study

A case study was carried out to empirically estimate outage costs for different types of electricity consumers in Cascavel, Brazil—a city of about 100,000 persons (in 1977), experiencing rapid growth and industrialization.

For residential consumers, during an outage, electricity-dependent housekeeping chores could be effectively rescheduled without much inconvenience, and cooking activity was not disrupted since it is done almost entirely by gas in Cascavel. However, leisure activities were significantly affected by outages since the enjoyment of leisure in most households was constrained to occur over a relatively fixed period of time in the evening, especially for wage earners, and the use of electricity could be considered essential to the enjoyment of certain leisure activities (e.g., TV watching, reading, dining, etc.) during these night time hours.

The results of a survey of residential consumers and an analysis of their outage costs confirmed the result of a detailed theoretical model presented elsewhere,[19] and showed that: (i) the chief impact of unexpected outages on electricity using households was the loss of a critical 90 minute period of leisure during the evening hours when electricity was considered essential, whereas domestic activities that were interrupted during the daytime could be rescheduled with relatively little inconvenience; and (ii) over this 1½ hour period, the monetary value of lost leisure could be measured in terms of the net wage or income earning rate of affected households, as confirmed by their short-term willingness-to-pay to avoid outages. Estimated residential outage costs in 1977 were in the range of US\$1.30 to 2.00 per kilowatt-hour lost or not consumed due to outages. The principal advantage of this method for estimating the leisure costs of outages to residential consumers was its reliance on relatively easy-to-obtain income data.

In general industrial consumers suffer outage costs because materials and

---

[19]M. Munasinghe, *J. Consumer Research, op. cit.*

Table 13.3. Industrial Outage Cost Functions in Brazil (1977).

INDUSTRY	OC FUNCTION TYPE[2]	ESTIMATED COEFFICIENTS[1]		REMARKS[3]
		$(\times 10^{-4})^1$	$(\times 10^{-4})^2$	
1. Mechanical and metallurgy	linear	0.0013	3.9	D
2. Nonmetallic minerals	linear	0.077	3.5	D
3. Wood	linear	0.94	3.9	D
4. Vegetable oils	piecewise-linear	0.037	2.1	$d < 0.5$ hours, A
		3.6	1.9	$d \geq 0.5$ hours, A
5. Food and beverages	piecewise-linear	1.22	2.30	$d < 0.5$ hours, D
		7.07	2.02	$d \geq 0.5$ hours, D
	linear	0.66	0.72	N
6. Other	log	7.2	2.9	$d \geq 0.01$ hours, D
7. Telephone	linear	0.002	1.1	A
8. Water treatment	linear	-0.27	1.1	$d \geq 2$ hours, D

[1] $\overline{R}^2 > 0.95$ for all estimations.
[2] Linear and log refer to functions of the type $C/Q = (a + b \cdot d)$ and $C/Q = (A + b \cdot \log_{10}d)$, respectively, where C = outage cost, Q = annual value added, and d = outage duration (hours).
[3] D = daytime operation; N = night-time operation; A = 24-hour operation.

products are spoiled, and normal production cannot take place; the disrupted production results in an opportunity cost in the form of idle capital and labor, both during the outage and any restart period following the outage. If there is slack capacity, some of the lost value added may be recovered by using this productive capacity more intensively during normal working hours. In addition, the firm may operate overtime to make up lost production. Based on these considerations, a survey of the 20 principal industrial users of electricity in Cascavel was made to determine outage costs for outages of various durations (i.e., 1 minute to 5 hours). Mathematical functions were estimated giving the relationship between outage cost per unit of value added, and outage duration, as summarized in Table 13.3.[20] The results of the analysis indicated that there were wide variations in the effects of outages on industrial consumers, e.g.,

[20]For further details, see M. Munasinghe and M. Gellerson.

US$1–7 per kWh lost, depending on the type of industry, the duration of the outage, and the time of day during which it occurred. This approach was helpful to rank industries in terms of sensitivity to outages, e.g., for emergency load shedding purposes.

An outage which affects public illumination imposes a cost in the form of foregone community benefits such as security, improved motoring safety, etc. One can argue that foregone benefits are worth at least as much as the net supply cost which the community would have incurred for public illumination during the outage periods, e.g., the annuitized value of capital equipment and routine maintenance expenditures—electricity costs are not included since they are not incurred during outages. Two hospitals (80 beds and 200 beds) were surveyed to estimate the opportunity costs of both productive factors which are made idle (e.g., electricity using equipment, labor, etc.) and intermediate products, such as blood and medicines, which might spoil because of outages. The principal outage costs of US¢5.5 per hospital bed per hour of outage were found to occur during the night period (i.e., 1900–0600 hours), due to idle labor and capital. Estimating the outage costs resulting from possible loss of life is a task exceeding the scope of this study, and therefore, such costs are not considered here. The existence of standby batteries for the intensive care and surgical equipment suggests that death will be avoided in most cases; the cost of these batteries is very small.

Outage costs for government offices and commercial customers were found to be minimal, because in most cases reliance on electricity using equipment such as calculators and xerox machines was small, thus permitting work to continue by daylight. Furthermore, there was sufficient slack during the normal hours of work for jobs delayed by any outage to be made up. Supermarkets and hotels reported minor amounts of spoilage for long outages, i.e., over five hours; however, such outages are extremely rare. Rural consumers in the vicinity of Cascavel could be neglected since their energy consumption was less than 2% of the total.

The residential outage costs of US$1.3 to 2.0 kWh lost (depending on the duration) in Cascavel may be compared with corresponding results of other studies: US$0.4 to 0.7 (Sweden, 1948); 0.7 to 1.5 (Sweden, 1969); and 0.5 to 1.5 (England, 1975). Similarly Cascavel industrial outage costs of US$1 to 7 per kWh lost correspond to values of US$1 to 2 (Sweden, 1948); 0.1 to 3 (Sweden, 1969); 0.2 to 8 (Chile, 1973); and 1 to 9 (Canada, 1976). (All values in 1977 dollars.)[21]

---

[21]For details, see M. Munasinghe, *The Economics of Power System Reliability and Planning,* Johns Hopkins Press, Baltimore, 1979, Appendix E.

ANNEX TO CHAPTER 13
THEORETICAL ELECTRICITY DEMAND MODELS

RESIDENTIAL ELECTRICITY DEMAND

Consider a residential consumer of electricity.

Let $U = U(B,N)$ be the consumer's utility function

where B = quantity consumed of all goods and services except energy

and N = total quantity of energy services consumed, that is, from both electric
energy services, E, and substitute energy services, S.

In general $N = N(E,S)$ which says that energy services are a function of electricity
services and substitute energy services.

The consumer's budget constraint may be written as

$$Y = p_b B + p_e E + p_s S$$

where

Y = income

p_b = price of all goods and services except energy

p_e = effective price of electric energy services

p_s = effective price of substitute energy services

The consumer's utility maximizing problem can thus be written as

$$\text{Max. } L = U[B,N(E,S)] + \lambda(Y - p_b B - p_e E - p_s S)$$

where

λ = an appropriate Lagrange multiplier.

The first order conditions reduce to

$$\frac{(\partial U/\partial B)}{p_b} = \frac{(\partial U/\partial N) \cdot (\partial N/\partial S)}{p_s} \tag{A.13.1}$$

$$\frac{(\partial N/\partial S)}{(\partial N/\partial E)} = \frac{p_s}{p_e} \tag{A.13.2}$$

Suppose the consumer's utility function can be written in the log-linear form $U = B^{\alpha_1} N^{\alpha_2}$ where α_1 and α_2 are constants.[1] Condition (A.13.1) then reduces to $N/(\partial N/\partial S) = (\alpha_2/\alpha_1) \cdot (p_b B/p_s)$.

---

[1]This form of utility function usually has certain restrictive implications for the income and price elasticities of the terms that appear in the function. Specifically, both B and N should have income elasticities of 1 and price elasticities of -1. In this instance, however, due to the two tier nature of the formulation it can be shown that this is no longer necessary.

If expenditure on energy forms a relatively small share of the consumer's budget p_bB is approximately equal to Y.

Therefore

$$N/(\partial N/\partial S) = (\alpha_2/\alpha_1) \cdot (Y/p_s) \qquad (A.13.3)$$

Suppose $N = \exp(S^\beta E^\gamma)$ where β and γ are parameters at our disposal. This function has sufficient generality while still allowing to proceed with the mathematical analysis. Substituting for the partial derivatives of N in equations (A.13.2) and (A.13.3) and eliminating S gives

$$E = [(\alpha_1/\alpha_2) \cdot (1/\beta) \cdot (\gamma/\beta)^{\beta-1}]^{1/\gamma+\beta-1} p_s^{\beta/\gamma+\beta-1} p_e^{1-\beta/\gamma+\beta-1} Y^{-1/\gamma+\beta-1}$$

i.e.,

$$E = K p_s^{t_1} p_e^{t_2} Y^{t_3} \qquad (A.13.4)$$

where K, t_1, t_2, and t_3 depend on α_1, α_2, β, and γ.

Note that $t_1 + t_2 + t_3 = 0$, which is a linear restriction on the coefficients. t_1 and t_2 are price elasticities of demand for electricity services, while t_3 is the income elasticity of demand for electricity services. Assuming no money illusion, or that E is homogeneous of degree zero in p_s, p_e, and Y, the condition $t_1 + t_2 + t_3 = 0$ is the familiar elasticity condition derived in consumer theory by a straightforward application of Euler's theorem to a general demand function for E.

E and S have been defined as the quantities of electricity and substitute energy services, respectively. For substitute energy we will simply assume that the quantity of service is directly proportional to the quantity of such energy consumed. The constant of proportionality may be set to one without loss of generality. With this assumption, p_s can be reinterpreted as simply the price of substitute energy. For electricity services, however, the notion of quality of supply enters into the argument. Assume that electricity services depend on the actual quantity of electricity consumed, X_{er} (measured in kilowatt hours), and on the quality of electricity supply, R. The subscripts e and r refer to electricity and residential consumers, respectively. The simplest reasonable form of E is $E = f(R) \cdot X_{er}$, which says that electricity services are directly proportional to the quantity of electricity consumed and that the constant of proportionality depends on the quality of supply.

Variations in R affect the effective price of electricity services P_e according to the relationship $p_e E = p_x X_{er} = p_x \cdot [E/f(R)]$, where p_x is the price of electricity (per kilowatt hour). Solving for p_e and substituting for p_e and E in (A.13.4) we have

$$X_{er} = = h(R) p_s^{t_1} p_x^{t_2} y^{t_3} \qquad \text{where } h(R) = K f(R)^{-1-t_2}$$

This equation relates demand and the quality of supply. As discussed in the introduction, when demand changes with changes in quality, consumption benefits will also change. In order to estimate these consumption benefits it is necessary to calculate the area under the (compensated) demand curve for electricity up to the point p_x, the actual price of electricity. In this range, p_x will vary from p_x to infinity. This reveals a weakness in the above demand function. If $-1 < t_2 < 0$ the area representing total benefits becomes infinite.

One possible way to avoid this weakness, while making the demand function more realistic at the same time, is to allow the own price elasticity of demand (t_2) to vary.[2] It is unreasonable to assume that t_2 remains constant over a wide range of price variations. One simple assumption about this elasticity may be represented by $t_2 = \delta_0 + \delta_1 \ln p_x$. Since we expect $t_2 < 0$, and since demand for electricity is likely to become more elastic as p_x increases, we expect $\delta_0, \delta_1 < 0$.

Substituting for t_2 in the expression for X_{er} we have

$$X_{er} = h(R)p_s^{t_1}p_x^{\delta_0 + \delta_1 \ln p_x}Y^{+3}$$

If $g(R) = \ln [h(R)]$, taking logarithms of both sides we have

$$\ln X_{er} = g(R) + t_1 \ln p_s + \delta_0 \ln p_x + \delta_1(\ln p_x)^2 + t_3 \ln Y \quad \text{(A.13.5)}$$

One difficulty associated with allowing the price elasticity of demand to vary in the double-log formulation is that the demand equation is no longer homogenous of degree zero. The homogeneity of the equation depends on the condition $t_1 + t_2 + t_3 = 0$ or in this case $t_1 + \delta_0 + \delta_1 \ln p_x + t_3 = 0$. Since p_x can vary for any estimated values of the coefficients t_1, δ_0, δ_1, and t_3, the left hand side of this equation cannot always be zero unless $\delta_1 = 0$. If $\delta_1 = 0$, the price elasticity would no longer be variable. Postulating that some other variable besides p_x causes the variation in price elasticity will not eliminate the inconsistency. The left hand side of the equation cannot be zero for different values of this variable unless the coefficient associated with this variable is zero and $t_1 + \delta_0 + t_3 = 0$.

Although a demand function estimated using the functional form (A.13.5) is not homogeneous of degree zero in income and prices it is still possible to make this condition hold at least approximately. This may be done during estimation by placing the linear restriction $t_1 + \delta_0 + \delta_1 \ln \mu_{px} + t_3 = 0$ on the coefficients of the model, where μ_{px} is the mean price of electricity in the sample. This restriction constrains the demand function to be homogeneous of degree zero at the mean price.

INDUSTRIAL AND COMMERCIAL ELECTRICITY DEMAND

Consider an industrial or commercial firm consuming electricity and other types of energy. Let $O = O(J,N)$ be the firm's production function where N is the total quantity of energy services consumed, from both electric energy services, E, and substitute energy services, S, and J is the quantity consumed of all other inputs.

---

[2]The use of variable elasticities in the double log formulation of demand has been suggested by other writers. For example, see, Roger R. Betancourt, "An Econometric Analysis of Peak Electricity Demand in the Short Run," *Energy Economics,* January 1981, pp. 14–29.

As for residential consumers, suppose $N = N(E,S)$.

The producers' optimizing problem can be stated as one of minimizing the cost of producing some quantity of output $\overline{O}$ and may be written in Lagrangian form as

$$\text{Min.} L = p_e E + p_s S + p_j J + \lambda [\overline{O} - O\{J,N(E,S)\}]$$

where
p_e = effective price of electric energy services
p_s = effective price of substitute energy services
p_j = price of other inputs
λ = a Lagrange multiplier
The first order conditions reduce to

$$\frac{(\partial O/\partial J)}{p_j} = \frac{(\partial O/\partial N) \cdot (\partial N/\partial S)}{p_s} \qquad \text{(A.13.6)}$$

and

$$\frac{(\partial N/\partial S)}{(\partial N/\partial E)} = \frac{p_s}{p_e} \qquad \text{(A.13.7)}$$

Suppose that the production function can be written in Cobb-Douglas form $0 = J^{\beta_1} N^{\beta_2}$ where β_1 and β_2 are constants.

Condition (A.13.6) then reduces to $N/(\partial N/\partial S) = (\beta_2/(\beta_1) \cdot (p_j J/p_s)$. Define $V_i = p_j J$ as the total returns to all factors except energy.
Therefore

$$N/(\partial N/\partial S) = (\beta_2/\beta_1) \cdot (V_i/p_s) \qquad \text{(A.13.8)}$$

Just as for the case of residential consumers suppose $N = \exp(S^\theta E^\phi)$ where θ and ϕ are parameters at our disposal.

Equations (A.13.7) and (A.13.8) are identical in form to (A.13.2) and (A.13.3), respectively, except for different constants and V_i instead of Y. Proceeding exactly as for the case of residential consumers leads to an equation of the form of (A.13.5) with V_i in place of Y.
Therefore

$$\ln X_{ei} = h(R) + r_1 \ln p_s + r_2 \ln p_x + r_3 (\ln p_x)^2 + r_4 \ln V_i \qquad \text{(A.13.9)}$$

where X_{ei} is the firm's electricity demand.

EMPIRICAL ESTIMATION OF ELECTRICITY DEMAND FUNCTIONS

One common characteristic of the demand functions (A.13.5) and (A.13.9) is the existence of a term of the form $g(R)$ representing the effect of the quality of electricity supply on demand.

Two problems arise when attempting to estimate a function with such a term. First, the exact functional form $g(R)$ is unknown. Second, as discussed earlier the quality of electricity supply, R, encompasses a variety of different aspects of electricity and it will be very difficult to quantify such a concept. Both problems may be mitigated if, rather than a continuous range of values for R we consider a set of discrete levels R_0, $R_1, \ldots, R_p$ where R_0 refers to the lowest quality of supply and $R_1, R_2, \ldots$ are successively higher qualities of supply, and we introduce a set of dummy variables D_1, $D_2, \ldots, D_p$ where

$$D_1 = D_2 = D_3 = \ldots = D_p = 0 \qquad \text{when } R = R_0$$
$$D_1 = 1, D_2 = D_3 = \ldots = D_p = 0 \qquad \text{when } R = R_1, \text{ and so on.}$$

Introducing an error term u into (A.13.5) we can then write

$$\ln X_{er} = (C + a_1 D_1 + a_2 D_2 + \ldots + a_p D_p) + t_1 \ln p_s$$
$$+ \delta_0 \ln p_x + \delta_1 (\ln p_x)^2 + t_3 \ln Y + u \qquad \text{(A.13.10)}$$

where
C is a constant
a_1 represents the difference between $g(R_0)$ and $g(R_1)$
a_2 represents the difference between $g(R_0)$ and $g(R_2)$, and so on.
Using the regression model (A.13.10) we can estimate $C, a_1, a_2, \ldots, a_p, t_1, \delta_0, \delta_1$, and t_3.

From equation (A.13.5) the fractional change in quantity demanded when the quality of supply is changed from R_i to R_j is given by $[\exp \{g(R_j) - g(R_i)\} - 1]$. Using (A.13.10) this expression may now be represented by $[\exp (a_j - a_i) - 1]$. Though the presentation here has been for the residential consumer category an identical approach may be applied industrial and commercial consumers.

COMPARISON WITH OTHER DEMAND STUDIES

It is interesting to compare these demand functions derived earlier with others used in the economic literature to estimate electricity demand.[3] In these studies residential electricity demand has received by far the widest attention.[4] The most commonly used residential demand functions are either linear or double-log in form. The dependent variable is usually the per capita demand for electricity (or its logarithm) while the

---

[3] For a detailed summary and evaluation of the existing econometric research on the demand for electricity see, Lester D. Taylor, "The Demand for Electricity: A Survey," *Bell Journal of Economics,* Spring 1975, pp. 74–110.
[4] See, for example, H. S. Houthaker, "Some Calculations of Electricity Consumption in Great Britain," *Journal of the Royal Statistical Society,* Ser. A, 114, 1951, pp. 351–71; K. P. Anderson, "Residential Energy Use: An Econometric Analysis," Report no. R-1297-NSF, The Rand Corporation, October 1973; J. W. Wilson, "Residential Demand for Electricity," *Quarterly Review of Economics and Business,* Vol. 11, No. 1, Spring 1971, pp. 7–22; Betancourt, "An Econometric Analysis of Peak Electricity Demand in the Short Run," *Energy Economics,* January 1981, pp. 14–29.

independent variables are usually the ex post average price of electricity, per capita income, and a variety of other variables including the price of substitute forms of energy and the stock of electricity using equipment.

For some constant level of the quality of supply the present demand function (A.13.5) takes a double-log form identical to that used in many of the earlier studies. The independent variables price of electricity and per capita income are present as in all previous studies using the double-log form. The price of substitute energy is also present as in many previous studies, while the price elasticity is now no longer constrained to be constant.

Many of the earlier residential demand studies suffer from the shortcoming that the estimated demand functions are not based on utility maximizing theory. (For example, see the studies cited by Anderson, Houthaker, and Wilson.)

Sato has shown that by using additive utility functions it is possible to derive a system of double-log demand functions with variable elasticities.[5] Betancourt uses this result to argue that the demand functions employed in his estimations are consistent with utility mazimizing theory. Nevertheless Sato's derivation has the shortcoming that the additive nature of the utility function does not allow for strong substitutability between goods or services. If the price of substitute energy is to be included in the double-log form, as is done in many estimates of electricity demand, Sato's derivation is no longer relevant and it is not possible to argue that the double-log demand function is consistent with utility maximizing theory.

The demand functions used in this study contain the price of substitute energy and in contrast with earlier studies are based on explicit maximizing theory. In fact, the approach used to derive demand functions in this chapter could be used to derive similar functions for other types of goods and services, besides electricity. These functions would allow for strong substitutability and would have the empirically convenient double-log form.

CHANGES IN THE TOTAL BENEFITS OF CONSUMPTION

Equation (A.13.5), which represents the demand for electricity, may be rewritten as $X_{er}^L = e^{g(R_L)} p_s^{t_1} p_x^{t_1} p_x^{\delta_0 + \delta_1 \ln p_x} Y^{t_3}$ for a system providing electricity of low quality, R_L. Using this equation (and Figure 13.1), the total benefits (TB_L) due to consumption of $\overline{X}_{er}^L$ units of electricity may be written as

$$TB_L = \overline{p}\,\overline{X}_{er}^L + e^{g(R_L)} \int_{\overline{p}}^{\infty} p_s^{t_1} p_x^{\delta_0 + \delta_1 \ln p_x} Y^{t_3} dp_x \qquad (A.13.11)$$

A similar equation can also be written for the total benefits (TB_M) when the quality of supply is medium and the quantity consumed is $\overline{X}_{er}^M$ at the same price p. If R_M refers to medium supply quality

[5]K. Sato, "Additive Utility Functions with Double-Log Consumer Demand Functions," *Journal of Political Economy*, 1982, pp. 102–24.

$$TB_M = \overline{p}\overline{X}_{er}^M + e^{g(RM)} \int_{\overline{p}}^{\infty} p_s^{t_1} p_x^{\delta_0 + \delta_1 \ln\, p_x} Y^{t_3} dp_x \qquad \text{(A.13.12)}$$

Furthermore, $\overline{X}_{er}^M / \overline{X}_{er}^L = e^{g(RM)} / e^{g(RL)}$ from (2.5). This expression along with (A.13.11) and (A.13.12) yields
i.e.,

$$TB_M = \frac{e^{g(RM)}}{e^{g(RL)}} \cdot TB_L$$

If TR_M and TR_L are the total revenues for the medium and low quality plans, respectively, $TR_M = \overline{p}\overline{X}_{er}^M$ and $TR_L = \overline{p}\overline{X}_{er}^L$.
Therefore,

$$TR_M = (\overline{X}_{er}^M / \overline{X}_{er}^L) TR_L = \frac{e^{g(RM)}}{e^{g(RL)}} TR_L$$

Thus, both revenues and total benefits change by the same percentage when moving from the low to medium quality plans. This would in turn imply that consumer surplus also increases by the factor $[e^{g(RM)} / e^{g(RL)}]$ when moving from the low to the medium quality plan.

Index